Springer Proceedings in Mathematics & Statistics

Volume 239

Springer Proceedings in Mathematics & Statistics

This book series features volumes composed of selected contributions from workshops and conferences in all areas of current research in mathematics and statistics, including operation research and optimization. In addition to an overall evaluation of the interest, scientific quality, and timeliness of each proposal at the hands of the publisher, individual contributions are all refereed to the high quality standards of leading journals in the field. Thus, this series provides the research community with well-edited, authoritative reports on developments in the most exciting areas of mathematical and statistical research today.

More information about this series at http://www.springer.com/series/10533

Adriano Polpo · Julio Stern
Francisco Louzada · Rafael Izbicki
Hellinton Takada
Editors

Bayesian Inference and Maximum Entropy Methods in Science and Engineering

MaxEnt 37, Jarinu, Brazil, July 09–14, 2017

Springer

Editors
Adriano Polpo
Department of Statistics
Federal University of São Carlos
São Carlos, São Paulo
Brazil

Rafael Izbicki
Department of Statistics
Federal University of São Carlos
São Carlos, São Paulo
Brazil

Julio Stern
Applied Mathematics
University of São Paulo
São Paulo, São Paulo
Brazil

Hellinton Takada
Itaú Asset Management
Banco Itaú-Unibanco
São Paulo, São Paulo
Brazil

Francisco Louzada
Institute of Mathematical Sciences and
Computing
University of São Paulo
São Carlos, São Paulo
Brazil

ISSN 2194-1009 ISSN 2194-1017 (electronic)
Springer Proceedings in Mathematics & Statistics
ISBN 978-3-030-08186-7 ISBN 978-3-319-91143-4 (eBook)
https://doi.org/10.1007/978-3-319-91143-4

Mathematics Subject Classification (2010): 60G35, 62-06, 62A01, 62F99, 65C40, 65C05, 81P05, 82B31, 82B41, 85A35

Printed on acid-free paper

This Springer imprint is published by the registered company Springer International Publishing AG part of Springer Nature
The registered company address is: Gewerbestrasse 11, 6330 Cham, Switzerland

Preface

MaxEnt 2017

This book brings the contributed works of MaxEnt 2017—37th International Workshop on Bayesian Inference and Maximum Entropy Methods in Science and Engineering (http://www.gis.des.ufscar.br/meetings/2017maxent/).

The 37th MaxEnt happened in Jarinu-SP from 9 to 14 July 2017 and the event aimed at:

- strengthening research and presenting contributions in all aspects of Bayesian methods and maximum entropy, as well as extending its applications in problems from different scientific communities and in areas such as astronomy, reliability, cosmology, econometrics, engineering, quantum mechanics, stochastic processes, survival, etc.;
- providing an environment in which researchers could interact, presenting recent developments and discussing related issues;
- allowing graduate students to have contact with senior researchers to discuss their work and to start possible contacts for future doctorate studies and post-doctorate projects.

The main objective of this workshop was the discussion about Bayesian computational techniques such as Monte Carlo Markov Chain and approximate inferential methods, questions about foundation of probability, and information theory. It also considered the presentation and discussion of new applications of inference foundations of physical theories.

The event has had excellent talks given by researchers of international reputation, whose works are currently highlighted in applied mathematics and statistics.

The MaxEnt 2017 was organized by: Inductive Statistics Group (GIS), Department of Statistics of the Federal University of São Carlos (DEs/UFSCar), Institute of Mathematics and Statistics of the University of São Paulo (IME/USP), Mathematics and Computer Science, University of São Paulo (ICMC/USP), Postgraduate Program in Statistics of UFSCar/ICMC-USP, and CeMEAI.

The Conference Chairing Committee was

- Adriano Polpo (UFSCar, Brazil)
- Julio M. Stern (USP, Brazil)

The members of the Organizing Committee were

- Adriano Polpo (UFSCar, Brazil)
- Carlos Alberto de Bragança Pereira (IME-USP, Brazil)
- Francisco Louzada Neto (ICMC-USP, Brazil)
- Hellinton H. Takada (Itaú Asset Management, Brazil)
- Juliana Cobre (ICMC-USP, Brazil)
- Julio Stern (IME-USP, Brazil)
- Katiane Conceição (ICMC-USP, Brazil)
- Márcio Alves Diniz (UFSCar, Brazil)
- Nestor Caticha (IF-USP, Brazil)
- Rafael Bassi Stern (UFSCar, Brazil)
- Rafael Izbicki (UFSCar, Brazil)
- Teresa Cristina Martins Dias (UFSCar, Brazil)
- Victor Fossaluza (IME-USP, Brazil)

The members of the Scientific Committee were

- Adriano Polpo (UFSCar, Brazil)
- Ali Mohammad-Djafari (CNRS, France)
- Ariel Caticha (Univ. at Albany, USA)
- Carlos Alberto de Bragança Pereira (IME-USP, Brazil)
- Francisco Louzada Neto (ICMC-USP, Brazil)
- Hellinton H. Takada (Itaú Asset Management, Brazil)
- Julian Center (Autonomous Exploration Inc., USA)
- Julio Stern (IME-USP, Brazil)
- Kevin Knuth (Univ. at Albany, USA)
- Paul M. Goggans (Univ. of Mississippi, USA)
- Rafael Izbicki (UFSCar, Brazil)
- Robert Niven (UNSW, Australia)

The members of the Executive Committee were

- Luana A. Takahashi
- Rita Volckov
- Sylvia Regina A. Takahashi

The workshop included nine invited speakers, whose names and respective institutions are listed below

- Ariel Caticha (University at Albany, USA)
- Flávio B. Gonçalves (UFMG, Brazil)
- Udo von Toussaint (Max-Planck-Institut fuer Plasmaphysik, Germany)
- Karim Anaya-Izquierdo (University of Bath, United Kingdom)

- Estevam R. Hruschka Jr. (UFSCar, Brazil)
- John Skilling (Maximum Entropy Data Consultants, Ireland)
- Thais Fonseca (UFRJ, Brazil)
- Rubens Sampaio (PUC-RJ, Brazil)
- Kevin Knuth (University at Albany, USA)

Also, four tutorial sessions have been given and were presented by

- Adriano Polpo (UFSCar, Brazil)
- Ali Mohammad-Djafari (Centre National de la Recherche Scientifique, France)
- Hellinton H. Takada (Itaú Asset Management, Brazil)
- Rafael B. Stern (UFSCar, Brazil)

It is worth noting the significant participation of students (29), most of whom presented their works orally or in posters. This is an important indicator of a promising future for the areas covered by the event in the Brazilian scientific community. The workshop has been attended by 20 researchers from foreign institutions. The photo of the participants is shown in Fig. 1

We point out that all oral sessions were recorded. They are available to the entire community at https://goo.gl/p3KVu5.

The MaxEnt 2017 Organizing Committee is grateful for the support received from the following agencies and institutions: FAPESP, CNPq, CAPES, Interinstitutional Program of Postgraduate Studies in Statistics UFSCar/ICMC-USP, Program

Fig. 1 Group photo of MaxEnt 2017 participants

of Postgraduation in Applied Mathematics IME-USP, Postgraduate Program in Probability and Statistics IME-USP, Springer, Entropy Journal, and Boise State University.

São Carlos, Brazil
July 2018

Adriano Polpo
Julio Stern
Francisco Louzada
Rafael Izbicki
Hellinton Takada

Contents

Contributors

Markus Abel Ambrosys GmbH/University of Potsdam, Potsdam, Germany

Mariane B. Alves Department of Statistics, Centro de Tecnologia, Federal University of Rio de Janeiro, Rio de Janeiro, Brazil

Camila B. Martins Federal University of São Paulo, São Paulo, Brazil

Junior Barrera Instituto de Matemática e Estatística—USP, São Paulo, Brazil

Troy Broderick University at Albany, Albany, NY, USA

Nicholas Carrara Department of Physics, University at Albany–SUNY, Albany, NY, USA

Ariel Caticha Department of Physics, University at Albany–SUNY, Albany, NY, USA

Nestor Caticha Instituto de Física, Universidade de São Paulo, São Paulo, SP, Brazil

Camille Chapdelaine Laboratoire des signaux et systèmes, CNRS, CentraleSupélec, Université Paris-Saclay, Gif sur Yvette, France

Oswaldo L. V. Costa Polytechnic School, University of São Paulo, São Paulo, Brazil

Rafael de C. Ceregatti Federal University of São Carlos, São Carlos, São Paulo, Brazil

Celma de O. Ribeiro Polytechnic School, University of São Paulo, São Paulo, Brazil

Marcio A. Diniz Federal University of São Carlos, São Carlos, Brazil

Mircea Dumitru Laboratoire des signaux et systèmes, CNRS, CentraleSupélec, Université Paris-Saclay, Gif sur Yvette, France

Keith A. Earle University at Albany, Albany, NY, USA

Rafael S. Erbisti Department of Statistics, Centro de Tecnologia, Federal University of Rio de Janeiro, Rio de Janeiro, Brazil

M. Faitsch Max-Planck-Institute for Plasma Physics, Garching, Germany

Leandro A Ferreira Departament of Statistics, University of São Paulo, São Paulo, Brazil

Thais C. O. Fonseca Department of Statistics, Centro de Tecnologia, Federal University of Rio de Janeiro, Rio de Janeiro, Brazil

Victor Fossaluza Departament of Statistics, University of São Paulo, São Paulo, Brazil; IME-USP, São Paulo, Brazil

Nicolas Gac Laboratoire des signaux et systèmes, Gif sur Yvette, Rue Joliot-Curie, France

Sandro Gallo Federal University of São Carlos, São Carlos, Brazil

Paul M. Goggans University of Mississippi, Oxford, Mississippi, USA

Roger Guimera Rovira i Virgili University, Tarragona, Spain

R. Wesley Henderson University of Mississippi, Oxford, Mississippi, USA

Paulo Hubert IME-USP, São Paulo, Brazil

Rafael Izbicki Federal University of São Carlos, São Carlos, São Paulo, Brazil

Oleks Kazakov University at Albany, Albany, NY, USA

Roberta Lima PUC-Rio, Department of Mechanical Engineering, Gávea, Brazil

Diego Marcondes Instituto de Matemática e Estatística—USP, São Paulo, Brazil

Patricio Maturana Russel Department of Statistics, University of Auckland, Auckland, New Zealand

Ali Mohammad-Djafari Laboratoire des signaux et systèmes, CNRS – CentraleSupélec – Université Paris-Saclay, Gif sur Yvette, France

Nathália Demetrio Vasconcelos Moura University of São Paulo, São Paulo, Brazil

D. Nille Max-Planck-Institute for Plasma Physics, Garching, Germany

Robert K. Niven School of Engineering and Information Technology, The University of New South Wales, Canberra, NSW, Australia

Natalia Lombardi Oliveira Federal University of São Carlos, São Carlos, Brazil

Linilson Padovese EP-USP, São Paulo, Brazil

Carlos A. de B. Pereira University of São Paulo, São Paulo, Brazil

Jony Arrais Pinto Junior Instituto de Matemática e Estatística (UFF), Niterói, RJ, Brazil

Adriano Polpo Federal University of São Carlos, São Carlos, Brazil

R. Preuss Max-Planck-Institut für Plasmaphysik, Garching, Germany

Rafael Catoia Pulgrossi Federal University of São Carlos, São Carlos, Brazil

Luis Ernesto B. Salasar Federal University of São Carlos, São Carlos, São Paulo, Brazil

Rubens Sampaio PUC-Rio, Department of Mechanical Engineering, Gávea, Brazil

Michael Schlegel Technische Universität Berlin, Berlin, Germany

B. Sieglin Max-Planck-Institute for Plasma Physics, Garching, Germany

Adilson Simonis Instituto de Matemática e Estatística—USP, São Paulo, Brazil

Lucas Silva Simões Instituto de Física, Universidade de São Paulo, São Paulo, SP, Brazil

Antal Spector-Zabusky Department of Computer and Information Science, University of Pennsylvania, Philadelphia, PA, USA

Donald Spector Department of Physics, Hobart and William Smith Colleges, Geneva, NY, USA

Julio M. Stern IME-USP, São Paulo, Brazil; Institute of Mathematics and Statistics, University of São Paulo, São Paulo, Brazil

Rafael Bassi Stern Federal University of São Carlos, São Carlos, SP, Brazil

Hellinton H. Takada Quantitative Research, Itaú Asset Management, São Paulo, Brazil

Milene Vaiano Farhat IME-USP, São Paulo, Brazil

Afonso Vaz Federal University of São Carlos, São Carlos, SP, Brazil

Geert Verdoolaege Department of Applied Physics, Ghent University, Ghent, Belgium; Laboratory for Plasma Physics, Royal Military Academy, Brussels, Belgium

Patrícia Viana da Silva Faculdade de Matematica (UFU), Uberlândia, Brazil

U. von Toussaint Max-Planck-Institut für Plasmaphysik, Garching, Germany

Nicholas Wagner Eugenio IME-USP, São Paulo, Brazil

Steven H. Waldrip School of Engineering and Information Technology, The University of New South Wales, Canberra, NSW, Australia

Li Wang Laboratoire des signaux et systèmes, CNRS, CentraleSupélec, Université Paris-Saclay, Gif sur Yvette, France

Sergio Wechsler University of São Paulo, São Paulo, Brazil

Quantum Phases in Entropic Dynamics

Nicholas Carrara and Ariel Caticha

Abstract In the Entropic Dynamics framework, the dynamics is driven by maximizing entropy subject to appropriate constraints. In this work, we bring Entropic Dynamics one-step closer to full equivalence with quantum theory by identifying constraints that lead to wave functions that remain single-valued even for multi-valued phases by recognizing the intimate relation between quantum phases, gauge symmetry, and charge quantization.

Keywords Entropic Dynamics · Quantum phases · Charge quantization

1 Introduction

In the Entropic Dynamics (ED) framework, the Schrödinger equation is derived as an application of entropic methods of inference[1] and, as always with inference, the first and most crucial step is to be clear about what we want to infer. What microstates are we talking about? This defines the ontology of the model. Once that choice is made the dynamics is driven by entropy subject to information expressed by constraints [2–5].

ED takes the epistemic view of the wave function Ψ to its logical conclusion. Within an inferential framework, it is not sufficient to just state that the probability

[1]The principle of maximum entropy as a method for inference can be traced to the pioneering work of E. T. Jaynes. For a pedagogical overview of Bayesian and entropic inference and further references see [1].

Presented at MaxEnt 2017, the 37th International Workshop on Bayesian Inference and Maximum Entropy Methods in Science and Engineering (July 9–14, 2017, Jarinu, Brazil).

N. Carrara (✉) · A. Caticha (✉)
Department of Physics, University at Albany–SUNY, Albany, NY 12222, USA
e-mail: ncarrara@albany.edu

A. Caticha
e-mail: acaticha@albany.edu

A. Polpo et al. (eds.), *Bayesian Inference and Maximum Entropy Methods in Science and Engineering*, Springer Proceedings in Mathematics & Statistics 239, https://doi.org/10.1007/978-3-319-91143-4_1

$|\Psi|^2$ reflects a state of knowledge; it is also necessary to demand that the phase receive an epistemic interpretation, and that all changes in Ψ be dictated by the maximum entropy and Bayesian updating rules. Thus, the ED framework is very restrictive: it must account for *both* the unitary time evolution described by the Schrödinger equation *and* the collapse of the wave function during measurement.

But even after ED succeeds in accomplishing these tasks a challenge still remains: is ED fully equivalent to quantum mechanics (QM) or does it merely reproduce a subset of its solutions? Problems of this kind were pointed out long ago by Takabayasi [6] in the context of the hydrodynamical interpretation of QM, and later revived by Wallstrom [7, 8] in the context of Nelson's stochastic mechanics. Wallstrom's objection is that stochastic mechanics leads to phases and wave functions that are either both multi-valued or both single-valued. Both alternatives are unsatisfactory because on the one hand, QM requires single-valued wave functions, while on the other hand single-valued phases exclude states that are physically relevant (e.g., states with nonzero angular momentum).

In previous work, the constraints that drive the dynamics were introduced in two different ways, either by postulating some extra variables [2, 9], or by the explicit introduction of a "drift" potential [3–5, 10]. One of the goals of this paper is to show that these two types of constraint can be imposed simultaneously which lends the theory greater flexibility and expands the range of future potential applications. We identify constraints that lead to single-valued wave functions, but nevertheless allow for multi-valued phases[2] and naturally lead to the local gauge symmetry required for electromagnetic interactions. Our argument involves two ingredients. The first is the recognition that a deeper understanding of the phase of the wave function must consider the intimate relation between quantum phases and gauge symmetry. The second ingredient is the recognition that in order for ED to agree with experiment it is necessary that the dynamics be linear. ED differs from standard QM in many crucial ways but its demand for linearity is not one of them. The demand that the linear and the probabilistic structures be compatible with each other implies that ED constraints must lead to single-valued wave functions [12].

Here, we will focus on the derivation of the Schrödinger equation but the ED approach has been applied to a variety of other topics including the quantum measurement problem [13, 14]; momentum and uncertainty relations [15]; the Bohmian limit [10, 16] and the classical limit [17]; the extensions to curved spaces [18] and to relativistic fields [19, 20].

---

[2] A hint towards a satisfactory resolution of Wallstrom's objection is found in Takabayasi's later work which incorporates spin into his hydrodynamical approach [11]. Although here we focus on non-spinning particles our choice of constraints can be generalized to particles with spin $1/2$ — a project to be addressed in a future publication.

2 Entropic Dynamics — A Brief Review

The statistical model– We consider N particles living in a flat Euclidean space $\mathbf{X}$ with metric δ_{ab}. The first important assumption is that position plays a distinguished role: it defines the ontic state of the system. The fact that at all times particles have *definite* positions deviates from the standard Copenhagen interpretation according to which definite values are created by measurement.[3] In ED, positions are in general *unknown*; they are the quantities to be inferred.

The position of each particle will be denoted by x_n^a where the index $n = 1 \ldots N$ labels the particle and $a = 1, 2, 3$ its spatial coordinates. The position of the system in configuration space $\mathbf{X}_N = \mathbf{X} \times \ldots \times \mathbf{X}$ will be denoted either by x or by the components x^A where $A = (n, a)$, and the corresponding volume element is $d^{3N}x = dx$.

The second assumption is that in addition to the particles there also exist some other variables denoted y [2, 9]. This assumption is not unreasonable: the world does contain stuff beyond the N particles of interest. It is also most fortunate that we need not be too specific about these y variables. It turns out that the relevant information is conveyed by their entropy,

$$S(x) = -\int dy\, p(y|x) \log \frac{p(y|x)}{q(y)}\,, \tag{1}$$

where we assume that the probability distribution $p(y|x)$ depends on the location x of the particles and $q(y)$ is some unspecified underlying measure.

Having identified the microstates $(x, y) \in \mathbf{X}_N \times \mathbf{Y}$ we tackle the dynamics. The goal is to find the probability density $P(x'|x)$ for the transition from an initial x to a new x'. Since both x' and the corresponding y' are unknown the relevant space is not just $\mathbf{X}_N$ but $\mathbf{X}_N \times \mathbf{Y}$. The distribution we seek is the joint distribution $P(x', y'|x, y)$. It is found by maximizing the appropriate entropy,

$$\mathscr{S}[P, Q] = -\int dx'dy'\, P(x', y'|x, y) \log \frac{P(x', y'|x, y)}{Q(x', y'|x, y)}\,, \tag{2}$$

relative to a joint prior $Q(x', y'|x, y)$ and subject to the appropriate constraints.
The prior– We adopt a prior $Q(x', y'|x, y)$ that represents a state of extreme ignorance: knowledge of x' tells us nothing about y' and vice versa. This is a product, $Q(x', y'|x, y) = Q(x'|x, y)Q(y'|x, y)$, in which $Q(x'|x, y)dx'$ and $Q(y'|x, y)dy'$ are uniform,[4] that is, proportional to the respective volume elements, $d^{3N}x = dx$ and

[3] On the other hand, in ED — just as in the Copenhagen interpretation — other observables such as energy or momentum do not, in general, have definite values; their values are created by the act of measurement. These other quantities are epistemic in that they do not reflect properties of the particles but of the wave function.

[4] Strictly uniform non-normalizable priors can be mathematically problematic but here no such difficulties arise. By "uniform" we actually mean any distribution that is essentially flat over the support of the posterior which in our case will be infinitesimally narrow.

$q(y)dy$. Since proportionality constants have no effect on the entropy maximization, the joint prior is

$$Q(x', y'|x, y) = q(y'). \tag{3}$$

The constraints– We first write the posterior as a product,

$$P(x', y'|x, y) = P(x'|x, y)P(y'|x', x, y) . \tag{4}$$

We require that the new x' depends only on x so we set $P(x'|x, y) = P(x'|x)$. We also require that the uncertainty in y' depends only on x', $P(y'|x', x, y) = p(y'|x')$. Therefore, the first constraint is

$$P(x', y'|x, y) = P(x'|x)p(y'|x'). \tag{5}$$

To implement it substitute (3) and (5) into (2),

$$\mathscr{S}[P, Q] = -\int dx'\, P(x'|x) \log P(x'|x) + \int dx'\, P(x'|x)S(x') , \tag{6}$$

where $S(x)$ is given in Eq. (1). Next, the continuity of the motion is enforced by requiring that the steps Δx_n^a from x_n^a to $x_n'^a = x_n^a + \Delta x_n^a$ taken by each individual particle be infinitesimally short. This is implemented by imposing N independent constraints,

$$\int dx'\, P(x'|x)\Delta x_n^a \Delta x_n^b \delta_{ab} = \langle \Delta x_n^a \Delta x_n^b \rangle \delta_{ab} = \kappa_n , \qquad (n = 1 \ldots N) . \tag{7}$$

where repeated indices are summed over and we eventually take the limit $\kappa_n \to 0$. The κ_n's are chosen to be constant to reflect the translational symmetry of the space **X** and they are n-dependent in order to accommodate non-identical particles.

The transition probability– Varying $P(x'|x)$ to maximize (6) subject to (7) and normalization gives

$$P(x'|x) = \frac{1}{\zeta} \exp\left[S(x') - \frac{1}{2}\textstyle\sum_n \alpha_n \delta_{ab} \Delta x_n^a \Delta x_n^b \right] , \tag{8}$$

where ζ is a normalization constant and the Lagrange multipliers α_n are chosen to implement the constraints Eq. (7). In Eq. (8) it is clear that the infinitesimally short steps are obtained in the limit of large α_n. It is therefore useful to Taylor expand,

$$S(x') = S(x) + \textstyle\sum_n \Delta x_n^a \frac{\partial S}{\partial x_n^a} + \ldots \tag{9}$$

and rewrite $P(x'|x)$ as

$$P(x'|x) = \frac{1}{Z}\exp\left[-\frac{1}{2}\sum\nolimits_{n}\alpha_{n}\,\delta_{ab}\left(\Delta x_{n}^{a}-\langle\Delta x_{n}^{a}\rangle\right)\left(\Delta x_{n}^{b}-\langle\Delta x_{n}^{b}\rangle\right)\right], \tag{10}$$

where Z is a new normalization constant and Δx_{n}^{a} is given by Eq. (12) below.

To find how these short steps accumulate, we introduce time as a book-keeping device. As discussed in [2–5] entropic time is measured by the fluctuations themselves (see Eq. (14) below) which leads to the choice

$$\alpha_{n} = \frac{m_{n}}{\eta\Delta t}, \tag{11}$$

where Δt is the time taken by the short step, the m_n are particle-specific constants that will be called "masses," and η is a constant that fixes the units of time relative to those of length and mass. A generic displacement is then expressed as an expected drift plus a fluctuation,

$$\Delta x_{n}^{a} = \Delta x^{A} = b^{A}\Delta t + \Delta w^{A}, \tag{12}$$

where $b^{A}(x)$ is the drift velocity,

$$\langle\Delta x^{A}\rangle = b^{A}\Delta t \quad \text{with} \quad b^{A} = \frac{\eta}{m_{n}}\delta^{AB}\partial_{B}S = \eta m^{AB}\partial_{B}S, \tag{13}$$

and $\partial_{A} = \partial/\partial x_{n}^{a}$; $m_{AB} = m_{n}\delta_{AB}$ is the "mass" tensor and $m^{AB} = \delta^{AB}/m_{n}$ is its inverse. The fluctuations Δw^{A} satisfy,

$$\langle\Delta w^{A}\rangle = 0 \quad \text{and} \quad \langle\Delta w^{A}\Delta w^{B}\rangle = \frac{\eta}{m_{n}}\delta^{AB}\Delta t = \eta m^{AB}\Delta t. \tag{14}$$

Thus ED leads to the non-differentiable trajectories that are characteristic of a Brownian motion.

The Fokker–Planck equation– Once the probability for a single short step is found, Eq. (10), the accumulation of many short steps leads to a probability distribution $\rho(x,t)$ in configuration space that obeys a Fokker–Planck equation (FP), [1–3]

$$\partial_{t}\rho = -\sum\nolimits_{n}\partial_{na}\left(\rho v_{n}^{a}\right) = -\partial_{A}\left(\rho v^{A}\right), \tag{15}$$

where v^{A} is the velocity of the probability flow in configuration space or *current velocity*. It is given by

$$v^{A} = m^{AB}\partial_{B}\Phi_{0} \quad \text{and} \quad \Phi_{0} = \eta S - \eta\log\rho^{1/2} \tag{16}$$

where Φ_{0} will be called the *phase*.

Hamiltonian entropic dynamics– The FP equation (15) describes a *standard diffusion* of a single dynamical field, $\rho(x)$, that evolves in response to a non-dynamical field given by the entropy $S(x)$. In contrast, a *quantum dynamics* includes a second dynamical field, the phase of the wave function. In ED, this evolving phase is introduced by continuously updating the constraint (5) which allows the entropy $S(x)$, or equivalently the phase $\Phi_0(x)$, to become dynamical.

First, we note that without loss of generality we can always find a functional $\tilde{H}[\rho, \Phi_0]$ so that $\partial_t \rho = \delta \tilde{H}/\delta \Phi_0$ reproduces the FP equation (15). The specific updating rule for S or Φ_0 is inspired by an idea of Nelson's [21]: requiring that Φ_0 be updated in such a way that the functional $\tilde{H}[\rho, \Phi_0]$ be conserved leads to Hamilton's equations [4],

$$\partial_t \rho = \frac{\delta \tilde{H}}{\delta \Phi_0} \quad \text{and} \quad \partial_t \Phi_0 = -\frac{\delta \tilde{H}}{\delta \rho} . \tag{17}$$

$\tilde{H}[\rho, \Phi_0]$ is the "ensemble" Hamiltonian. The second equation in (17) is a Hamilton–Jacobi equation (HJ). Additional arguments from information geometry [4] can then be invoked to suggest that the natural choice of $\tilde{H}$ is

$$\tilde{H}[\rho, \Phi_0] = \int dx\, \rho \left[\frac{1}{2} m^{AB} \partial_A \Phi_0 \partial_B \Phi_0 + V + \xi m^{AB} \frac{1}{\rho^2} \partial_A \rho \partial_B \rho \right] . \tag{18}$$

The first term in the integrand is the "kinetic" term that reproduces the FP equation (15). The second term represents the simplest nontrivial interaction and introduces the standard potential $V(x)$. The third term, motivated by information geometry, is the trace of the Fisher information and is called the "quantum" potential. The parameter ξ controls the relative contributions of the two potentials: $\xi = 0$ leads to a stochastic classical mechanics; $\xi > 0$ leads to quantum theory — in fact, ξ *defines* Planck's constant as $\hbar = (8\xi)^{1/2}$.

The Schrödinger equation– To conclude this brief review of ED, we note that at this point the dynamics is fully specified by Eqs. (17) and (18). We can combine ρ and Φ_0 into a single complex function, $\Psi_0 = \rho^{1/2} \exp(i\Phi_0/\hbar)$. Then, the pair of Hamilton's equations (17) can be rewritten as a single complex Schrödinger equation that is explicitly linear,

$$i\hbar \partial_t \Psi_0 = -\frac{\hbar^2}{2} m^{AB} \partial_A \partial_B \Psi_0 + V \Psi_0 . \tag{19}$$

However, even though Eq. (17) can be written in the form (19), this does not mean that they are equivalent to the full quantum theory. The problem is that Eq. (17) only reproduce a subset of all the wave functions required by quantum mechanics. More specifically, since both $S(x)$ and $\rho(x)$ are single-valued — the total change as one moves in a closed path vanishes,

$$\Delta S = \oint_\Gamma d\ell^A \partial_A S = 0 \quad \text{and} \quad \Delta \rho = \oint_\Gamma d\ell^A \partial_A \rho = 0 , \tag{20}$$

so that both Φ_0 and Ψ_0 are single-valued too. The single-valuedness of Ψ_0 is precisely what we want, but the single-valuedness of Φ_0 is too restrictive. It excludes, for example, eigenstates of angular momentum that have manifestly multi-valued phases ($\Psi \propto e^{im\phi}$, where ϕ is the azimuthal angle and m is an integer).

3 Gauge Symmetry and Multi-Valued Phases

A minimal ED was derived in the previous section. A richer dynamics that allows additional interactions can be achieved by imposing additional constraints.

Additional constraints– We assume that the motion of each particle is affected by an additional potential field $\varphi(x)$ where $x \in \mathbf{X}$ is a point in 3D space with the topological properties of an angle ($\varphi(x)$ and $\varphi(x) + 2\pi$ describe the same angle). We further assume that these angles can be redefined by different amounts $\chi(x)$ at different places, that is, the origin from which these angles are measured can be set independently at each x. This is a local gauge symmetry and it immediately raises the question of how can one compare angles at different locations in order to define derivatives. The answer is well known: introduce a *connection* field, a vector potential $A_a(x)$ that defines which angle at $x + \Delta(x)$ is the "same" as the angle at x. This is implemented by imposing that as $\varphi \to \varphi + \chi$ then the connection transforms as $A_a \to A_a + \partial_a \chi$ so that the corrected derivative $\partial_a \varphi - A_a$ remains invariant.[5]

To derive an ED that incorporates gauge invariant interactions with these potentials, in addition to (7) and normalization, for each particle we impose the constraint

$$\langle \Delta x_n^a \rangle [\partial_a \varphi(x_n) - A_a(x_n)] = \kappa_n'(x_n) , \qquad (n = 1 \dots N) \tag{21}$$

where $\kappa_n'(x_n)$ are functions to be specified below.

The transition probability $P(x'|x)$ that maximizes the entropy $\mathscr{S}[P, Q]$ in (6) is

$$P(x'|x) = \frac{1}{\zeta} \exp\left[S(x') - \sum\nolimits_n \left(\frac{\alpha_n}{2}\delta_{ab}\Delta x_n^a \Delta x_n^b - \beta_n \left(\partial_{na}\varphi(x_n) - A_a(x_n)\right) \Delta x_n^a\right)\right] \tag{22}$$

where $\partial_{na} = \partial/\partial x_n^a$, α_n and β_n are Lagrange multipliers, and ζ is a normalization constant. For large α_n Taylor expand $S(x')$ about x, and use Eq. (11), then, as in Eq. (12) a generic displacement Δx^A can be expressed in terms of an expected drift plus a fluctuation, $\Delta x^A = b^A \Delta t + \Delta w^A$, but the drift velocity (13) now includes a new term,

$$b_n^a = \frac{\eta}{m_n}\delta^{ab}\left[\partial_{nb}\{S(x) + \beta_n \varphi(x_n)\} - \beta_n A_b(x_n)\right] , \tag{23}$$

---

[5]Note that since φ is dimensionless the vector potential A_a has units of inverse length and this implicitly defines the units of electric charge. These are not the units conventionally adopted in electromagnetism.

while the fluctuations Δw^A, Eq. (14), remain unchanged.

Hamilton's equations– As before, the accumulation of many short steps leads to the FP equation (15), but now the current velocity $v^A = v_n^a$ must be suitably modified,

$$v^A = m^{AB}\left(\partial_B \Phi - \bar{A}_B\right) \quad \text{with} \quad \Phi = \eta(S + \bar{\varphi} - \log \rho^{1/2}) , \tag{24}$$

where we introduced the configuration space quantities,

$$\bar{A}_A(x) = \eta \beta_n A_a(x_n)\, and\, \bar{\varphi}(x) = \Sigma_n \beta_n \varphi(x_n). \tag{25}$$

where $A = (n, a)$. Note that v^A is gauge invariant. The new ensemble Hamiltonian $\tilde{H}$, Eq. (18), is

$$\tilde{H}[\rho, \Phi] = \int dx \left[\frac{1}{2}\rho m^{AB}\left(\partial_A \Phi - \bar{A}_A\right)\left(\partial_B \Phi - \bar{A}_B\right) + \rho V + \frac{\hbar^2}{8\rho} m^{AB} \partial_A \rho \partial_B \rho\right] , \tag{26}$$

and the new FP equation now reads,

$$\partial_t \rho = -\partial_A \left[\rho m^{AB}\left(\partial_B \Phi - \bar{A}_B\right)\right] = \frac{\delta \tilde{H}}{\delta \Phi} . \tag{27}$$

As before, the requirement that $\tilde{H}$ be conserved for arbitrary initial conditions amounts to imposing the conjugate Hamilton equation, Eq. (17), which leads to the Hamilton–Jacobi equation,

$$\partial_t \Phi = -\frac{\delta \tilde{H}}{\delta \rho} = -\frac{1}{2} m^{AB}\left(\partial_A \Phi - \bar{A}_A\right)\left(\partial_B \Phi - \bar{A}_B\right) - V + \frac{\hbar^2}{2} m^{AB} \frac{\partial_A \partial_B \rho^{1/2}}{\rho^{1/2}} . \tag{28}$$

Finally, we combine ρ and Φ into a single wave function, $\Psi = \rho^{1/2} \exp(i\Phi/\hbar)$, to obtain the linear Schrödinger equation,

$$i\hbar \partial_t \Psi = -\sum\nolimits_n \frac{\hbar^2}{2m_n} \delta^{ab} \left(\frac{\partial}{\partial x_n^a} - \frac{i}{\hbar} \eta \beta_n A_a(x_n)\right)\left(\frac{\partial}{\partial x_n^b} - \frac{i}{\hbar} \eta \beta_n A_b(x_n)\right) \Psi + V\Psi. \tag{29}$$

4 Discussion

Electric charges are Lagrange multipliers– Recalling the standard expression for covariant derivatives,

$$\frac{\partial}{\partial x_n^a} - \frac{i q_n}{\hbar c} A_a(x_n) , \tag{30}$$

(q_n is the electric charge of particle n and c is the speed of light) shows that (29) is indeed the Schrödinger equation provided the multipliers β_n are chosen to be particle-dependent *constants* that are related to electric charges by

$$\beta_n = \frac{q_n}{\eta c} \quad \text{or} \quad q_n = c\eta\beta_n . \tag{31}$$

Thus, in ED *electric charges are Lagrange multipliers* that measure the strength of the particles' coupling to the φ_n and A_a potentials.

Single-valued wave functions, quantized circulation, and quantized charges– The success of any framework for inference such as ED depends on identifying the correct constraints. The choice of constraints in Sect. 2 succeeds in reproducing many of the features of quantum theory including a linear Schrödinger equation but is ultimately unsatisfactory because it leads to single-valued wave functions with single-valued phases that fail to include all quantum states.

The choice of constraints adopted in Sect. 3 represent an improvement because they take into account the relation between quantum phases and gauge symmetry. However, the wave functions Ψ obtained for generic choices of the multipliers β_n are also problematic in that they give multi-valued phases Φ, Eq. (24), that lead to multi-valued wave functions. Indeed, since φ is an angle the integral over a closed loop Γ_n in which all particles except n are kept fixed gives

$$\Delta\varphi = \oint_{\Gamma_n} d\ell_n^a \partial_{na}\varphi = 2\pi\nu(\Gamma_n) , \tag{32}$$

where $\nu(\Gamma_n)$ is an integer that depends on the loop Γ_n. Since S and $\log\rho$ are single-valued, from (24), we have

$$\Delta\frac{\Phi}{\hbar} = \oint_{\Gamma_n} d\ell_n^a \partial_{na}\frac{\Phi}{\hbar} = \frac{\eta\beta_n}{\hbar}\oint_{\Gamma_n} d\ell_n^a \partial_{na}\varphi = \frac{\eta\beta_n}{\hbar}2\pi\nu(\Gamma_n) , \tag{33}$$

so that Ψ is not single-valued.

Unfortunately, this means that even though (29) is linear, its linearity is in conflict with the underlying probabilistic structure. To see the problem consider two multi-valued ED solutions of (29), Ψ_1 and Ψ_2. Their magnitudes $|\Psi_1|^2 = \rho_1$ and $|\Psi_2|^2 = \rho_2$ are single-valued because they are probability densities. However, even though $\alpha_1\Psi_1 + \alpha_2\Psi_2 = \Psi_3$ is also a solution, it turns out that its magnitude $|\Psi_3|^2$ will in general turn out to be multi-valued which precludes a probabilistic interpretation [11]. Mere linearity is not enough. The condition for the linear and probabilistic structures to be compatible with each other is that wave functions be single-valued.

Inspection of Eq. (33) for arbitrary loops shows that the choice of constraint (21) — that is, the choice of β_n— that leads to single-valued wave functions is

$$\frac{\eta\beta_n}{\hbar} = \mu \tag{34}$$

where μ is an integer.

Equation (31) then shows that electric charges must be quantized in units of a basic charge $q = \hbar c$

$$\frac{\eta \beta_n}{\hbar} = \frac{q_n}{\hbar c} = \mu \quad \text{or}\, q_n = \mu q . \tag{35}$$

Changing to conventional units for charges and potentials is straightforward; just rescale $\lambda q_n = q_n'$ and $A_a/\lambda = A_a'$ so that $q_n A_a = q_n' A_a'$.

Conclusion– The equivalence of ED and quantum mechanics with a wave functions that remain single-valued even for multi-valued phases is achieved by imposing constraints that recognize the intimate relation between quantum phases and gauge symmetry. The condition for compatibility between the probabilistic and linear structures is that charges be quantized.

Acknowledgements We would like to thank M. Abedi, D. Bartolomeo, C. Cafaro, N. Caticha, S. DiFranzo, A. Giffin, S. Ipek, D.T. Johnson, K. Knuth, S. Nawaz, M. Reginatto, C. Rodríguez, K. Vanslette, for many discussions on entropy, inference, and quantum mechanics.

References

1. For a pedagogical review see Caticha, A.: Entropic Inference and the Foundations of Physics (11th Brazilian Meeting on Bayesian Statistics – EBEB-2012. http://www.albany.edu/physics/ACaticha-EIFP-book.pdf
2. Caticha, A.: Entropic dynamics, time and quantum theory. J. Phys. A: Math. Theor. **44**, 225303 (2011). arXiv.org:1005.2357
3. Caticha, A.: Entropic dynamics: an inference approach to quantum theory, time and measurement. J. Phys.: Conf. Ser. **504**, 012009 (2014). arXiv.org:1403.3822
4. Caticha, A., Bartolomeo, D., Reginatto, M.: Entropic dynamics: from entropy and information geometry to hamiltonians and quantum mechanics. AIP Conf. Proc. **1641**, 155 (2015). arXiv.org:1412.5629
5. Caticha, A.: Entropic dynamics. Entropy **17**, 6110 (2015). arXiv.org:1509.03222
6. Takabayasi, T.: On the formulation of quantum mechanics associated with classical pictures. Prog. Theor. Phys. **8**, 143 (1952)
7. Wallstrom, T.C.: On the derivation of the Schrödinger equation from stochastic mechanics. Found. Phys. Lett. **2**, 113 (1989)
8. Wallstrom, T.C.: Inequivalence between the Schrödinger equation and the Madelung hydrodynamic equations. Phys. Rev. A **49**, 1613 (1994)
9. Caticha, A.: Entropic dynamics: mechanics without mechanism. arXiv.org:1704.0266
10. Bartolomeo, D., Caticha, A.: Entropic dynamics: the Schroedinger equation and its Bohmian limit. AIP Conf. Proc. **1757**, 030002 (2016). arXiv.org:1512.09084
11. Takabayasi, T.: Vortex, spin and triad for quantum mechanics of spinning particle I: general theory. Prog. Theor. Phys. **70**, 1–17 (1983)
12. Merzbacher, E.: Single valuedness of wave functions. Am. J. Phys. **30**, 237 (1962)
13. Johnson, D.T., Caticha, A.: Entropic dynamics and the quantum measurement problem. AIP Conf. Proc. **1443**, 104 (2012). arXiv:1108.2550
14. Vanslette, K., Caticha, A.: Quantum measurement and weak values in entropic dynamics. AIP Conf. Proc. **1853**, 090003 (2017). arXiv:1701.00781

15. Nawaz, S., Caticha, A.: Momentum and uncertainty relations in the entropic approach to quantum theory. AIP Conf. Proc. **1443**, 112 (2012). arXiv:1108.2629
16. Bartolomeo, D., Caticha, A.: Trading drift and fluctuations in entropic dynamics: quantum dynamics as an emergent universality class. J. Phys.: Conf. Ser. **701**, 012009 (2016). arXiv.org:1603.08469
17. Demme, A., Caticha, A.: The classical limit of entropic quantum dynamics. AIP Conf. Proc. **1853**, 090001 (2017). arXiv.org:1612.01905
18. Nawaz, S., Abedi, M., Caticha, A.: Entropic dynamics on curved spaces. AIP Conf. Proc. **1757**, 030004 (2016). arXiv.org:1601.01708
19. Ipek, S., Caticha, A.: Entropic quantization of scalar fields. arXiv.org:1412.5637
20. Ipek, S., Abedi, M., Caticha, A.: A covariant approach to entropic dynamics. AIP Conf. Proc. **1853**, 090002 (2017)
21. Nelson, E.: Lect. Notes in Phys. **100**, 168 (Springer, Berlin, 1979)

Bayesian Approach to Variable Splitting Forward Models

Ali Mohammad-Djafari, Mircea Dumitru, Camille Chapdelaine and Li Wang

Abstract Classical single additive noise forward model can be extended to account for different uncertainties by variable splitting models. For example, one can distinguish between observation noise and forward model uncertainty or even to account for other forward model uncertainties. In this paper, we consider different cases and propose to use the Bayesian approach to handle them. As a by-product, we see that when MAP estimator is used we can find the same kind of optimization algorithms as Alternating Direction Method of Multipliers (ADMM) or Iterative Shrinkage Thresholding Algorithm (ISTA) optimization ones. However, the Bayesian approach gives us the tools to go further by estimating the hyperparameters of the inversion problems which are often crucial in real applications.

Keywords Inverse problems · Variable splitting forward models · Bayesian MAP estimate · ADMM optimization

1 Introduction

The classical forward model for linear inverse problems is:

$$g = \boldsymbol{H}\boldsymbol{f} + \boldsymbol{\varepsilon}, \tag{1}$$

A. Mohammad-Djafari (✉) · M. Dumitru · C. Chapdelaine · L. Wang
Laboratoire des signaux et systèmes, CNRS – CentraleSupélec – Université Paris-Saclay, 3, Rue Joliot-Curie, 91192 Gif sur Yvette, France
e-mail: Mohammad-Djafari@lss.supelec.fr

M. Dumitru
e-mail: Dumitru@lss.supelec.fr

C. Chapdelaine
e-mail: Chapdelaine@lss.supelec.fr

L. Wang
e-mail: Wang@lss.supelec.fr

A. Polpo et al. (eds.), *Bayesian Inference and Maximum Entropy Methods in Science and Engineering*, Springer Proceedings in Mathematics & Statistics 239, https://doi.org/10.1007/978-3-319-91143-4_2

where $\boldsymbol{f}$, a vector of length N represents the unknown, $\boldsymbol{g}$, a vector of length M represents the data, $\boldsymbol{H}$ is the forward model matrix and all the uncertainties are summarized by the vector $\boldsymbol{\varepsilon}$. For this simple model, nowadays, almost everything has been told, starting by Least Squares (LS), then Quadratic Regularization (QR), L_1 regularization, Bayesian approach with simple Gaussian models for the noise and Gaussian prior model, Double Exponential (DE) prior, Student-t prior [1] to much more sophisticated Hierarchical models [2–4]. However, in many real applications, it is needed to propose forward models which can account for other sources of uncertainties. For example, if we want to distinguish between the measurement noise and the forward model uncertainties, we can propose the following variable splitting model: $\boldsymbol{g} = \boldsymbol{H}\boldsymbol{f} + \boldsymbol{\xi} + \boldsymbol{\varepsilon}$ which can also be written as:

$$\begin{cases} \boldsymbol{g} = \boldsymbol{g}_0 + \boldsymbol{\varepsilon}, \\ \boldsymbol{g}_0 = \boldsymbol{H}\boldsymbol{f} + \boldsymbol{\xi}, \end{cases} \tag{2}$$

where $\boldsymbol{\varepsilon}$ represents the measurement noise and $\boldsymbol{\xi}$ can represent the modeling errors [5]. However, now, we have two quantities to infer on: $\boldsymbol{f}$ and $\boldsymbol{g}_0$. Interestingly, as we will see in Sect. 3, the Maximum A Posteriori (MAP) estimate of $\boldsymbol{f}$ and $\boldsymbol{g}_0$ results in an algorithm which looks like the ADMM. However, the Bayesian approach has many advantages over the deterministic regularization methods. For example, we may assign different probability laws to $\boldsymbol{\varepsilon}$ and $\boldsymbol{\xi}$. For example, we can assign as usual a Gaussian model to $\boldsymbol{\varepsilon}$ and a more heavy tailed model for $\boldsymbol{\xi}$. See the details in [6, 7].

We can also go further in details by considering the following models:

$$\begin{cases} \boldsymbol{g} = \boldsymbol{g}_0 + \boldsymbol{\varepsilon}, \\ \boldsymbol{g}_0 = \boldsymbol{H}(\boldsymbol{f} + \boldsymbol{f}_0) + \boldsymbol{\xi}, \end{cases} \tag{3}$$

where now, we have three unknowns $\boldsymbol{f}$, $\boldsymbol{g}_0$ and $\boldsymbol{f}_0$, where this last one can model some unknown background in the input space.

This model can also be written differently as:

$$\begin{cases} \boldsymbol{g} = \boldsymbol{g}_0 + \boldsymbol{\varepsilon}, \\ \boldsymbol{g}_0 = \boldsymbol{H}\boldsymbol{f} + \boldsymbol{u} + \boldsymbol{\xi}, \end{cases} \tag{4}$$

where $\boldsymbol{u} = \boldsymbol{H}\boldsymbol{f}_0$. We may also rewrite this:

$$\begin{cases} \boldsymbol{g} = \boldsymbol{g}_0 + \boldsymbol{u} + \boldsymbol{\varepsilon}, \\ \boldsymbol{g}_0 = \boldsymbol{H}\boldsymbol{f} + \boldsymbol{\xi}, \end{cases} \tag{5}$$

where, this time, $\boldsymbol{u}$ may represent a background in the measurement space. Finally, we can consider

$$\begin{cases} \boldsymbol{g} = \boldsymbol{g}_0 + \boldsymbol{\varepsilon}, \\ \boldsymbol{g}_0 = (\boldsymbol{H} + \delta\boldsymbol{H})\boldsymbol{f} + \boldsymbol{\xi}, \end{cases} \tag{6}$$

where $\delta \boldsymbol{H}$ represents some error in the elements of the forward model, which again, can be written either as in (4) or (5) with $\boldsymbol{u} = \delta \boldsymbol{H} \boldsymbol{f}$ or

$$\begin{cases} \boldsymbol{g} = \boldsymbol{g}_0 + \boldsymbol{\varepsilon}, \\ \boldsymbol{g}_0 = \boldsymbol{H}\boldsymbol{f} + \boldsymbol{u} + \boldsymbol{\xi}, \\ \boldsymbol{u} = \delta \boldsymbol{H}\, \boldsymbol{f} + \boldsymbol{\zeta}. \end{cases} \tag{7}$$

In the following, we are going to consider the different forward models (1)–(7) and give more insights for each case. In particular, we compare the classical optimization algorithms with Bayesian MAP, but also, we enhance the other advantages of the Bayesian approach to handle these models. For some cases, we mention and refer to the real application of these models in Computed Tomography (CT) [1, 7].

2 Forward Model 1

For the case of the forward model (1), if we consider the classical regularization methods, the solution is defined as the optimizer of

$$J(\boldsymbol{f}) = \|\boldsymbol{g} - \boldsymbol{H}\boldsymbol{f}\|_2^2 + \lambda R(\boldsymbol{f}) \tag{8}$$

where the regularization term $R(\boldsymbol{f})$ has been, classically chosen between any of the following:

$$R(\boldsymbol{f}) = \left\{ \|\boldsymbol{f}\|_2^2, \|\boldsymbol{f}\|_1, \|\boldsymbol{f}\|_\beta^\beta, \|\boldsymbol{D}\boldsymbol{f}\|_2^2, \|\boldsymbol{D}\boldsymbol{f}\|_1, \|\boldsymbol{D}\boldsymbol{f}\|_\beta^\beta \right\} \tag{9}$$

where $\boldsymbol{D}$ is a linear sparsifying operator (gradient, Laplacien, Wavelet Transform, etc.). There are also other more general expressions for it.

$$p(\boldsymbol{f}) \propto \sum_j \phi(f_j) \text{ or } p(\boldsymbol{f}) \propto \sum_j \phi(f_j - f_{j-1}) \tag{10}$$

where $\phi(.)$ is in general a potential function. The particular cases of $\phi(f_j) = |f_j|^2$, $\phi(f_j) = |f_j - f_{j-1}|^2$ or $\phi(f_j) = |f_j|$, $\phi(f_j) = |f_j - f_{j-1}|$ are the classical ones.

In the Bayesian framework, the classical Gaussian priors for the forward model 1 is written as follows:

$$\begin{cases} p(\boldsymbol{g}|\boldsymbol{f}, v_\epsilon) = \mathcal{N}(\boldsymbol{g}|\boldsymbol{H}\boldsymbol{f}, v_\epsilon \boldsymbol{I}), \\ p(\boldsymbol{f}|v_f) = \mathcal{N}(\boldsymbol{f}|\boldsymbol{0}, v_f \boldsymbol{I}), \end{cases} \tag{11}$$

which results in $p(\boldsymbol{f}|\boldsymbol{g}, v_\epsilon, v_f) = \mathcal{N}(\boldsymbol{f}|\widehat{\boldsymbol{f}}, \widehat{\boldsymbol{\Sigma}})$ with

$$\begin{cases} \widehat{f} = [H'H + \lambda I]^{-1} H' g, \\ \widehat{\Sigma} = v_\epsilon [H'H + \lambda I]^{-1}, \quad \lambda = \frac{v_\epsilon}{v_f}. \end{cases} \tag{12}$$

Interestingly, in this simple case, $\widehat{f}$ can also be written as the optimizer of:

$$J(f) = \|g - Hf\|_2^2 + \lambda \|f\|_2^2, \tag{13}$$

which makes the link with QR, for which many optimization algorithms, such as gradient-based algorithms have been proposed. One of the simplest one (simple gradient descent) can be written as:

$$f^{(k+1)} = f^{(k)} + \alpha^{(k)} [H'(g - Hf^{(k)}) - \lambda f^{(k)}]. \tag{14}$$

Many other optimization algorithms have been proposed for more efficient and faster implementation of this optimization problem, in particular for great dimensional applications.

Many other prior prior models have also been proposed, in particular, Double Exponential (DE) $p(f) \propto \exp[-\alpha \|f\|_1]$ which makes the link with L_1 regularization criterion: $J(f) = \|g - Hf\|_2^2 + \lambda \|f\|_1$ with $\lambda = \alpha v_\epsilon$.

We may also mention the Student-t and other hierarchical models such as:

$$\begin{cases} g = Hf + \epsilon, \\ f = Dz + \zeta, \end{cases} \tag{15}$$

where z here may represent the coefficients of any linear transformation whose probability law is modeled by any heavy tailed distribution such as DE [8] or Student-t [9].

Coming back to the regularization-based methods, and in particular, for the cases of L_1 or Total Variation (TV) regularization, special purpose algorithms have been developed based on the Augmented Lagragian (AL) [10, 11] and Bregman Duality (BD) [12] which can be summarized as trying to solve the following optimization problems:

- Analysis criterion and AL:

$$\text{minimize } J(f) = \|g - Hf\|_2^2 + \lambda R(Df) \text{ s.t. } Hf = g, \tag{16}$$

for which the solution is obtained as the stationary point of the AL:

$$\mathscr{L}(f, \mu) = \|g - Hf\|_2^2 + \lambda R(Df) + \mu'(Hf - g), \tag{17}$$

which gives the following algorithm:

$$\begin{cases} f^{(k+1)} = f^{(k)} + \alpha_1^{(k)} [2H'(g - Hf^{(k)}) - \lambda D' \nabla R(f^{(k)}) - H'\mu] \\ \mu^{(k+1)} = \mu^{(k)} + \alpha_2^{(k)} (Hf^{(k)} - g). \end{cases} \tag{18}$$

• Synthesis criterion and AL:

$$\min J(\boldsymbol{z}) = \|\boldsymbol{g} - \boldsymbol{H}\boldsymbol{D}\boldsymbol{z}\|_2^2 + \lambda R(\boldsymbol{z}) \text{ s.t. } \boldsymbol{H}\boldsymbol{D}\boldsymbol{z} = \boldsymbol{g}, \tag{19}$$

for which the solution is obtained as the stationary point of the AL:

$$\mathscr{L}(\boldsymbol{z}, \boldsymbol{\mu}) = \|\boldsymbol{g} - \boldsymbol{H}\boldsymbol{D}\boldsymbol{z}\|_2^2 + \lambda R(\boldsymbol{z}) + \boldsymbol{\mu}'(\boldsymbol{H}\boldsymbol{D}\boldsymbol{z} - \boldsymbol{g}), \tag{20}$$

which gives the following algorithm:

$$\begin{cases} \boldsymbol{z}^{(k+1)} = \boldsymbol{z}^{(k)} + \alpha_1^{(k)}[\boldsymbol{D}'\boldsymbol{H}'(\boldsymbol{g} - \boldsymbol{H}\boldsymbol{D}\boldsymbol{z}^{(k)}) - \lambda\nabla R(\boldsymbol{z}^{(k)}) - \boldsymbol{D}'\boldsymbol{H}'\boldsymbol{\mu}] \\ \boldsymbol{\mu}^{(k+1)} = \boldsymbol{\mu}^{(k)} + \alpha_2^{(k)}(\boldsymbol{H}\boldsymbol{D}\boldsymbol{f}^{(k)} - \boldsymbol{g}). \end{cases} \tag{21}$$

At the end of the iterations, we can compute $\widehat{\boldsymbol{f}} = \boldsymbol{D}\widehat{\boldsymbol{z}}$. Some other alternative optimization algorithms based on Bregman iterations are also proposed [10, 11, 13–25].

The main difficulties in these regularization methods are twofold: how to determine λ and how to quantify the remaining uncertainty on the obtained solution. The Bayesian approach gives these possibilities. First, it is possible to include the hyperparameters $\boldsymbol{\theta}$ in the estimation process by looking at the joint posterior:

$$p(\boldsymbol{f}, \boldsymbol{\theta}|g) \propto p(\boldsymbol{g}|\boldsymbol{f}, \boldsymbol{\theta}_1)p(\boldsymbol{f}|\boldsymbol{\theta}_2)p(\boldsymbol{\theta}) \tag{22}$$

with $\boldsymbol{\theta} = (\boldsymbol{\theta}_1, \boldsymbol{\theta}_2)$ [26, 27]. Second, we can use this joint posterior to quantify the uncertainty on the solution. This can be done, for example, by computing the posterior covariance. However, often, the exact computation of the posterior mean and covariance may not be easy or may be too costly. Hopefully, solutions exist. For example, we can use the variational Bayesian Approximation (VBA) methods to approximate $p(\boldsymbol{f}, \boldsymbol{\theta}|g)$ by $q_1(\boldsymbol{f})q_2(\boldsymbol{\theta})$ by choosing appropriate families for q_1 and q_2. For more details, see [2, 28, 29].

3 Forward Model 2

For the case of forward model (2), if we consider the classical regularization methods, the solution is defined as the optimizer of

$$J(\boldsymbol{f}, \boldsymbol{g}_0) = \|\boldsymbol{g} - \boldsymbol{g}_0\|_2^2 + \|\boldsymbol{g}_0 - \boldsymbol{H}\boldsymbol{f}\|_2^2 + \lambda R(\boldsymbol{f}), \tag{23}$$

where, classically, λ has been chosen in an ad hoc way. A simple alternate optimization algorithm can be:

$$\begin{cases} \boldsymbol{f}^{(k+1)} = \boldsymbol{f}^{(k)} + \alpha_1^{(k)}\left[2\boldsymbol{H}'(\boldsymbol{g}_0 - \boldsymbol{H}\boldsymbol{f}^{(k)}) - \lambda\nabla R(\boldsymbol{f}^{(k)})\right], \\ \boldsymbol{g}_0^{(k+1)} = \boldsymbol{g}_0^{(k)} + \alpha_2^{(k)}\left[(\boldsymbol{g} - \boldsymbol{H}\boldsymbol{f}^{(k)}\right], \end{cases} \tag{24}$$

where we can compare it with ADMM like of (18) where g_0 plays the role of $\boldsymbol{\mu}$.

For the particular cases of L_1 or Total Variation (TV) regularization-based methods, special purpose algorithms have been developed, such as ISTA and FISTA [11, 30, 31].

In the Bayesian framework, the classical Gaussian model is written as follows:

$$\begin{cases} p(\boldsymbol{g}|\boldsymbol{g}_0, v_\epsilon) = \mathcal{N}(\boldsymbol{g}|\boldsymbol{g}_0, v_\epsilon \boldsymbol{I}), \\ p(\boldsymbol{g}_0|\boldsymbol{f}, v_\xi) = \mathcal{N}(\boldsymbol{g}_0|\boldsymbol{H}\boldsymbol{f}, v_\xi \boldsymbol{I}), \\ p(\boldsymbol{f}|v_f) = \mathcal{N}(\boldsymbol{f}|\boldsymbol{0}, v_f \boldsymbol{I}), \end{cases} \tag{25}$$

which results to $p(\boldsymbol{f}, \boldsymbol{g}_0|\boldsymbol{g}, v_\epsilon, v_\xi, v_f) \propto \exp[-J(\boldsymbol{f}, \boldsymbol{g}_0)]$ with

$$J(\boldsymbol{f}, \boldsymbol{g}_0) = \frac{1}{2v_\epsilon}\|\boldsymbol{g} - \boldsymbol{g}_0\|_2^2 + \frac{1}{2v_\xi}\|\boldsymbol{g}_0 - \boldsymbol{H}\boldsymbol{f}\|_2^2 + \frac{1}{2v_f}\|\boldsymbol{f}\|_2^2 \tag{26}$$

where we can see the similarity with (23).

Indeed, in this approach we have access to the joint posterior law $p(\boldsymbol{f}, \boldsymbol{g}_0|\boldsymbol{g})$ and we can quantify the uncertainties. We can also estimate v_ϵ, v_f and $\mathbf{r}_\xi$ at the same time with $\boldsymbol{f}$ and $\boldsymbol{g}_0$. For this, we have to assign them appropriate prior laws, for example, conjugate priors (like Inverse Gamma or Generalized Inverse Gaussian) or the Jeffreys prior.

Another extension is to choose a heavy tailed prior for $\boldsymbol{\xi}$ to enforce its sparsity, a property that can be understood if we want to satisfy $\boldsymbol{H}\boldsymbol{f} = \boldsymbol{g}_0$ as much as possible. For this, we may choose the Double Exponential or Student-t distribution, which is considered in this paper. The main interest of Student-t is that we can model it as the marginal of a Normal-Inverse Gamma. In summary:

$$\begin{cases} p(\boldsymbol{g}|\boldsymbol{g}_0, v_\epsilon) = \mathcal{N}(\boldsymbol{g}|\boldsymbol{g}_0, v_\epsilon \boldsymbol{I}), \quad p(v_\epsilon) = \mathcal{IG}(v_\epsilon|\alpha_{\epsilon_0}, \beta_{\epsilon_0}), \\ p(\boldsymbol{g}_0|\boldsymbol{f}, \mathbf{r}_\xi) = \mathcal{N}(\boldsymbol{g}_0|\boldsymbol{H}\boldsymbol{f}, \boldsymbol{V}_\xi), \quad \boldsymbol{V}_\xi = \operatorname{diag}\left[\boldsymbol{v}_\xi\right], \\ p(\boldsymbol{v}_\xi) = \prod_{i=1}^M p(\mathbf{r}_{\xi_i}) = \prod_{i=1}^M \mathcal{IG}(\mathbf{r}_{\xi_i}|\alpha_{\xi_0}, \beta_{\xi_0}), \\ p(\boldsymbol{f}|v_f) = \mathcal{N}(\boldsymbol{f}|\boldsymbol{0}, v_f \boldsymbol{I}), \quad p(v_f) = \mathcal{IG}(v_f|\alpha_{f_0}, \beta_{f_0}). \end{cases} \tag{27}$$

This gives: $p(\boldsymbol{f}, \boldsymbol{g}_0, \boldsymbol{v}_\xi, v_f, v_\epsilon|\boldsymbol{g}) \propto \exp\left[-J(\boldsymbol{f}, \boldsymbol{g}_0, \boldsymbol{v}_\xi, v_f, v_\epsilon)\right]$ with

$$\begin{aligned} J(\boldsymbol{f}, \boldsymbol{g}_0, \boldsymbol{v}_\xi, v_f, v_\epsilon) &= \tfrac{1}{2v_\epsilon}\|\boldsymbol{g} - \boldsymbol{g}_0\|_2^2 + \tfrac{1}{2}\|\boldsymbol{V}_\xi^{-1/2}(\boldsymbol{g}_0 - \boldsymbol{H}\boldsymbol{f})\|_2^2 + \tfrac{1}{2v_f}\|\boldsymbol{f}\|_2^2 \\ &+ (\alpha_{\epsilon_0} + 1)\ln v_\epsilon + \tfrac{\beta_{\epsilon_0}}{v_\epsilon} + (\alpha_{f_0} + 1)\ln v_f + \tfrac{\beta_{f_0}}{v_f} + \sum_{i=1}^M \left[(\alpha_{\xi_0} + 1)\ln v_{\xi_i} + \tfrac{\beta_{\xi_0}}{v_{\xi_i}}\right] \end{aligned} \tag{28}$$

Again, the alternate optimization of J with respect to its arguments results in an ADMM-like iterative algorithm with the advantage of updating the hyperparameters.

4 Forward Model 3

For the case of forward model (3), we give directly the Bayesian framework:

$$\begin{cases} p(\boldsymbol{g}|\boldsymbol{g}_0, v_\epsilon) = \mathcal{N}(\boldsymbol{g}|\boldsymbol{g}_0, v_\epsilon \boldsymbol{I}), \;\; p(v_\epsilon) = \mathcal{IG}(v_\epsilon|\alpha_{\epsilon_0}, \beta_{\epsilon_0}), \\ p(\boldsymbol{g}_0|\boldsymbol{f}, \boldsymbol{f}_0, \boldsymbol{v}_\xi) = \mathcal{N}(\boldsymbol{g}_0|\boldsymbol{H}(\boldsymbol{f}+\boldsymbol{f}_0), \boldsymbol{V}_\xi), \quad \boldsymbol{V}_\xi = \text{diag}\left[\boldsymbol{v}_\xi\right], \\ p(\boldsymbol{v}_\xi) = \prod_{i=1}^M p(\mathbf{r}_{\xi_i}) = \prod_{i=1}^M \mathcal{IG}(\mathbf{r}_{\xi_i}|\alpha_{\xi_0}, \beta_{\xi_0}), \\ p(\boldsymbol{f}|\boldsymbol{v}_f) = \mathcal{N}(\boldsymbol{f}|\boldsymbol{0}, \boldsymbol{V}_f), \quad \boldsymbol{V}_f = \text{diag}\left[\boldsymbol{v}_f\right] \\ p(\boldsymbol{v}_f) = \prod_{j=1}^N p(\mathbf{r}_{f_j}) = \prod_{j=1}^N \mathcal{IG}(v_{f_j}|\alpha_{f_0}, \beta_{f_0}), \\ p(\boldsymbol{f}_0) = \mathcal{N}(\boldsymbol{f}_0|\boldsymbol{0}, v_u \boldsymbol{I}), \end{cases} \tag{29}$$

which results in: $p(\boldsymbol{f}, \boldsymbol{g}_0, \boldsymbol{f}_0, v_\epsilon, \boldsymbol{v}_\xi, \boldsymbol{v}_f|\boldsymbol{g}) \propto \exp\left[-J(\boldsymbol{f}, \boldsymbol{g}_0, \boldsymbol{f}_0, v_\epsilon, \boldsymbol{v}_\xi, \boldsymbol{v}_f)\right]$ with

$$\begin{aligned} & J(\boldsymbol{f}, \boldsymbol{g}_0, \boldsymbol{f}_0, v_\epsilon, \boldsymbol{v}_\xi, \boldsymbol{v}_f, \mathbf{r}_u) = \\ & \tfrac{1}{2v_\epsilon}\|\boldsymbol{g}-\boldsymbol{g}_0\|_2^2 + \tfrac{1}{2}\|\boldsymbol{V}_\xi^{-1/2}(\boldsymbol{g}_0 - \boldsymbol{H}(\boldsymbol{f}+\boldsymbol{f}_0))\|_2^2 \\ & + \tfrac{1}{2}\|\boldsymbol{V}_f^{-1/2}\boldsymbol{f}\|_2^2 + \tfrac{1}{2v_u}\|\boldsymbol{f}_0\|_2^2 + (\alpha_{\epsilon_0}+1)\ln v_\epsilon + \tfrac{\beta_{\epsilon_0}}{v_\epsilon} \\ & + \sum_{i=1}^M \left[(\alpha_{\xi_0}+1)\ln v_{\xi_i} + \tfrac{\beta_{\xi_0}}{v_{\xi_i}}\right] + \sum_{j=1}^N \left[(\alpha_{f_0}+1)\ln \mathbf{r}_{f_j} + \tfrac{\beta_{f_0}}{\mathbf{r}_{f_j}}\right] \end{aligned} \tag{30}$$

Alternate optimization of this criterion with respect to its arguments results in an iterative algorithm where, again, we can see a certain relation with ADMM type algorithms. The main advantage here is that the hyperparameters are also estimated. However, it is not easy to study its convergency. In practical situations, we could observe local convergency to solutions with desired properties.

5 Forward Models 4 and 5

For the cases of forward models (4) and (5), we can always choose appropriate prior laws and obtain the expressions of the posterior law and try to obtain the Joint Maximum A Posteriori (JMAP) solution by alternate optimization or any other algorithms.

The case of the models 4 and 5 are very similar. A particular use of the forward model 5 with an extra relation $\boldsymbol{f} = \boldsymbol{D}\boldsymbol{z}$ where $\boldsymbol{D}$ represents a wavelet transform (and $\boldsymbol{z}$ its representation in that domain) with application in Computed Tomography (CT) is in preparation in [9]. The main idea again is to use a sparsity enforcing prior for $\boldsymbol{z}$. In summary:

$$\begin{cases} p(\boldsymbol{g}|\boldsymbol{g}_0, \boldsymbol{u}, v_\epsilon) = \mathcal{N}(\boldsymbol{g}|\boldsymbol{g}_0 + \boldsymbol{u}, v_\epsilon \boldsymbol{I}), \\ p(v_\epsilon) = \mathcal{IG}(v_\epsilon|\alpha_{\epsilon_0}, \beta_{\epsilon_0}), \\ p(\boldsymbol{g}_0|\boldsymbol{f}, \boldsymbol{v}_\xi) = \mathcal{N}(\boldsymbol{g}_0|\boldsymbol{H}\boldsymbol{f}, \boldsymbol{V}_\xi), \boldsymbol{V}_\xi = \mathrm{diag}\left[\boldsymbol{v}_\xi\right], \\ p(\boldsymbol{v}_\xi) = \prod_{i=1}^M p(\mathbf{r}_{\xi_i}) = \prod_{i=1}^M \mathcal{IG}(\mathbf{r}_{\xi_i}|\alpha_{\xi_0}, \beta_{\xi_0}), \\ \boldsymbol{f} = \boldsymbol{D}\boldsymbol{z} \quad \text{and} \quad \boldsymbol{z} = \boldsymbol{D}'\boldsymbol{f}, \\ p(\boldsymbol{z}|\boldsymbol{v}_z) = \mathcal{N}(\boldsymbol{z}|\boldsymbol{0}, \boldsymbol{V}_z), \quad \boldsymbol{V}_z = \mathrm{diag}\left[\boldsymbol{v}_z\right], \\ p(\boldsymbol{v}_z) = \prod_{j=1}^N p(\mathbf{r}_{z_j}) = \prod_{j=1}^N \mathcal{IG}(\mathbf{r}_{z_j}|\alpha_{z_0}, \beta_{z_0}), \\ p(\boldsymbol{u}|v_u) = \mathcal{N}(\boldsymbol{u}|\boldsymbol{0}, v_u \boldsymbol{I}), \\ p(v_u) = \mathcal{IG}(v_u|\alpha_{u0}, \beta_{u0}). \end{cases} \tag{31}$$

This gives: $p(\boldsymbol{z}, \boldsymbol{g}_0, \boldsymbol{u}, v_\epsilon, \boldsymbol{v}_\xi, \boldsymbol{v}_z, v_u|\boldsymbol{g}) \propto \exp\left[-J(\boldsymbol{z}, \boldsymbol{g}_0, \boldsymbol{u}, v_\epsilon, \boldsymbol{v}_\xi, \boldsymbol{v}_z, v_u)\right]$ with

$$\begin{aligned} &J(\boldsymbol{z}, \boldsymbol{g}_0, \boldsymbol{u}, v_\epsilon, \boldsymbol{v}_\xi, \boldsymbol{v}_z, v_u) = \tfrac{1}{2v_\epsilon}\|\boldsymbol{g} - \boldsymbol{g}_0 - \boldsymbol{u}\|_2^2 + \tfrac{1}{2}\|\boldsymbol{V}_\xi^{-1/2}(\boldsymbol{g}_0 - \boldsymbol{H}\boldsymbol{D}\boldsymbol{z})\|_2^2 \\ &+\tfrac{1}{2}\|\boldsymbol{V}_z^{-1/2}\boldsymbol{z}\|_2^2 + \tfrac{1}{2v_u}\|\boldsymbol{u}\|_2^2 + (\alpha_{\epsilon 0} + 1)\ln v_\epsilon + \tfrac{\beta_{\epsilon 0}}{v_\epsilon} \\ &+\textstyle\sum_{i=1}^M \left[(\alpha_{\xi_0} + 1)\ln v_{\xi_i} + \tfrac{\beta_{\xi 0}}{\mathbf{r}_{z_j}}\right] + \sum_{j=1}^N \left[(\alpha_{z_0} + 1)\ln v_{z_j} + \tfrac{\beta_{z0}}{\mathbf{r}_{z_j}}\right] + (\alpha_{u0} + 1)\ln v_u + \tfrac{\beta_{u0}}{v_u} \end{aligned} \tag{32}$$

This time again, an alternate optimization algorithm can be used to obtain $\widehat{\boldsymbol{z}}$ from which we can obtain $\widehat{\boldsymbol{f}} = \boldsymbol{D}'\widehat{\boldsymbol{z}}$.

6 Forward Models 6 and 7

For the cases of forward models (6) and (7) too, we present directly the Bayesian framework. In particular, for the forward model (6):

$$\begin{cases} p(\boldsymbol{g}|\boldsymbol{g}_0, v_\epsilon) = \mathcal{N}(\boldsymbol{g}|\boldsymbol{g}_0, v_\epsilon \boldsymbol{I}), \\ p(v_\epsilon) = \mathcal{IG}(v_\epsilon|\alpha_{\epsilon_0}, \beta_{\epsilon_0}), \\ p(\boldsymbol{g}_0|\boldsymbol{f}, \delta\boldsymbol{H}_0, \boldsymbol{v}_\xi) = \mathcal{N}(\boldsymbol{g}_0|(\boldsymbol{H} + \delta\boldsymbol{H}_0)\boldsymbol{f}, \boldsymbol{V}_\xi), \quad \boldsymbol{V}_\xi = \mathrm{diag}\left[\boldsymbol{v}_\xi\right], \\ p(\boldsymbol{v}_\xi) = \prod_{i=1}^M p(\mathbf{r}_{\xi_i}) = \prod_{i=1}^M \mathcal{IG}(\mathbf{r}_{\xi_i}|\alpha_{\xi_0}, \beta_{\xi_0}), \\ p(\boldsymbol{f}|\boldsymbol{v}_f) = \mathcal{N}(\boldsymbol{f}|\boldsymbol{0}, \boldsymbol{V}_f), \quad \boldsymbol{V}_f = \mathrm{diag}\left[\boldsymbol{v}_f\right] \\ p(\boldsymbol{v}_f) = \prod_{j=1}^N p(\mathbf{r}_{f_j}) = \prod_{j=1}^N \mathcal{IG}(\mathbf{r}_{f_j}|\alpha_{f_0}, \beta_{f_0}), \\ p(\delta\boldsymbol{H}_0|v_h) = \mathcal{N}(\delta\boldsymbol{H}_0|\boldsymbol{0}, v_h \boldsymbol{I}). \end{cases} \tag{33}$$

which results to: $p(\boldsymbol{f}, \boldsymbol{g}_0, \delta\boldsymbol{H}_0, v_\epsilon, \boldsymbol{v}_\xi, \boldsymbol{v}_f|\boldsymbol{g}, v_h) \propto \exp\left[-J(\boldsymbol{f}, \boldsymbol{g}_0, \delta\boldsymbol{H}_0, v_\epsilon, \boldsymbol{v}_\xi, \boldsymbol{v}_f)\right]$ with

$$\begin{aligned} &J(\boldsymbol{f}, \boldsymbol{g}_0, \delta\boldsymbol{H}_0, v_\epsilon, \boldsymbol{v}_\xi, \boldsymbol{v}_f) = \tfrac{1}{2v_\epsilon}\|\boldsymbol{g} - \boldsymbol{g}_0\|_2^2 + \tfrac{1}{2}\|\boldsymbol{V}_\xi^{-1/2}(\boldsymbol{g}_0 - (\boldsymbol{H} + \delta\boldsymbol{H}_0)\boldsymbol{f})\|_2^2 \\ &+ \tfrac{1}{2}\|\boldsymbol{V}_f^{-1/2}\boldsymbol{f}\|_2^2 + \tfrac{1}{2v_h}\|\delta\boldsymbol{H}_0\|_2^2 + (\alpha_{\epsilon_0} + 1)\ln v_\epsilon + \tfrac{\beta_{\epsilon_0}}{v_\epsilon} \\ &+ \textstyle\sum_{i=1}^M \left[(\alpha_{\xi_0} + 1)\ln v_{\xi_i} + \tfrac{\beta_{\xi_0}}{v_{\xi_i}}\right] + \sum_{j=1}^N \left[(\alpha_{f_0} + 1)\ln \mathbf{r}_{f_j} + \tfrac{\beta_{f_0}}{\mathbf{r}_{f_j}}\right] \end{aligned} \tag{34}$$

Here, the expression $p(\delta \boldsymbol{H}_0) = \mathcal{N}(\delta \boldsymbol{H}_0 | \boldsymbol{0}, v_h \boldsymbol{I})$ has to be understood as the product of

$$p([\delta \boldsymbol{H}_0]_{ij}) = \mathcal{N}([\delta \boldsymbol{H}_0]_{ij} | 0, v_h). \tag{35}$$

The same kind of equations can be written for the case 7.

7 Conclusions

In this work, we extended the classical single additive noise, linear forward model to account for different uncertainties by variable splitting techniques. In the first step, we splitted the error term in two parts to distinguish between observation noise and forward model uncertainty. Then, we accounted for unknown background in the input or at the output. For any of these forward models, we examined in parallel the deterministic regularization and the Bayesian MAP approach focusing more on the second. As a by-product, we could see the links between an alternate optimization algorithm for the MAP estimator and the optimization algorithms, such as Alternate Descent Minimization Maximization (ADMM) or ISTA or its fast version FISTA for the particular case of double exponential prior. In a second step, we extended this to also account for the input or output background modeling errors and the errors on the elements of the linear forward model. This last one can occur in Blind deconvolution problems.

The main advantage of the Bayesian approach is giving the tools to go further by estimating the hyperparameters and being able to quantify the uncertainties of the solutions of the inversion problems. These two points are often crucial in real applications. However, one may be careful about choosing appropriate hyperparameters, the order of the optimization, and the convergency of the algorithms. As the problems are in general very ill-posed, we have to carefully choose the priors to guarantee, at least, the local convergency of the algorithms. But this problem is a very general task for any inverse problem.

One main question is still open: finding the reason for better performances of these variable splitting algorithms. This goes in the opposite direction of the regularization idea of restricting the space of the solution to obtain a regularized solution. As a final conclusion, we may try to answer the following question: *Is it better to restrict the space and define a criterion with a global minimum or, in the opposite, increase the dimension of the unknown space, define a criterion which may have many minima and looking for a local minimum of it?*

References

1. Wang, G., Schultz, L., Qi, J.: IEEE Trans. Nucl. Sci. **56**(4), 2480 (2009)
2. Mohammad-Djafari, A. In: International Workshop on Systems, Signal Processing and Applications (WOSSPA 2013) Proceedings, (2013) [Tutorial]

3. Dumitru, M., Mohammad-Djafari, A., Sain, B.S.: EURASIP Journal on Bioinformatics and Systems Biology, vol. 3. Springer, Berlin (2016). https://doi.org/10.1186/s13637-015-0033-6
4. Arhab, S., Ayasso, H., Duchêne, B., Mohammad-Djafari, A.: In: 4th International Workshop on New Computational Methods for Inverse Problems, pp. 6. Cachan, France (2014). http://hal.univ-grenoble-alpes.fr/hal-01011188
5. Afonso, M.V., Bioucas-Dias, J.M., Figueiredo, M.A.T.: IEEE Trans. Image Process. **19**(9), 2345 (2010). https://doi.org/10.1109/TIP.2010.2047910
6. Marvasti, F., Mohammad-Djafari, A., Chambers, J.: Special issue on sparse signal processing. EURASIP J. Adv. Signal Process (2012). 10.1186/1687. http://asp.eurasipjournals.com/content/pdf/1687-6180-2012-90.pdf
7. Mohammad-Djafari, A.: Special issue on sparse signal processing. EURASIP J. Adv. Signal Process. **2012**:52 (2012). http://asp.eurasipjournals.com/content/pdf/1687-6180-2012-52.pdf
8. Dobigeon, N., Hero, A.O., Tourneret, J.Y.: IEEE Trans. Image Process. **18**(9), 2059 (2009)
9. Wang, L., Mohammad-Djafari, A., Gac, N., Dumitru, M.: In: 2016 IEEE International Conference on Acoustics, Speech and Signal Processing (ICASSP), pp. 883–887. IEEE (2016)
10. Weller, D.S., Ramani, S., Fessler, J.A.: IEEE Trans. Med. Imagin. **33**(2), 351 (2014). https://doi.org/10.1109/TMI.2013.2285046
11. He, C., Hu, C., Zhang, W., Shi, B.: IEEE Transactions on Image Process. **23**(12), 4954 (2014). https://doi.org/10.1109/TIP.2014.2360133
12. Duan, J., Liu, Y., Zhang, L.: IEEE Signal Process. Lett. **20**(8), 831 (2013). https://doi.org/10.1109/LSP.2013.2268206
13. Elad, M., Milanfar, P., Rubinstein, R.: Inverse Probl. **23**(3), 947 (2007)
14. Zhao, N., Wei, Q., Basarab, A., Dobigeon, N., Kouamé, D., Tourneret, J.Y.: IEEE Trans. Image Process. **25**(8), 3683 (2016). https://doi.org/10.1109/TIP.2016.2567075
15. Chun, I.Y., Adcock, B., Talavage, T.M.: IEEE Trans. Med. Imaging **35**(1), 354 (2016). https://doi.org/10.1109/TMI.2015.2474383
16. Chen, S., Liu, H., Hu, Z., Zhang, H., Shi, P., Chen, Y.: IEEE Trans. Biomed. Eng. **62**(7), 1784 (2015). https://doi.org/10.1109/TBME.2015.2404296
17. Zhang, H., Wu, C., Zhang, J., Deng, J.: IEEE Trans. Vis. Comput. Graph. **21**(7), 873 (2015). https://doi.org/10.1109/TVCG.2015.2398432
18. Muckley, M.J., Noll, D.C., Fessler, J.A.: IEEE Trans. Med. Imaging **34**(2), 578 (2015). https://doi.org/10.1109/TMI.2014.2363034
19. Allison, M.J., Ramani, S., Fessler, J.A.: IEEE Trans. Med. Imaging **32**(3), 556 (2013). https://doi.org/10.1109/TMI.2012.2229711
20. Iordache, M.D., Bioucas-Dias, J.M., Plaza, A.: IEEE Trans. Geosci. Remote Sens. **50**(11), 4484 (2012). https://doi.org/10.1109/TGRS.2012.2191590
21. Xie, S., Rahardja, S.: IEEE Trans. Image Process. **21**(11), 4557 (2012). https://doi.org/10.1109/TIP.2012.2206043
22. Ramani, S., Fessler, J.A.: IEEE Trans. Med. Imaging **30**(3), 694 (2011). https://doi.org/10.1109/TMI.2010.2093536
23. Chen, D., Cheng, L.: IET Image Process. **4**(5), 353 (2010). https://doi.org/10.1049/iet-ipr.2009.0186
24. Chun, S.Y., Dewaraja, Y.K., Fessler, J.A.: IEEE Trans. Med. Imaging **33**(10), 1960 (2014). https://doi.org/10.1109/TMI.2014.2328660
25. Sroubek, F., Milanfar, P.: IEEE Trans. Image Process. **21**(4), 1687 (2012). https://doi.org/10.1109/TIP.2011.2175740
26. Mohammad-Djafari, A.: In: 1993 IEEE International Conference on Acoustics, Speech, and Signal Processing, ICASSP-93., vol. 5 (1993), vol. 5, pp. 495–498. https://doi.org/10.1109/ICASSP.1993.319857
27. Mohammad-Djafari, A.: In: Maximum Entropy and Bayesian Methods, Kluwer Academic Publisher, (1996), pp. 135–143. http://djafari.free.fr/pdf/
28. Mohammad-Djafari, A.: Comput. J. **52**(1), 126 (2009)
29. Chu, N., Mohammad-Djafari, A., Gac, N., Picheral, J.: In: 2014 13th Workshop on Information Optics (WIO) (2014), pp. 1–4. https://doi.org/10.1109/WIO.2014.6933297

30. Charbonnier, P., Blanc-Féraud, L., Aubert, G., Barlaud, M.: IEEE Trans. Image Process. **6**(2), 298 (1997)
31. Chan, T.F., Golub, G.H., Mulet, P.: SIAM J. Sci. Comput. **20**(6), 1964 (1999)

Prior Shift Using the Ratio Estimator

Afonso Vaz, Rafael Izbicki and Rafael Bassi Stern

Abstract Several machine learning applications use classifiers as a way of quantifying the prevalence of positive class labels in a target dataset, a task named quantification. For instance, a naive a way of determining what proportion of people like a given product with no labeled reviews is to (i) train a classifier based on the Google Shopping reviews to predict whether a user likes a product given its review, and then (ii) apply this classifier to Facebook/Google+ posts about that product. It is well known that such a two-step approach, named Classify and Count, fails because of dataset shift, and thus, several improvements have been recently proposed under an assumption named prior shift. Unfortunately, these methods only explore the relationship between the covariates and the response via classifiers. Moreover, the literature lacks in the theoretical foundation to improve these techniques. We propose a new family of estimators named Ratio Estimator which is able to explore the relationship between the cov ariates and the response using any function $g : \mathscr{X} \rightarrow \mathbb{R}$ and not only classifiers. We show that for some choices of g, our estimator matches standard estimators used in the literature. We also explore alternative ways of constructing functions g that lead to estimators with good performance, and compare them using real datasets. Finally, we provide a theoretical analysis of the method.

Keywords Quantification · Prior shift · Machine learning

A. Vaz (✉) · R. Izbicki · R. B. Stern
Federal University of São Carlos, Rod. Washington Luís km 235,
310 São Carlos, SP, Brazil
e-mail: afonsofvaz@gmail.com

R. Izbicki
e-mail: rafaelizbicki@gmail.com

R. B. Stern
e-mail: rbstern@gmail.com

A. Polpo et al. (eds.), *Bayesian Inference and Maximum Entropy Methods in Science and Engineering*, Springer Proceedings in Mathematics & Statistics 239, https://doi.org/10.1007/978-3-319-91143-4_3

1 Introduction

In many statistical learning applications, we use classifiers as a way of estimating the proportion (or distribution, or prevalence) of the positive labels in a target sample where we do not observe the labels. This task is named quantification and its definition is usually attributed to Forman [1]. For instance, one company may be interested in estimating the proportion of positive reviews in a page such as Facebook or Twitter about a given product but without to label it because is very expensive. A naive (and often used) approach to deal with this problem is: (i) Estimate a classifier using a labeled training sample such as Google Shopping reviews, (ii) apply this classifier to Facebook or Twitter reviews about the target product, and (iii) use the positive classification number to estimating the proportion. This approach is called classify and count. Classify and count may lead to wrong results because it does not consider that the distribution in the training and target samples may be substantially different. This fact is known as dataset shift (or dataset drift) [2, 3]. In order to deal with this problem, [1] proposes an adjustment to the classify and count methodology motivated by the prior probability shift assumption [4]. Although such adjustment improves the estimates, this estimator only explores the relationship between the covariates and the response via classifiers, and there is almost no theoretical study in order to find its properties. Here, we propose a new method, named Ratio Estimator which is a generalization of the method from [1]. We study its theoretical properties (and hence, the theoretical properties from [1]), explore a version of this estimator based on the Reproducing Kernel Hilbert Spaces, and compare all approaches real datasets.

In Sect. 2, we present the notation of the paper and formally state the goal of quantification methods. In Sect. 3, we review the classify and count method and prove it is inconsistent. We also introduce the ratio estimator and study its theoretical properties. In Sect. 4, we present applications to four datasets. Our concluding remarks are shown in Sect. 5. Additional proofs may be found in the supplementary material at www.small.ufscar.br/priorShift, and codes to perform the experiments can be found at https://github.com/afonsofvaz/PriorShiftQuantification.

2 Setting and Goals

Consider two different populations

$$(Y^{tr}, \mathbf{X}^{tr}), (Y^{tg}, \mathbf{X}^{tg}) \in \{0, 1\} \times \mathscr{X}$$

which we call "training" and "target" population, respectively. Let $(Y_i^{tr}, \mathbf{X}_i^{tr})$, $i = 1, \ldots, n_{tr}$, be an iid sample from $(Y^{tr}, \mathbf{X}^{tr})$, which we call "training sample", and $(Y_i^{tg}, \mathbf{X}_i^{tg})$, $i = 1, \ldots, n_{tg}$, be an iid sample from $(Y^{tg}, \mathbf{X}^{tg})$, which we call "target sample". Our goal is to estimate the probability of the positive class in the target population, which we denote by $\theta^{tg} := \mathbf{P}(Y^{tg} = 1)$ without having access to $Y_1^{tg}, \ldots, Y_{n_{tg}}^{tg}$. We also define $\theta^{tr} := \mathbf{P}(Y^{tr} = 1)$.

3 Quantification Methods

3.1 The Classify and Count Estimator (CCE)

We start by reviewing the Classify and Count estimator, which is often used in practice.

Definition 3.1 [*CCE*] Let $f : \mathscr{X} \longrightarrow \{0, 1\}$ be a classifier trained using a training sample. Given a target sample, the classify and count estimator is defined by

$$\widehat{\theta}_{cc}^{tg} := \frac{\sum_{i=1}^{n_{tg}} \mathbb{i}_P(i)}{n_{tg}} \tag{1}$$

where $P = \left\{i : f(\mathbf{X}_i^{tg}) = 1\right\}$.

In words, CCE consists in (i) training a classifier using a training sample, (ii) applying it to the target sample and, (iii) counting the number of positives classifications in the target set. Unfortunatelly, although intuitive, CCE may be not consistent because classifiers typically also make mistakes. We prove this in the next section.

3.1.1 Estimator Properties

Our formulation for CCE (Eq. (1), allow us to study its properties, which are presented in the following Theorem.

Theorem 1 *Consider the CCE defined in Eq. 1. Then, if* $\mathbf{P}\left(f(\mathbf{X}^{tg}) = 1\right) \neq \theta^{tg}$, *the estimator is not consistent for* θ^{tg} *in the sense that* $\lim_{n_{tg} \to \infty} \widehat{\theta}_{cc}^{tg} \neq \theta^{tg}$ *even if there is an infinite amount of training samples.*

Proof Note that we may rewrite the sums of the indicators variables involvided in Eq. 1 as $\sum_{i=1}^{n_{tg}} \mathbb{i}_P(i) \sim \text{Binomial}\left(n_{tg}, \mathbf{P}\left(f(\mathbf{X}^{tg}) = 1\right)\right)$. Therefore,

$$\mathbf{E}\left[\widehat{\theta}_{cc}^{tg}\right] = \mathbf{P}\left(f(\mathbf{X}^{tg}) = 1\right)$$

It follows from the law of large numbers that $\lim_{n_{tg} \to \infty} \widehat{\theta}_{cc}^{tg} = \mathbf{P}\left(f(\mathbf{X}^{tg}) = 1\right)$ almost surely. Finally, notice that this holds even if f is the Bayes classifier. Thus, the conclusion remains true even if there is an infinite amount of training samples. □

3.2 The Ratio Estimator (RE)

Now, we present our approach to the quantification problem. As in the case of the estimator from [5], the method relies on the prior shift assumption:

Assumption 1

$$\mathbf{X}^{tr}|Y^{tr} \stackrel{\mathscr{D}}{=} \mathbf{X}^{tg}|Y^{tg} \tag{2}$$

where "$\stackrel{\mathscr{D}}{=}$" denotes equality in distribution.

Let $g : \mathscr{X} \longrightarrow \mathbb{R}$ be any real function of the covariates. Moreover, let f_{tr} and f_{tg} be the probability density function of $\mathbf{X}^{tr}$ and $\mathbf{X}^{tg}$ respectively. Then, by the total probability theorem, it holds that under Assumption 1,

$$f_{tg}(\mathbf{x}) = f_{tg}(\mathbf{x}|Y^{tg} = 1)\mathbf{P}(Y^{tg} = 1) + f_{tg}(\mathbf{x}|Y^{tg} = 0)\mathbf{P}(Y^{tg} = 0) \ \forall\, \mathbf{x} \in \mathscr{X}. \tag{3}$$

Multiplying both sides of Eq. 3 by $g(\mathbf{x})$ and integrating them with respect to $\mathbf{x}$, it follows that

$$\int_{\mathscr{X}} g(\mathbf{x}) f_{tg}(\mathbf{x}) d\mathbf{x} = \sum_{j \in \{0,1\}} \mathbf{P}(Y^{tg} = j) \int_{\mathscr{X}} g(\mathbf{x}) f_{tg}(\mathbf{x}|Y^{tg} = j) d\mathbf{x}$$

and so

$$\mathbf{E}\left[g(\mathbf{X}^{tr})\right] = \theta^{tg}\mathbf{E}\left[g(\mathbf{X}^{tg})|Y^{tg} = 1\right] + (1 - \theta^{tg})\mathbf{E}\left[g(\mathbf{X}^{tg})|Y^{tg} = 0\right]. \tag{4}$$

We conclude that

$$\begin{aligned} \theta^{tg} &= \frac{\mathbf{E}[g(\mathbf{X}^{tg})] - \mathbf{E}[g(\mathbf{X}^{tg})|Y^{tg} = 0]}{\mathbf{E}[g(\mathbf{X}^{tg})|Y^{tg} = 1] - \mathbf{E}[g(\mathbf{X}^{tg})|Y^{tg} = 0]} \\ &\stackrel{A1}{=} \frac{\mathbf{E}[g(\mathbf{X}^{tg})] - \mathbf{E}[g(\mathbf{X}^{tr})|Y^{tr} = 0]}{\mathbf{E}[g(\mathbf{X}^{tr})|Y^{tr} = 1] - \mathbf{E}[g(\mathbf{X}^{tr})|Y^{tr} = 0]} \end{aligned} \tag{5}$$

Our estimator for θ^{tg} consists in replacing the quantities involved in Eq. 5 by their respective sample estimates:

Definition 3.2 [*RE*] Let

$$\widehat{\mu}^{tg} = \frac{1}{n_{tg}} \sum_{i=1}^{n_{tg}} g(\mathbf{X}_i^{tg}) \text{ and } \widehat{\mu}_j^{tr} = \frac{1}{n_{tr}^j} \sum_{i=1}^{n_{tr}} \mathbb{i}_{A_j}(i) g(\mathbf{X}_i^{tr})$$

where

$$A_j = \{i : Y_i^{tr} = j\} \text{ and } n_{tr}^j = \sum_{i=1}^{n_{tr}} \mathbb{i}_{A_j}(i)$$

$i = 1, 2, \ldots, n_{tr}$ and $j = 0, 1$. The ratio estimator (given a function g) is defined by

$$\widehat{\theta}_R^{tg} = \frac{\widehat{\mu}^{tg} - \widehat{\mu}_0^{tr}}{\widehat{\mu}_1^{tr} - \widehat{\mu}_0^{tr}}.$$

Note that if f is a classifier estimated using a training sample and $g(\mathbf{x}) = \mathrm{i}(f(\mathbf{x}) = 1)$, then $\widehat{\theta}_R^{tg}$ coincides with the Adjusted Classify and Count Estimator (ACCE) proposed by Forman [5].

3.2.1 Estimator Properties

Lemma 1 *Let* $\mu_j^{tr} := \mathbf{E}\left[g(\mathbf{X}^{tr})|Y^{tr} = j\right]$ *and* $\mu_j^{tg} := \mathbf{E}\left[g(\mathbf{X}^{tg})|Y^{tg} = j\right]$, $\xi_j^{tr} := \mathbf{Var}\left[g(\mathbf{X}^{tr})|Y^{tr} = j\right]$ *and* $\xi_j^{tg} := \mathbf{Var}\left[g(\mathbf{X}^{tg})|Y^{tg} = j\right]$ *for* $j = 0, 1$. *Under Assumption 1, it holds that*

$$\mathbf{E}[g(\mathbf{X}^{tg})] = \theta^{tg}\mu_1^{tr} + (1 - \theta^{tg})\mu_0^{tr}$$

and

$$\mathbf{Var}[g(\mathbf{X}^{tg})] = (\mu_1^{tr} - \mu_1^{tr})^2\theta^{tg}(1 - \theta^{tg}) + \xi_1^{tr}\theta^{tg} + \xi_0^{tr}(1 - \theta^{tg}).$$

Proof By the conditional expectation properties we have that

$$\begin{aligned}\mathbf{E}[g(\mathbf{X}^{tg})] &= \mathbf{E}\left[\mathbf{E}[g(\mathbf{X}^{tg})|Y^{tg}]\right] \overset{A1}{=} \mathbf{E}\left[\mathbf{E}[g(\mathbf{X}^{tr})|Y^{tr}]\right] \\ &= \mathbf{E}[g(\mathbf{X}^{tr})|Y^{tr} = 1]\theta^{tg} + \mathbf{E}[g(\mathbf{X}^{tr})|Y^{tr} = 0](1 - \theta^{tg}) \\ &= \theta^{tg}\mu_1^{tr} + (1 - \theta^{tg})\mu_0^{tr}.\end{aligned}$$

Similarly, by conditional variance properties,

$$\begin{aligned}\mathbf{Var}[g(\mathbf{X}^{tg})] &= \mathbf{Var}[\mathbf{E}[g(\mathbf{X}^{tg})|Y^{tg}]] + \mathbf{E}[\mathbf{Var}[g(\mathbf{X}^{tg})|Y^{tg}]] \\ &\overset{A1}{=} \mathbf{Var}[\mathbf{E}[g(\mathbf{X}^{tr})|Y^{tr}]] + \mathbf{E}[\mathbf{Var}[g(\mathbf{X}^{tr})|Y^{tr}]] \\ &= (\mu_1^{tr} - \mu_0^{tr})^2\theta^{tg}(1 - \theta^{tg}) + \xi_1^{tr}\theta^{tg} + \xi_0^{tr}(1 - \theta^{tg}).\end{aligned}$$

□

Lemma 2 *Under Assumption 1 and using the notation from Lemma 1, it holds that*

$$\mathbf{E}\left[\widehat{\mu}_j^{tr}\right] = \mu_j^{tr};$$

$$\mathbf{Var}\left[\widehat{\mu}_0^{tr}\right] \approx \frac{\xi_0^{tr}}{n_{tr}(1 - \theta^{tr})} \quad \textit{and} \quad \mathbf{Var}\left[\widehat{\mu}_1^{tr}\right] \approx \frac{\xi_1^{tr}}{n_{tr}\theta^{tr}};$$

$$\mathbf{Cov}\left[\widehat{\mu}_0^{tr}, \widehat{\mu}_1^{tr}\right] = 0;$$

$$\mathbf{Cov}\left[\widehat{\mu}^{tg} - \widehat{\mu}_0^{tr}, \widehat{\mu}_1^{tr} - \widehat{\mu}_0^{tr}\right] = \mathbf{Var}\left[\widehat{\mu}_0^{tr}\right];$$

$$\mathbf{Var}\left[\widehat{\mu}_1^{tr} - \widehat{\mu}_0^{tr}\right] \approx \frac{\xi_0^{tr}}{n_{tr}(1-\theta^{tr})} + \frac{\xi_1^{tr}}{n_{tr}\theta^{tr}}.$$

The proof of the lemma 2 may be found in the supplementary material.

Theorem 2 *The RE is approximately unbiased for θ^{tg}, that is*

$$\mathbf{E}\left[\widehat{\theta}_R^{tg}\right] \approx \theta^{tg}.$$

Proof Using the results found in Lemma 2 and by the delta method [6], the expectation of the ratio estimator may be approximed as

$$\begin{aligned}
\mathbf{E}\left[\widehat{\theta}_R^{tg}\right] &= \mathbf{E}\left[\frac{\widehat{\mu}^{tg} - \widehat{\mu}_0^{tr}}{\widehat{\mu}_1^{tr} - \widehat{\mu}_0^{tr}}\right] \approx \frac{\mathbf{E}\left[\widehat{\mu}^{tg} - \widehat{\mu}_1^{tr}\right]}{\mathbf{E}\left[\widehat{\mu}_1^{tr} - \widehat{\mu}_0^{tr}\right]} = \frac{\mathbf{E}\left[\widehat{\mu}^{tg}\right] - \mathbf{E}\left[\widehat{\mu}_0^{tr}\right]}{\mathbf{E}\left[\widehat{\mu}^{tr}\right] - \mathbf{E}\left[\widehat{\mu}_0^{tr}\right]} = \frac{\mathbf{E}[g(\mathbf{X}^{tg})] - \mu_0^{tr}}{\mu_1^{tr} - \mu_0^{tr}} \\
&= \frac{\mathbf{E}[g(\mathbf{X}^{tg})|Y^{tg}=0]\mathbf{P}(Y^{tg}=0) + \mathbf{E}[g(\mathbf{X}^{tg})|Y^{tg}=1]\mathbf{P}(Y^{tg}=1) - \mu_0^{tr}}{\mu_1^{tr} - \mu_0^{tr}} \\
&= \frac{\mathbf{E}[g(\mathbf{X}^{tr})|Y^{tr}=0](1-\theta^{tg}) + \mathbf{E}[g(\mathbf{X}^{tr})|Y^{tr}=1]\theta^{tg} - \mu_0^{tr}}{\mu_1^{tr} - \mu_0^{tr}} \\
&= \frac{\mu_0^{tr}(1-\theta^{tg}) + \mu_1^{tr}\theta^{tg} - \mu_0^{tr}}{\mu_1^{tr} - \mu_0^{tr}} = \frac{\theta^{tg}(\mu_1^{tr} - \mu_0^{tr})}{\mu_1^{tr} - \mu_0^{tr}} = \theta^{tg}.
\end{aligned}$$

□

Theorem 3 *The mean squared error of the RE is approximately given by*

$$\begin{aligned}
\frac{1}{\left(\mu_1^{tr} - \mu_0^{tr}\right)^2} &\left[\frac{\left(\mu_1^{tr} - \mu_0^{tr}\right)^2 \theta^{tg}\left(1-\theta^{tg}\right) + \xi_1^{tr}\theta^{tg} + \xi_0^{tr}\left(1-\theta^{tg}\right)}{n_{tg}}\right. \\
&\left. + \frac{\xi_0^{tr}}{n_{tr}(1-\theta^{tr})}\left(1-\theta^{tg}\right)^2 + \frac{\xi_1^{tr}}{n_{tr}\theta^{tr}}(\theta^{tg})^2\right] \approx \mathbf{Var}\left[\widehat{\theta}_R^{tg}\right] \quad (6)
\end{aligned}$$

Proof By Lemma 2, we know that the bias is approximately zero. Therefore, $\mathbf{MSE}\left[\widehat{\theta}_R^{tg}\right] \approx \mathbf{Var}\left[\widehat{\theta}_R^{tg}\right]$. Again, by delta method [6], the variance of the ratio estimator is approximately

$$\begin{aligned}
\left(\frac{\mathbf{E}\left[\widehat{\mu}^{tg} - \widehat{\mu}_0^{tr}\right]}{\mathbf{E}\left[\widehat{\mu}_1^{tr} - \widehat{\mu}_0^{tr}\right]}\right)^2 &\left[\frac{\mathbf{Var}\left[\widehat{\mu}^{tg} - \widehat{\mu}_0^{tr}\right]}{\mathbf{E}^2\left[\widehat{\mu}^{tg} - \widehat{\mu}_0^{tr}\right]} + \frac{\mathbf{Var}\left[\widehat{\mu}_1^{tr} - \widehat{\mu}_0^{tr}\right]}{\mathbf{E}^2\left[\widehat{\mu}_1^{tr} - \widehat{\mu}_0^{tr}\right]}\right. \\
&\left. -2\frac{\mathbf{Cov}\left[\widehat{\mu}^{tg} - \widehat{\mu}_0^{tr}, \widehat{\mu}_1^{tr} - \widehat{\mu}_0^{tr}\right]}{\mathbf{E}\left[\widehat{\mu}^{tg} - \widehat{\mu}_0^{tr}\right]\mathbf{E}\left[\widehat{\mu}_1^{tr} - \widehat{\mu}_0^{tr}\right]}\right]. \quad (7)
\end{aligned}$$

The conclusion follows by replacing the quantities found by Lemma 2 in expression 7.

□

It follows from Theorem 3 that

Corollary 1 *The RE converges in probability to θ^{tr} as $n_{tr}, n_{tg} \longrightarrow \infty$. Hence, this estimator is consistent.*

Theorem 3 gives us some insights on how to choose good functions g to use in the ratio estimator. For instance, assuming that $n_{tg} \gg n_{tr}$, we have

Corollary 2 *If $n_{tg} \gg n_{tr}$, the MSE of the ratio estimator is approximately*

$$\frac{1}{n_{tr}(\mu_1^{tr} - \mu_0^{tr})^2}\left[\frac{\xi_0^{tr}\left(1-\theta^{tg}\right)^2}{(1-\theta^{tr})} + \frac{\xi_1^{tr}\left(\theta^{tg}\right)^2}{\theta^{tr}}\right].$$

The MSE expression in Corollary 2 indicates that one should use functions g such that μ_1^{tr} is far from μ_0^{tr} (so that $\mu_1^{tr} - \mu_0^{tr}$ is large) and both variances ξ_0^{tr} and ξ_1^{tr} are small. This motivates us to use functions g that are associated to the labels. Roughly speaking, we want the distribution of $g(\mathbf{X}^{tr})|Y^{tr} = 1$ to be "far" from the distribution of $g(\mathbf{X}^{tr})|Y^{tr} = 0$. This motivates the use of $g(\mathbf{x}) = \mathrm{i}(\mathbf{x} \in A)$ and $A = \{\mathbf{x} \in \mathbb{R} : f(\mathbf{x}) = 1\}$, where f is a classifier build using the labeled data. This is precisely the solution given by [5] for the prior probability shift problem. Another function that can have good performance is $g(\mathbf{x}) = \widehat{P}(Y = 1|\mathbf{x})$, where $\widehat{P}(Y = 1|\mathbf{x})$ is estimated using the training set. In the next section, we provide another way of choosing g.

3.2.2 Choice Based on RKHS

A different approach to choose good functions g to use in the RE is by minimizing the approximation to the MSE given by Corollary 2 on some appropriate class of functions. In this section, we take this class to be a Reproducing Kernel Hilbert Space (RKHS; [7]). More precisely, let

$$\widehat{\mathrm{MSE}}(g) = \frac{1}{(\widehat{\mu}_1^{tr} - \widehat{\mu}_0^{tr})^2}\left[\frac{\widehat{\xi}_0^{tr}}{\left(1-\widehat{\theta}^{tr}\right)}(1-\widehat{\theta})^2 + \frac{\widehat{\xi}_1^{tr}\widehat{\theta}^2}{\widehat{\theta}^{tr}}\right],$$

where $\widehat{\theta}$ is an initial estimate on θ^{tr}, be an estimate of the mean squared error of the ratio estimator for a given function g assuming that n_{tg} is large. We will find a smooth function g such that $\widehat{\mathrm{MSE}}$ is small by finding the solution to

$$\arg\min_{g \in \mathscr{H}_K} \widehat{\mathrm{MSE}}(g) + \lambda ||g||^2_{\mathscr{H}_K},$$

where $\mathscr{H}_K$ is any RKHS.

Theorem 4 *Let K be a kernel, and let $\mathscr{H}_K$ be its corresponding reproducing kernel Hilbert space. For a given $\lambda > 0$, the solution to*

$$\arg\min_{g \in \mathscr{H}_K} \widehat{MSE}(g) + \lambda ||g||^2_{\mathscr{H}_K}$$

is given by

$$g(\mathbf{x}) = \sum_{i=1}^{n_{tr}} w_i K(\mathbf{x}, \mathbf{x}_i^{tr})$$

where $w = (w_1, \ldots, w_n)$ *is such that*

$$w = \arg\min_{w \in \Re^d} \frac{w^t N w}{w^t M w} + \lambda w^t \mathbb{K} w. \quad (8)$$

Here, $(\mathbb{K})_{i,j} = K(\mathbf{x}_i^{tr}, \mathbf{x}_j^{tr})$, $M = (\widehat{\mu}_1 - \widehat{\mu}_0)(\widehat{\mu}_1 - \widehat{\mu}_0)^t$ *and* $N = \frac{\widehat{\theta}^2}{\theta_{1|L}} \widehat{\Sigma}_1^l + \frac{(1-\widehat{\theta})^2}{\widehat{\theta}_{0|L}} \widehat{\Sigma}_0^1$, *where* $\widehat{\mu}_i$ *is the* $n_{tr} \times 1$ *vector with entrance* k *given by* $\frac{1}{n_{tr}^i} \sum_{j:y_j^{tr}=i} K(\mathbf{x}_j^{tr}, \mathbf{x}_k^{tr})$, *and* $\widehat{\Sigma}_i$ *is the* $n_{tr} \times n_{tr}$ *matrix with entrance* (k, l) *given by the covariance of the vectors* $(K(\mathbf{x}_j^{tr}, \mathbf{x}_k^{tr}))_{j:y_j^{tr}=1}$ *and* $(K(\mathbf{x}_j^{tr}, \mathbf{x}_l^{tr}))_{j:y_j^{tr}=1}$.

Proof The fact that $g(\mathbf{x})$ admits the representation $\sum_{i=1}^{n_{tr}} w_i K(\mathbf{x}, \mathbf{x}_i^{tr})$ follows directly from the Representer Theorem [7]. Let $g(\mathbf{x}) = \sum_{i=1}^{n_{tr}} w_i K(\mathbf{x}, \mathbf{x}_i^{tr})$. We have that

$$\widehat{\mu}_i = \frac{1}{n_{tr}^i} \sum_{j:y_j^{tr}=i} g(\mathbf{x}_j^{tr}) = \sum_{k=1}^{n_{tr}} w_k \frac{1}{n_{tr}^i} \sum_{j:y_j^{tr}=i} K(\mathbf{x}_j^{tr}, \mathbf{x}_k^{tr}) = w^t \widehat{\mu}_i,$$

Similarly,

$$\widehat{\xi}_i = \frac{1}{n_{tr}^i} \sum_{j:y_j^{tr}=i} (g(\mathbf{x}_j^{tr}) - \widehat{\mu}_i)^2 = w^t \widehat{\Sigma}_i w.$$

It follows that

$$\widehat{\text{MSE}}(g) + \lambda ||g||^2_{\mathscr{H}_K} = \frac{w^t(\widehat{\theta}\widehat{\Sigma}_1 + (1-\widehat{\theta})\widehat{\Sigma}_0)w}{w^t(\widehat{\mu}_1 - \widehat{\mu}_0)(\widehat{\mu}_1 - \widehat{\mu}_0)^t w} + \lambda w^t \mathbb{K} w.$$

□

Unfortunately, it is not trivial to minimize Eq. 8 [8]. Instead, we work with the case where $\lambda = 0$. In this case, the solution to Eq. 8 is the solution to the problem of finding the vector w associated to the largest eigenvalue λ^* of the generalized eigenvalue problem $Mw = \lambda^* N w$. If N is invertible, this problem reduces to the standard eigenvalue problem

$$N^{-1} M w = \lambda^* w,$$

i.e., one only needs to find the largest eigenvetor of $N^{-1}M$. which can be solved using any linear algebra package. If N is not invertible, we use $(N + \gamma I_{n_{tr}})^{-1} M$, where $I_{n_{tr}}$ is the identity matrix and γ is a small number that makes $N + \gamma I_{n_{tr}}$ invertible.

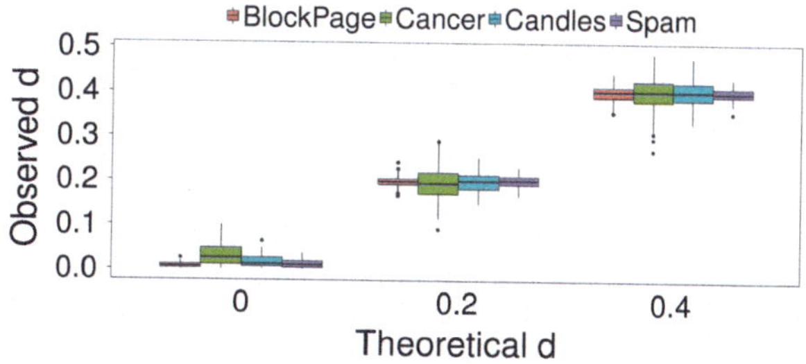

Fig. 1 Boxplots of the difference $d = |\theta^{tg} - \theta^{tr}|$ in each sample

Table 1 Methods compared with the experiments. The kernel for the RKHS approach was set to be a Gaussian kernel with bandwidth chosen by estimating the MSE via cross-validation; the value of $\gamma \in [0.00001, 0.1]$ was chosen in a similar fashion. The tuning parameters from all classifiers were also chosen by cross-validation

Methods	
Classify and count (CC)	Logistic regression (LR), k-NN, random forest (RF).
Ratio estimator	Logistic regression (LR), k-NN, random forest (RF), RKHS.

4 Experiments

In order to evaluate the quantification methods presented in Sect. 3, we consider Candles Dataset [9, 10], SPAM E-mail Database [11], Wisconsin Breast Cancer Database [11] and Blocks Classification [11]. We developed an algorithm (see the supplementary material) that partitions these datasets (in both training and target sample) so that we can simulate a probability prior shift scenario. Our algorithm allows one to control the average of the distance $d = |\theta^{tg} - \theta^{tr}|$, which is a measure of difficulty for the quantification problem. The simulations were performed setting $d \in \{0, 0.2, 0.4\}$ for each dataset, generating 100 different partitions. The results we present are averaged over these partitions.

The boxplots of d in each set are presented in Fig. 1, which shows that the observed distance is indeed close to the desired one in average. Moreover, the dispersion is inversely proportional to the number of samples in the dataset.

We evaluate the following methods in these datasets:

The results of the experiments are shown in Fig. 2. We can observe that, in most of settings, the ratio estimator has better performance than the classify and count approach. The only exception to these are in the cancer data (with $d = 0, 0.2$) and in the candles data (with $d = 0.2$). This may be justified by the fact that these are the datasets in which we have the smaller sample sizes. The figure also indicates that, in many settings, using RKHS to build the function g leads to better performance when compared to using classifiers in the ratio estimator (Table 1).

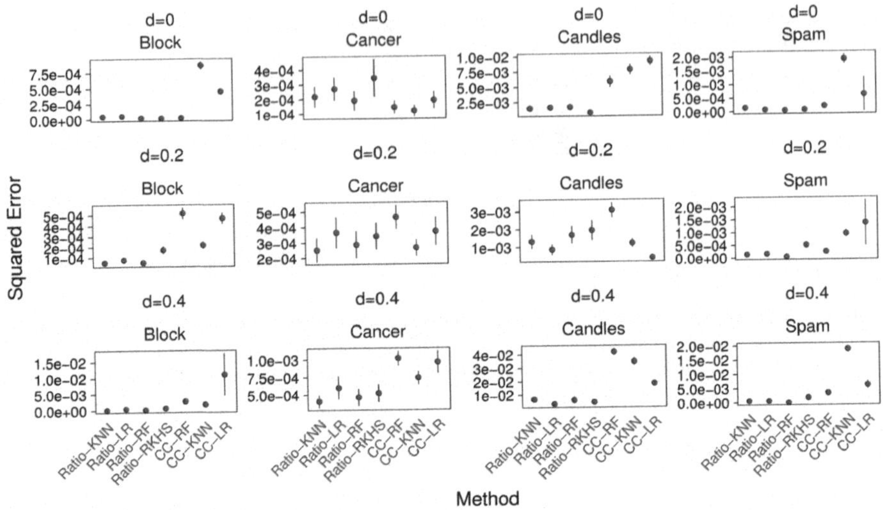

Fig. 2 Simulation results

5 Final Discussion

In this work, we show that the ratio estimator is a promising method to solve the quantification problem under the prior shift assumption. Moreover, we prove it is consistent.

We have noticed in our experiments that the tuning parameters used by the RKHS approach have a large influence on the results. However, because the cross-validation approach to choose such parameters is computationally intensive, we could not explore a wide variety of tuning parameter values. Thus, it is interesting to investigate alternative approaches to choose optimal values. In future works, we will also explore novel ways of constructing the function g.

Acknowledgements This work was partially supported by FAPESP grant 2017/03363-8 and CAPES.

References

1. Forman, G.: Quantifying trends accurately despite classifier error and class imbalance. In: Proceedings of the 12th ACM SIGKDD International Conference on Knowledge Discovery and Data Mining, pp. 157–166 (2006)
2. Quionero-Candela, J., Sugiyama, M., Schwaighofer, A., Lawrence, N.D.: Dataset Shift in Machine Learning. The MIT Press, Cambridge (2009)
3. Izbicki, R., Lee, A.B., Freeman, P.E.: Photo-z estimation: an example of nonparametric conditional density estimation under selection bias. Ann. Appl. Stat. **11**(2), 698–724 (2017)

4. Du Plessis, M.C., Sugiyama, M.: Semi-supervised learning of class balance under class-prior change by distribution matching. Neural Netw. **50**, 110–119 (2014)
5. Forman, G.: Quantifying counts and costs via classification. Data Min. Knowl. Discov. **17**, 164–206 (2008)
6. Lehmann, E.L.: Elements of Large-sample Theory. Springer Science & Business Media, Berlin (2004)
7. Scholkopf, B., Smola, A.J.: Learning With Kernels: Support Vector Machines, Regularization, Optimization, and Beyond. MIT press, Cambridge (2001)
8. Zhang, L.H.: On optimizing the sum of the Rayleigh quotient and the generalized Rayleigh quotient on the unit sphere. Comput. Optim. Appl. **54**(1), 111 (2013)
9. Freeman, P.E., Izbicki, R., Lee, A.B., Newman, J.A., Conselice, C.J., Koekemoer, A.M., Lotz, J.M., Mozena, M.: New image statistics for detecting disturbed galaxy morphologies at high redshift. Mon. Not. R. Astron. Soc. **434**(1), 282–295 (2013)
10. Izbicki, R., Stern, R.B.: Learning with many experts: model selection and sparsity. Mon. Not. R. Astron. Soc. **6**(6), 565–577 (2013)
11. Blake, C.L., Merz, C.J.: UCI Repository of machine learning databases http://www.ics.uci.edu/~mlearn/MLRepository.html. Irvine, CA: University of California. Department of Information and Computer Science, vol. 55, (1998)

Bayesian Meta-Analytic Measure

Camila B. Martins, Carlos A. de B. Pereira and Adriano Polpo

Abstract Meta-analysis is a procedure that combines results from studies (or experiments) with a common interest: inferences about an unknown parameter. We present a meta-analytic measure based on a combination of the posterior density functions obtained in each of the studies. Clearly, the point of view is from a Bayesian perspective. The measure preserves both the heterogeneity between and within the studies, and it is assumed that the all of the data from each study are available.

Keywords Case-by-case · Meta-analysis · Mixture models

1 Introduction

Meta-analysis refers to combining different studies that have the same objective, inferences about a common parameter. Meta-analyses can be based on either summarized results (e.g., means and variances) or the whole data set, i.e., case-by-case results, if available. Here, it is assumed that the case-by-case observations of all the studies are available. A convex combination of the studies' posterior densities functions is considered as the complete available information about the parameter of interest. The measure incorporates both types of heterogeneities, within and between studies, keeping the important information from each experiment. Most familiar meta-analysis methods work with summarized statistics, such as the means or proportions. These types of meta-analyses eliminate the heterogeneity within studies,

C. B. Martins
Federal University of São Paulo, São Paulo, Brazil
e-mail: cb.martins@unifesp.br

C. A. de B. Pereira
University of São Paulo, São Paulo, Brazil
e-mail: cpereira@ime.usp.br

A. Polpo (✉)
Federal University of São Carlos, São Carlos, Brazil
e-mail: polpo@ufscar.br

A. Polpo et al. (eds.), *Bayesian Inference and Maximum Entropy Methods in Science and Engineering*, Springer Proceedings in Mathematics & Statistics 239, https://doi.org/10.1007/978-3-319-91143-4_4

most likely losing relevant information. Hierarchical modeling, how Bayesians may approach meta-analysis, is questionable since its first level of uncertainty considers invisible independent observations of a leader random variable. However, different locations may have different environments, for instance. In this manner, the posterior distribution of the parameters of the leader random variable may not describe correctly the information sought.

The simple method proposed here is to combine the final Bayesian analyses obtained in each study. Considering a weighted average of the posterior densities, we obtain a density that may better represent the different nuances of the studies without being restricted to any family of distributions. The proposed posterior meta-analytic measure is presented in Sect. 2. Section 3 illustrates the measure with an example, and Sect. 4 presents some final remarks.

2 Meta-Analysis Measure

Consider that we have N different studies with the aim to understand some characteristic θ, $\theta \in \Theta$. Let $\boldsymbol{X} = \{\boldsymbol{X}_1, \ldots, \boldsymbol{X}_N\}$, where $\boldsymbol{X}_j = \{X_{j1}, \ldots, X_{jn_j}\}$ are the data from the jth study, $j = 1, \ldots, N$. Consider that X_{ji} are independent random variables with density function $f(x_{ji} \mid \theta)$, $i = 1, \ldots, n_j$. The small caps $\boldsymbol{x}$ and $\boldsymbol{x}_j$ are the observed values of $\boldsymbol{X}$ and $\boldsymbol{X}_j$, respectively. The likelihood function of the jth study is

$$L_j(\theta \mid \boldsymbol{x}_j) = \prod_{i=1}^{n_j} f(x_{ji} \mid \theta).$$

Given the prior density function $\pi(\theta)$, the posterior meta-analytic measure is defined by

$$\pi(\theta \mid \boldsymbol{x}) = \sum_{j=1}^{N} \omega_j \pi_j(\theta \mid \boldsymbol{x}_j), \tag{1}$$

where $\omega_j > 0, \forall j, \sum_{j=1}^{N} \omega_j = 1$, and

$$\pi_j(\theta \mid \boldsymbol{x}_j) = \frac{L_j(\theta \mid \boldsymbol{x}_j)\pi(\theta)}{\int_\Theta L_j(\theta \mid \boldsymbol{x}_j)\pi(\theta)\, d\theta}.$$

The constant ω_j is the weight of each study. If there is a reason to consider one study as more important than others, then it is possible to set a higher value for the weight of this study. We consider that the importance of each study is proportional to its sample size, that is, $\omega_j = n_j \Big/ \left(\sum_{i=1}^{N} n_j\right)$.

Note that there is only one prior, $\pi(\theta)$, and one parameter, θ. We do not assign a parameter to each study and then combine them. We can write the proposed measure as

Table 1 Use of SAMe

j	x_j	n_j	x_j/n_j	ω_j
1	20	20	1.00	0.133
2	4	10	0.40	0.067
3	11	16	0.69	0.107
4	10	19	0.53	0.127
5	5	14	0.36	0.093
6	36	46	0.78	0.306
7	9	10	0.90	0.067
8	7	9	0.78	0.060
9	4	6	0.67	0.030
Total	106	150	0.71	–

j is for the jth study;
x_j is the number of success;
n_j is the sample size of each study;
and $\omega_j = n_j/150$.

$$\pi(\theta \mid \boldsymbol{x}) = \pi(\theta)\left[\sum_{j=1}^{N} c_j L_j(\theta \mid \boldsymbol{x}_j)\right], \tag{2}$$

where

$$c_j = \frac{\omega_j}{\int_{\Theta} L_j(\theta \mid \boldsymbol{x}_j)\pi(\theta)\,d\theta}.$$

From Eq. (1), the proposed posterior measure, $\pi(\theta \mid \boldsymbol{x})$, is a convex combination of the posterior distributions of each study. On the other hand, from Eq. (2), the proposed measure is the prior multiplied by a mixture of the likelihood of each study, which is a fully Bayesian procedure (posterior = prior × model). Both cases result in the same posterior meta-analytic measure, $\pi(\theta \mid \boldsymbol{x})$, which is a probability density function of θ given the data from all available studies.

It is important to note that for proper priors, the proposed measure is always a probability density function of θ after observing the data, such as any Bayesian analysis procedure.

3 Example

Table 1 presented the results of nine studies about the success in the use of the SAMe (an antidepressant drug S-adenosylmethionine). The data are presented in [1, 2].

Let X_j be a random variable related to the number of success in the use of SAMe. We have that X_j given θ has a binomial distribution with parameters n_j, the number of trials, and θ, the success rate associated with the use of SAMe, $j = 1, \ldots, 9$. We assume that n_j are fixed constants, and then the only parameter is θ. In this case, the likelihood function of the jth study is

$$L_j(\theta \mid x_j, n_j) = \theta^{x_j}(1-\theta)^{n_j - x_j}.$$

Considering that $\pi(\theta) = 1$, $\theta \in (0, 1)$, that is, the prior of θ is a uniform distribution over (0, 1). The posterior of each study is

$$\pi_j(\theta \mid x_j, n_j) = \frac{\Gamma(n_j + 2)}{\Gamma(x_j + 1)\Gamma(n_j - x_j + 1)}\theta^{x_j}(1-\theta)^{n_j - x_j},$$

where Γ is the mathematical gamma function. In this case, $\theta \mid x_j, n_j$ has a beta distribution with parameters $x_j + 1$ and $n_j - x_j + 1$.

The meta-analytic measure is

$$\pi(\theta \mid \boldsymbol{x}, \boldsymbol{n}) = \sum_{j=1}^{9} \omega_j \frac{\Gamma(n_j + 2)}{\Gamma(y_j + 1)\Gamma(n_j - x_j + 1)}\theta^{x_j}(1-\theta)^{n_j - x_j},$$

where $\boldsymbol{x} = \{x_1, \ldots, x_9\}$, and $\boldsymbol{n} = \{n_1, \ldots, n_9\}$. In this example, the meta-analytic measure is a mixture of beta distributions.

In the hierarchical model, we cannot perform a direct analysis of θ as a population parameter. As discussed in the introduction, at the first prior level, we will need $\theta_1, \ldots, \theta_9$, the success rates of each study. At the second level, we will have a prior for the hyper-parameter of θ_j, which will not yield the value of θ. Reference [1] performed a hierarchical analysis of these data from the perspective of multicenter analysis; his main interest lies in the estimation of θ_j and the differences between them.

On the other hand, if we had considered that we have a unique study, our data should be 106 success in 150 trials. Considering, the same uniform prior for θ, our posterior $\theta \mid x = 106, n = 150$ would be a beta distribution with parameters $x + 1 = 107$ and $n - x + 1 = 45$. The meta-analytic measure and Beta(107, 45) distribution are shown in Fig. 1. As expected the meta-analytic measure preserves the characteristic of the data, and the Beta(107, 45) distribution does not represent the different results of the studies.

If our interest lies in comparing studies, we can draw the posterior of each study together with the meta-analytic measure (Fig. 2). From the meta-analytic measure, we have that the median is 0.72, the mode is 0.78, the mean is 0.69, and the 95% high posterior density credible interval is (0.31; 1.00]. We can observe that the meta-analytic measure has three modes, 0.50, 0.78, and 1.00. This may suggest that we have three groups in the studies.

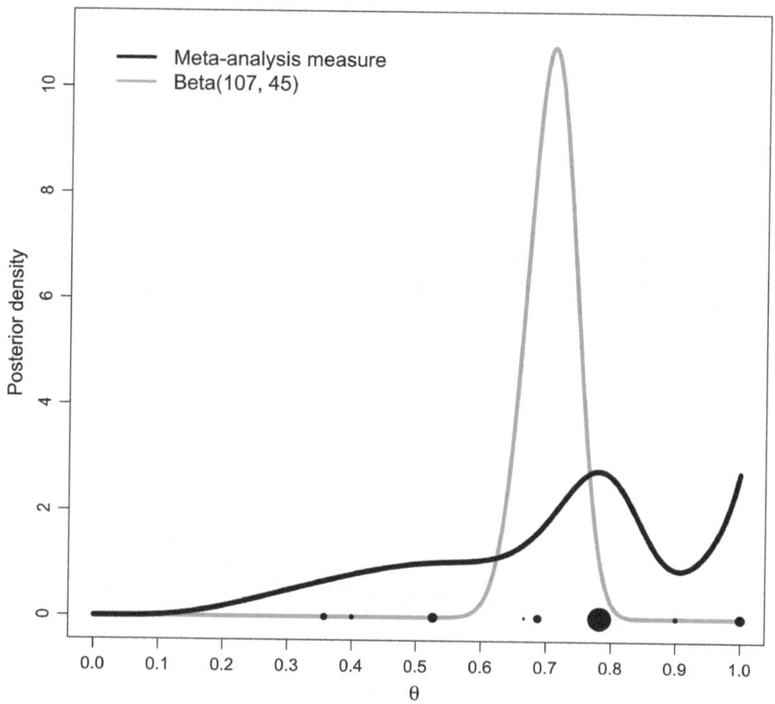

Fig. 1 Meta-analytic measure and posterior distribution considering all samples from a unique study, beta(107, 45); the dots represent the proportion of each study, and the size of the dot is proportional to the sample size of the study

4 Final Remarks

The proposed method is a posterior distribution, called the meta-analytic measure. The results show that we are not performing inference over means (the usual method for meta-analysis), and the proposed measure provides a complete inferential framework. We are able to evaluate the posterior mean, mode, median, variance, and credible interval, and we can even perform a hypothesis test. We have a measure that represents the observed data and the heterogeneity of the studies, and the analysis can be performed as with any traditional Bayesian method.

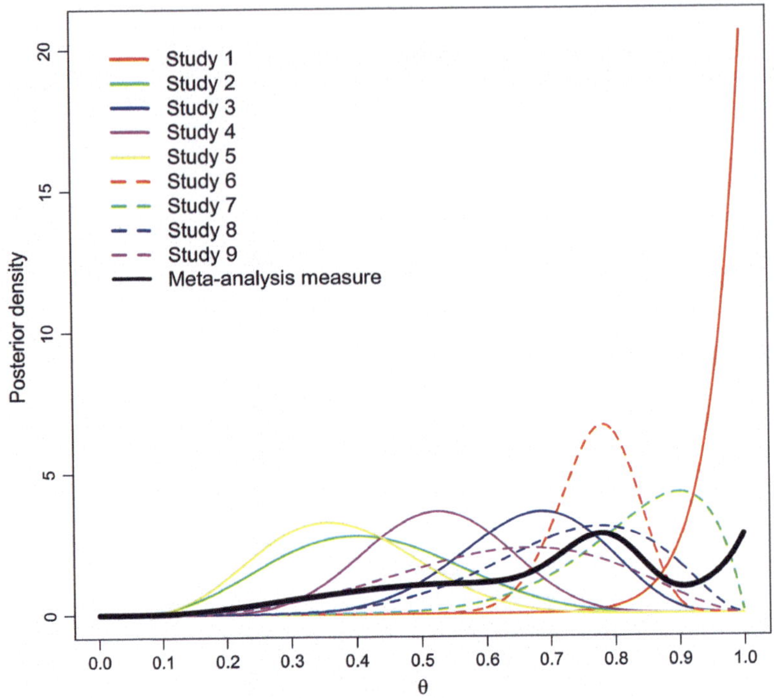

Fig. 2 Meta-analytic measure and posterior distributions of all studies

References

1. Berry, D.A.: A bayesian approach to multicenter trials and meta-analysis. Tech. Rep. ED325480, National Science Foundation, Washington, (1989)
2. Janicak, P., Lipinski, J., Davis, J., Coinaty, J., Waternaux, C., Cohen, B., Altman, E., Sharma, R.: S-adenosyl-methionine (same) in depresion: a literature review and preliminary data report. Ala. J. Med. Sci. **25**(3), 306–313 (1988)

Feature Selection from Local Lift Dependence-Based Partitions

Diego Marcondes, Adilson Simonis and Junior Barrera

Abstract The classical approach to feature selection consists in minimizing a cost function of the estimated joint distribution of the variable of interest and the feature vectors. However, in order to estimate the joint distribution, and therefore, the cost function, it is necessary to discretize the variables, so that feature selection algorithms are partition dependent, as they depend on the partitions in which the variables are discretized. In this framework, this paper aims to propose a systematic approach to the discretization of random vectors, which is based on the Local Lift Dependence. Our approach allows an interpretation of the local dependence between the variable of interest and the selected features, so that it is possible to outline the kind of dependence that exists between them. The proposed approach is applied to study the dependence between the performances on entrance exam subjects and on first semester courses of University of São Paulo Statistics and Computer Science undergraduate programs.

Keywords Feature selection · Local Lift Dependence · Mutual Information · Variable selection

1 Introduction

The Mutual Information $I(X, Y)$ between two random variables (X, Y), as proposed by [8], is a classical global dependence quantifier that, if the variables are absolutely continuous, is defined by

D. Marcondes (✉) · A. Simonis · J. Barrera
Instituto de Matemática e Estatística - USP, Rua do Matão, São Paulo 1010, Brazil
e-mail: dmarcondes@ime.usp.br

A. Simonis
e-mail: asimonis@ime.usp.br

J. Barrera
e-mail: jb@ime.usp.br

A. Polpo et al. (eds.), *Bayesian Inference and Maximum Entropy Methods in Science and Engineering*, Springer Proceedings in Mathematics & Statistics 239, https://doi.org/10.1007/978-3-319-91143-4_5

$$I(X, Y) = \iint\limits_{S_{X,Y}} \log\left(\frac{f(x, y)}{g(x)h(y)}\right) dF \tag{1}$$

in which f is the joint, and g and h are the respective marginal, probability densities of (X, Y), $S_{X,Y}$ is the support of (X, Y) and F is the joint distribution function of (X, Y). The value of $I(X, Y)$ represents the mass concentration variation on the distribution of Y due to the observation of X, and as the degree of dependence between the variables increases when mass concentration increases, $I(X, Y)$ also measures variable dependence.

The Mutual Information is also a classical cost function for feature selection algorithms, when we are interested in determining the variables (features) in $X = \{X_1, \ldots, X_n\}$ which are most related to a variable Y. However, in order to estimate $I(X, Y)$ given a random sample of $(X_i, Y), i = 1, \ldots, n$, it is necessary to discretize the variables and calculate the Mutual Information between the discrete random variables. Therefore, feature selection based on the Mutual Information is partition dependent, as different discretization processes may yield the selection of distinct features. In this paper, we propose a systematic manner of discretizing continuous variables, based on the Local Lift Dependence, in order to consistently select features and properly estimate cost functions as the Mutual Information.

2 Local Lift Dependence

In this section, we present the main concepts related to the Local Lift Dependence (LLD) and their implications on the Entropy and Mutual Information (MI) concepts [8]. The main concepts of this section are presented for continuous random variables X and Y defined on a support $S_{X,Y} = S_X \times S_Y$, although, with simple adaptations, the discrete case follows from it. We briefly present the MI and then discuss its relation to the LLD.

The MI, as defined in (1), is an index that measures the mass concentration of a joint probability density of two random variables. As greater the mass concentration on the joint probability density is, the more dependent the random variables are and greater is their MI. An useful property of the MI is that it may be expressed as

$$I(X, Y) = H(Y) - H(Y|X) \tag{2}$$

in which $H(Y)$ is the Entropy of Y and $H(Y|X)$ is the Conditional Entropy of Y given X. The form of the MI in (2) is useful because, if we fix a variable Y, and consider a set $\{X_1, \ldots, X_n\}$ of n random variables, we may determine which one of them is the most dependent with Y by observing only the Conditional Entropy of Y given each one of the random variables, as the variable that maximizes the MI is the same that minimizes the Conditional Entropy. Indeed, MI and Conditional Entropy are global and general measures of dependence, that summarize to an index a variety of dependence kinds that are expressed by mass concentration.

On the other hand, the LLD is a local and general measure of dependence that expands the global dependence measured by the MI into local indexes that enable local interpretation of the dependence between the variables. As the MI is an index that measures the dependence between random variables by measuring the mass concentration incurred in one variable by the observation of another, it may only give evidences about the existence of a dependence, but cannot assert what kind of dependence is being observed. Therefore, it is relevant to break down the MI by region, so that it can be interpreted in an useful manner and the kind of dependence outlined by it may be identified. The LLD is responsible for this break down, as it may be expressed by the Lift Function (LF) given by

$$L(x, y) = \frac{f(x, y)}{g(x)h(y)} := \frac{f(y|x)}{h(y)} \; \forall (x, y) \in S_{X,Y}. \tag{3}$$

in which $f(y|x) := \frac{f(x,y)}{g(x)}$ is the conditional density of Y given X.

Indeed, the MI is the expectation on (X, Y) of the LF, so that the LF presents locally the mass concentration measured by the MI. As the LF may be written as the ratio between the conditional probability density of Y given X and the marginal probability density of Y, the main interest in its behavior is in determining for which pairs $(x, y) \in S_{X,Y}$ $L(x, y) > 1$ and for which $L(x, y) < 1$. If $L(x, y) > 1$ then the fact of X being equal to x increases the density of Y being equal to y, as the conditional density is greater than the marginal one. Therefore, we say that the event $\{X = x\}$ lifts the event $\{Y = y\}$. In the same way, if $L(x, y) < 1$, we say that the event $\{X = x\}$ inhibits the event $\{Y = y\}$, as $f(y|x) < h(y)$. If $L(x, y) = 1, \forall (x, y) \in S_{X,Y}$, then the random variables are independent.

An important property of the LF is that it cannot be greater than one nor lesser than one for all pairs $(x, y) \in S_{X,Y}$. Therefore, if there are LF values greater than one, then there must be values lesser than one, what makes it clear that the values of the LF are dependent and that the *lift* is a pointwise characteristic of the joint probability density and not a global property of it. Thus, the study of the behavior of the LF may be accomplished by observing its level curves or *heatmap*, that presents the behavior of the LF by coloring the support $S_{X,Y}$ according to its values.

Although the LF, as defined in (3), presents a wide picture of the dependence between two random variables, it has some practical limitations. First of all, a great sample size may be necessary to estimate the LF via the kernel estimator. Second, the LF may assess the dependence between the random variables in too much detail, what may not be useful in practice, as it may be hard to interpret all the values of it. Finally, the LF treats only the dependence between two random variables, what narrows its application range. Thus, in order to solve those limitations, we propose that the random variables of interest be discretized in a manner so that their dependence may be interpreted by applying the LF in an useful way.

Therefore, this paper treats the scenario in which we have two groups of discrete and/or continuous random variables $Y = \{Y_1, \dots, Y_m\}$ and $X = \{X_1, \dots, X_n\}$ and want to measure the dependence between the variables in Y and X. We propose that such dependence study is made by assessing the local dependence between the discrete random variables $V(Y)$ and $U(X)$, constructed by the discretization of Y and X, respectively. The local dependence between $V(Y)$ and $U(X)$ is expressed by the discrete LF given, for all $(u, v) \in S_{U(X),V(Y)}$, by

$$L^*(u, v) = \frac{\mathbb{P}(U(X) = u, V(Y) = v)}{\mathbb{P}(U(X) = u)\mathbb{P}(V(Y) = v)} = \frac{\mathbb{P}(V(Y) = v|U(X) = u)}{\mathbb{P}(V(Y) = v)} \tag{4}$$

in which $S_{U(X),V(Y)} = S_{U(X)} \times S_{V(Y)}$ is the support of $(U(X), V(Y))$. The interpretation of L^* is similar to that of L, with the densities interchanged by their respective probability functions.

Even though the discrete LF (4) gives a wide view of the dependence between discrete random variables U and V, it may be of interest to summarize this dependence to an index. In order to do so, we use the normalized MI of V given U that is defined as

$$\eta(V|U) = \frac{\sum\limits_{(u,v)\in S_{U,V}} \mathbb{P}(U = u, V = v) \log L^*(u, v)}{-\sum_{v\in S_V} \mathbb{P}(V = v) \log \mathbb{P}(V = v)} = \frac{I(V, U)}{H(V)} \tag{5}$$

and is the ratio between the MI of (U, V) and the Entropy of V. We have that $0 \leq \eta(V|U) \leq 1$, that $\eta(V|U) = 0$ if and only if (V, U) are independent and that $\eta(V|U) = 1$ if, and only if, there exists a function $Q : S_U \to S_V$ such that $\mathbb{P}(V = Q(U)) = 1$. Note that the coefficient in (5) measures the influence of U in V and that it may differ from the coefficient $\eta(U|V)$, that measures the influence of V in U.

The local dependence measure given by the discrete LF, and the index given by η, may be useful in determining what subset of features is most related to a variable. Therefore, the LLD partition of random vectors into two discrete random variables may be a useful tool in feature selection, as it gives a systematic manner of measuring variable dependence. In the following sections, we present a classical approach to feature selection, that is the minimization of a cost function of an estimated joint distribution, having candidate parameter vectors in a Boolean lattice of feature vectors. However, we propose LLD-based partitions in which classic cost functions, as the η coefficient, may be applied to select features.

3 Feature Selection Algorithm from Local Lift Dependence-Based Partitions

In this section, we present the characteristics of a feature selection algorithm from LLD-based partitions. We first outline the classical approach to feature selection.

Then, we propose LLD-based partitions and cost functions that may be applied to feature selection. Lastly, we present stopping criteria for the proposed algorithm.

3.1 Classical Feature Selection Algorithm

Let Y and $X = \{X_1, \dots, X_n\}$ be random variables. We call the random variables in X features and note that the power set $\mathscr{P}(X)$ of X may be seen as a Boolean lattice of feature vectors, in which each vector represents a subset of features. Therefore, feature selection is given by the minimization, in the given Boolean lattice, of a cost function, defined as a function of the joint probability of the feature vectors and Y. In fact, the subset of features selected by this approach is given by

$$\chi = \underset{\chi^* \in \mathscr{P}(X)}{\arg\min}\, C(Y, \chi^*)$$

in which C is a cost function. The estimated error of a predictor Ψ as presented in [3, Chap. 2], for example, is a classical cost function.

In order to determine $C(Y, \chi^*)$ for a feature vector $\chi^* \in \mathscr{P}(X)$, we need to estimate the joint probability function of $(Y, \chi*)$. However, as Y may be a continuous variable, and $\chi*$ may contain continuous variables, it is convenient to discretize Y and $\chi*$, so that the joint probability distribution may be estimated when the sample size is not large. We propose that (Y, χ^*) be discretized into a LLD-based partition, i.e., into two discrete random variables.

3.2 Local Lift Dependence-Based Partitions

Although there are countless ways to discretize the random variables Y and $\chi^* \in \mathscr{P}(X)$, we present a manner of discretizing them into two random variables $(U(\chi^*), V(Y))$, what we call LLD-based partition, as the LLD is used to study the dependence between the discrete random variables. In order to propose LLD-based partitions, we separate the discretization process in three different types: when all the variables of χ^* are discrete, when all are continuous and when some of the variables in χ^* are continuous and others are discrete. However, we first treat the discretization of Y into $V(Y)$.

If Y is a discrete random variable, there is nothing to be done, as it is enough to take $V(Y) = Y$. However, if Y is continuous, then it may discretized in any usual manner. Nevertheless, an useful discretization process is that based on the sample percentiles of Y, as it enables a good interpretation of the categories of $V(Y)$.

In the same way, the discretization process of $\chi*$ when all of its features are discrete is pretty simple, as it is enough to define $U(\chi^*)$ as the discrete random variable whose support is the Cartesian Product of the supports of the variables in

$\chi*$. On the other hand, if all the features in $\chi*$ are continuous there is no trivial manner of discretizing them into a random variable $U(\chi^*)$. A possible approach would be to partition the support of $\chi*$ into equivalence classes delimited by hyper-rectangles, what is known as the histogram partition [3, Chap. 1], although it may not be practical if the dimension of χ^* is too high or if the support of $\chi*$ is too wide. Therefore, we propose a natural manner of discretizing $\chi*$ into $U(\chi^*)$ that is also based on the percentiles.

As the support of $\chi*$ is a subset of[1] $\mathbb{R}^k$, there is no trivial ordination of the $\chi*$ sample points as there was on the discretization process of the continuous variable Y, whose support was a subset of the real line $\mathbb{R}$. Therefore, we propose that a distance, defined in a statistical sense, be taken from each sample point of $\chi*$ and a fixed point $P \in \mathbb{R}^k$ and that the percentiles of this distance be used to discretize $\chi*$ into $U(\chi^*)$. If all the features in $\chi*$ are positive random variables, we may take, for example, $P = 0$ and then discretize the features $\chi*$ by the sample median (or any other percentiles) of the considered distance, so that we get a binary random variable $U(\chi^*)$ that may be easily interpreted: the observations on its first category are those with small values of the features in $\chi*$ jointly, while its second category contains the observations with great values of the features in $\chi*$ jointly.

If the Euclidean distance is used, then the support of χ^* is partitioned by hyper-spheres. On the other hand, if the Mahalanobis distance, as proposed by [2], is used instead, then the support of χ^* is partitioned by ellipsoids. Although any distance could be used, we propose the use of the Mahalanobis distance, as it consider the variances and covariances of the random variables within χ^*.

Finally, if there are some variables in χ^* that are continuous and others that are discrete, we may associate the methods presented above in order to discretize χ^* into $U(\chi^*)$. Indeed, we may take the distance between the continuous variables sample points and a fixed point P inside each category of the joint distribution of the discrete variables, i.e., the distances are calculated and the sample percentiles are determined considering only the observations inside each one of the categories of the discrete variables.

The discretization methods presented in this section may also be applied to study the dependence between two random vectors $Y = \{Y_1, \ldots, Y_m\}$ and $X = \{X_1, \ldots, X_n\}$, as it is enough to discretize Y into $V(Y)$ and X into $U(X)$ by one of the methods proposed above. In the same manner, it is possible to select the features $\chi \subset X$ which are most related to Y, as the relation between Y and a feature vector χ may be expressed by the LLD study of the dependence between $U(\chi)$ and $V(Y)$. With adaptations to the method, we may also find $\chi_X \subset X$ and $\chi_Y \subset Y$ that are most related.

[1] Supposing that $\chi*$ contains k features.

3.3 Cost Functions

Any cost function that can be applied to discrete variables may be used to select the features. However, in our approach, we use the η coefficient that is calculated for the discretized random variables. Therefore, the features $\chi \in \mathscr{P}(X)$ selected are those that maximizes the η coefficient of $V(Y)$ given $U(\chi)$.

3.4 Stopping Criteria for the Algorithm

As the Boolean lattice generated by the power set of $X = \{X_1, \ldots, X_n\}$ has cardinality $2^n - 1$, it is necessary to determine stopping criteria for the algorithm, as it is not computable for great values of n. A possible stopping approach is to employ an *U-curve* algorithm, so that it is not necessary to run an exhaustive search on all the power set of X. An *U-curve* algorithm, as presented in [5] and [6], searches $\mathscr{P}(X)$ for the subset that optimizes the cost function by visiting a *tree*, in which each *node* represents a subset of X. The algorithm penalizes the shortage of samples, which causes the algorithm not to visit many nodes of the tree: when the estimation error is greater than a fixed value, the algorithm *prune* the tree and do not search a group of nodes, what saves computational time.

However, the *U-curve* algorithm may not be applied if there are no discrete random variables on X. Therefore, we propose that an *U-curve* algorithm be used to treat the *nodes* of the Boolean lattice which contain discrete variables, while other stopping criteria may be used on nodes consisting only of continuous variables. A possible stopping criteria is to consider only the subsets of X whose dimension is lesser than a number $k < n$, so that the algorithm does not exhaustively search $\mathscr{P}(X)$.

4 Applications

In this section, we apply the methods proposed to study the relation between the performance on the courses of the first semester and the performance on the entrance exam test of students that enrolled in the Statistics and Computer Science undergraduate courses of the Institute of Mathematics and Statistics of the University of São Paulo between 2011 and 2016. There were considered only the students that had a mean grade on the first semester courses greater than five[2], which amounted to 129 students of Statistics and 251 students of Computer Science.

The entrance exam test is divided in eight subjects, namely, Mathematics, Physics, Biology, Chemistry, History, Geography, English, and Portuguese, besides an Essay. It is of great interest to know what are the entrance exam subjects that are most related to the performance on the courses of the first semester and, in order to determine the

[2]In a scale of 0 to 10.

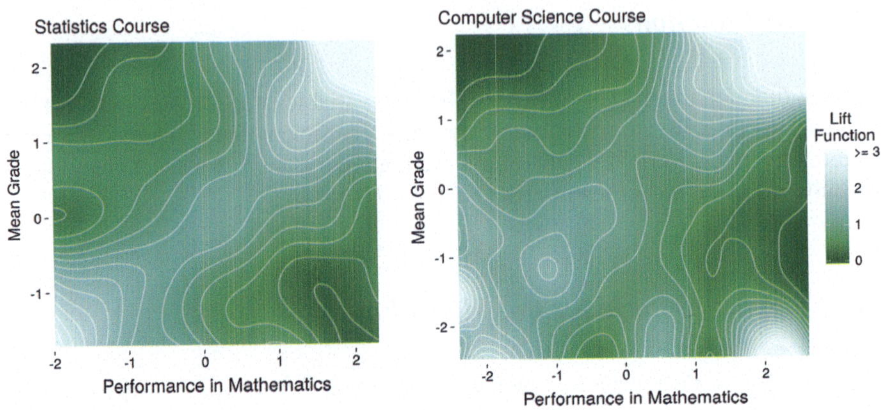

Fig. 1 Estimated *Heatmap* between the performance on mathematics and the mean grade on the first semester of the students of statistics and computer science, both standardized by year

subjects most related to these courses, we apply LLD methods. The performance on an exam subject is given by the percentage of correct answers on the subject.

First of all, we study the relation between the performance on Mathematics on the entrance exam and the mean grade on the first semester, that are both standardized by year, so that that they may be compared. The LF estimated *heatmap* between the mean grade and the performance on Mathematics for each course is presented in Fig. 1. The *heatmap* was estimated using the kernel estimated probability densities of the considered performances, based on [7, 9], and made by the functions *density* of the *Stats Package* [4] and *kde2d* from the *MASS Package* [10] of the *The R Project for Statistical Computing*. Analyzing Fig. 1, we see that high performance on Mathematics lifts high mean grade and that poor performance on Mathematics lifts poor mean grade, for both courses, as expected.

For the purpose of studying the relation between the joint performance on Mathematics and Physics and the mean grade, we may partition both the mean grade and the joint performance on Mathematics and Physics into two binary random variables according to the median of the mean grade within each year and the median of the Mahalanobis distance from zero, also within each year, of the joint performance on Mathematics and Physics. The Mahalanobis partition is presented in Fig. 2. The median distances and the median mean grades are taken within each year because it is known that the performance scale is not the same every year. Therefore, year may be seen as a block factor in our analysis [1, Chap. 5]. We observe that the support of the performances, i.e., $[0, 1]^2$, is partitioned by ellipses, so that the students inside the ellipse of their year are those with poor joint performance and the students outside it are those with high joint performance on Mathematics and Physics.

Table 1 presents the discrete LF for the performances considered above. We observe that having a performance on Mathematics and Physics above the median lifts the probability of having an above the median mean grade in both courses,

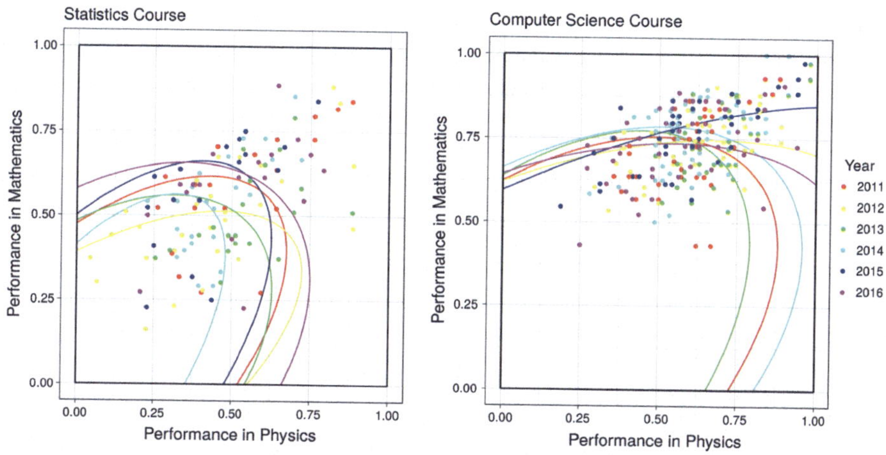

Fig. 2 Mahalanobis partition between the performances on physics and mathematics, within each year, by course

Table 1 Discrete LF between the mean grade and the joint performance on Mathematics and Physics given by the Mahalanobis partition by year. The numbers in parentheses represent sample sizes

Course	Mean grade	Mathematics and Physics		η
		Below median	Above median	
Statistics	Below median	1.31 (43)	0.682 (22)	0.07412
	Above median	0.682 (22)	1.32 (42)	
Computer science	Below median	1.17 (75)	0.829 (52)	0.02124
	Above median	0.829 (52)	1.18 (72)	

although the lift in the Statistic course is 32% while the lift in the Computer Science course is only 18%. On the same way, having a below median performance on Mathematics and Physics lifts the probability of having a below median mean grade, although the lift is greater in the Statistics course. Indeed, we observe from the η coefficients that the dependence between the performances on Mathematics and Physics and the mean grade is stronger in the Statistics course.

Finally, we apply the feature selection algorithm from LLD-based partitions to select the subjects that are most related to the mean grade using the η as a cost function. To that purpose, the mean grade is discretized by its median within each year. In the same way, the performance on a group of k subjects is discretized by the median of the Mahalanobis distance between the sample performances and zero, within each year. The subject selected by this method is Mathematics, for both the Statistics and Computer Science courses. Table 2 presents the discrete LF between the performance on Mathematics and the mean grade, both discretized within each year.

Table 2 Discrete LF between the mean grade and the performance on Mathematics for the Statistics and Computer Science courses. The numbers in parentheses represents sample sizes

Course	Mean grade	Mathematics		η
		Below median	Above median	
Statistics	Below median	1.33 (45)	0.662 (22)	0.08963
	Above median	0.64 (20)	1.37 (42)	
Computer science	Below median	1.23 (85)	0.768 (52)	0.04593
	Above median	0.728 (42)	1.28 (72)	

From Table 2 we see that the dependence between the performance on Mathematics and the mean grade is stronger in the Statistics course. On the one hand, if a Statistic student has an above the median performance on Mathematics, than his probability of having a mean grade above the median is lifted in 37%, when comparing with a student for which we do not know the performance on Mathematics. On the other hand, if a Computer Science student has an above the median performance on Mathematics, than his probability of having a mean grade above the median is lifted in only 28%, when comparing with a student for which we do not know the performance on Mathematics.

5 Final Remarks

The algorithm proposed in this paper has a couple of good qualities. First of all, it presents a systematic way of treating feature selection that can be applied to a variety of practical problems, as it may be adapted for continuous and discrete features. Second, the dependence between the selected features and Y may be interpreted by the use of LLD tools, as the LF and its *heatmap*. Therefore, the proposed method provides not only the features that are most related to Y in some sense, but also useful informations about the outlined relation.

On the other hand, the algorithm may not be computable if the number of continuous features is too great. However, some continuous features may be discretized beforehand, so that the majority of the variables in X are discrete and an *U-curve* algorithm may be applied. Nevertheless, we discourage the beforehand discretization of all continuous features, as important characteristics of the data may be lost in the discretization process. We believe there is much further work to be done in studying the Local Lift Dependence, and how it can be used to improve feature selection and the analysis of variable dependence.

References

1. Montgomery, D.C.: Design and Analysis of Experiments, 4th edn. John Wiley, (1997)
2. Mahalanobis, P.C.: On the generalized distance in statistics. Proc. Natl. Inst. Sci. (Calcutta) **2**, 49–55 (1936)
3. Braga Neto, U.M., Dougherty, E.R. Error Estimation for Pattern Recognition. Wiley (2015)
4. R Core Team. R: A Language and Environment for Statistical Computing. R Foundation for Statistical Computing, Vienna, Austria (2016)
5. Ris, M., Barrera, J., Martins, D.C.: U-curve: a branch-and-bound optimization algorithm for u-shaped cost functions on boolean lattices applied to the feature selection problem. Pattern Recognit. **43**(3), 557–568 (2010)
6. Reis, M.S.: Minimization of decomposable in U-shaped curves functions defined on poset chains–algorithms and applications. PhD thesis, Ph. D. thesis Institute of Mathematics and Statistics, University of São Paulo, Brazil (in Portuguese) (2012)
7. Scott, D.W.: Multivariate density estimation and visualization. In: Handbook of Computational Statistics, pp. 549–569. Springer, Berlin (2012)
8. Shannon, C.E., Weaver, W.: The Mathematical Theory of Communication. Urbana: University of Illinois Press, vol. 29, (1949)
9. Silverman, B.W.: Density Estimation for Statistics and Data Analysis, vol. 26. CRC press (1986)
10. Venables, W.N., Ripley, B.D.: Modern Applied Statistics with S, 4th edn. Springer, New York (2002). ISBN 0-387-95457-0

Probabilistic Inference of Surface Heat Flux Densities from Infrared Thermography

D. Nille, U. von Toussaint, B. Sieglin and M. Faitsch

Abstract In nuclear fusion research, based on the magnetic confinement, the determination of the heat flux density distribution onto the plasma facing components is important. The heat load poses the threat of damaging the components. The heat flux distribution is a footprint of the transport mechanisms in the plasma, which are still to be understood. Obtaining the heat flux density is an ill-posed problem. Most common is a measurement of the surface temperature by means of infrared thermography. Solving the heat diffusion equation in the target material with measured temperature information as boundary condition allows to determine the surface heat load distribution. A Bayesian analysis tool is developed as an alternative to deterministic tools, which aim for fast evaluation. The probabilistic evaluation uses adaptive kernels to model the heat flux distribution. They allow for self-consistent determination of the effective Degree of Freedom, depending on the quality of the measurement. This is beneficial, as the signal-to-noise ratio depends on the surface temperatures, ranging from room temperatures up to the melting point of tungsten.

Keywords Magnetically confined fusion · Power exhaust · Infrared thermography · Inverse problem

---

D. Nille (✉) · U. von Toussaint · B. Sieglin · M. Faitsch
Max-Planck-Institute for Plasma Physics, Boltzmannstrasse 2, 85748 Garching, Germany
e-mail: Dirk.Nille@ipp.mpg.de

U. von Toussaint
e-mail: Udo.v.Toussaint@ipp.mpg.de

B. Sieglin
e-mail: bernhard.sieglin@ipp.mpg.de

M. Faitsch
e-mail: Michael.Faitsch@ipp.mpg.de

A. Polpo et al. (eds.), *Bayesian Inference and Maximum Entropy Methods in Science and Engineering*, Springer Proceedings in Mathematics & Statistics 239, https://doi.org/10.1007/978-3-319-91143-4_6

1 Introduction

The shape and temporal evolution of the heat flux pattern of a magnetically confined plasma onto the first wall is of great interest for fusion research. Heat flux densities of several MW/m^2 pose a threat to the exposed material [1, 9]. The heat flux distribution is a footprint of the transport in the plasma edge [5, 6, 13]. Understanding the transport in the plasma edge is important to predict the behaviour of larger devices, aiming for a future fusion power plant. No direct measurement of the heat flux in the plasma is available. A method with sufficient spatial and temporal resolution to analyse many effects is to measure the thermal response of the target material, where the plasma deposits thermal energy. The impinging heat raises the temperature of the material, which itself transports the heat via conduction into the bulk. From the measured temporal evolution of the surface temperature, the heat flux into the material is deduced.

Contributions in the past years to improved infrared diagnostics in tokamaks, like [12], allow to deduce the heat flux density profile in great detail, by using a deterministic evaluation. However, a more precise deduction including proper error bars is desired. This work introduces a tool facilitating a probabilistic evaluation of the data. The forward model is based on the heat diffusion code called THEODOR (*THermal Energy Onto DivertOR*) [7]. It solves the heat diffusion equation in the target tile for the measured surface temperature as a boundary condition to deduce the heat flux onto the surface. This work presents a tool currently developed under the name BayTh (*Bayesian THEODOR*), using THEODOR to model the thermal response and adaptive kernel to describe the profile of interest.

The Bayesian approach allows to self-consistently deduce the optimal degree of freedom of the reconstruction. The experimental setup used at the tokamak ASDEX Upgrade in Garching is introduced in Sect. 2. The forward model is described in Sect. 3 including the model for the thermal response and the camera. Section 4 briefly introduces the adaptive kernel used to describe the heat flux profile. Section 5 outlines the implemented solvers to find the optimal solution. Section 6 explains how the statistical part is benchmarked in order to get a robust reconstruction. Section 7 shows an example of the developed tool BayTh applied to experimental data. Section 8 summarises this contribution.

2 The Measurement System

This section outlines the detailed description of the measurement system in [12]. Main component is an infrared camera, observing the surface of the divertor tile at a wavelength of 4.7 μm. Figure 1 shows a picture of the target tile taken by this camera. Assuming symmetry in toroidal direction in the device, it is sufficient to infer the 1D poloidal pattern from the 2D image. For the relevant direction, the spatial resolution is about 0.7 mm. The maximal fill level of the CMOS sensor corresponds to about 10^6 photons, which leads to a standard deviation of 10^3 photons for a normal distribution respective a signal-to-noise ratio of 10^3. A discretisation level of down to 46 photons

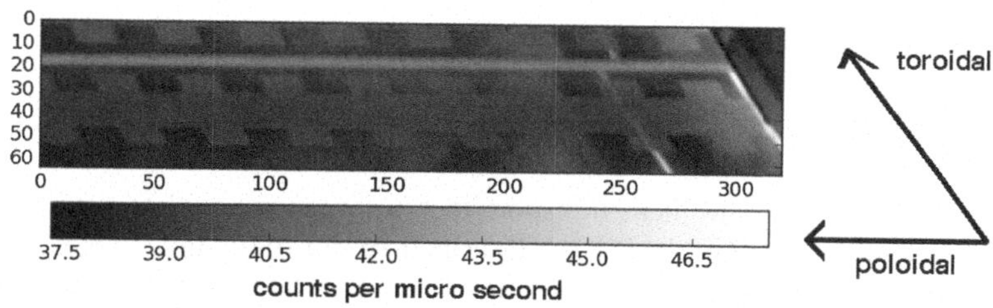

Fig. 1 IR camera image: the horizontal axis corresponds to the poloidal direction. Along the toroidal direction, the poloidal heat flux profile is about constant. The checkerboard like structure allows for movement correction

is achievable with the 15-bit ADC, which is mostly negligible compared to the photon noise. For the best SNR in all conditions the integration time is controlled by a real time system based on the histogram of the last image from 1 to 100 µs. An approximation of the expected noise level is derived from the heat diffusion equation, which is introduced in Sect. 3.1. The statistical noise in the photon flux translates to an effective temperature noise of about 30 mK. For an exemplary sample rate of 1 kHz —values between 800 and 3 kHz are typical —the effective noise in heat flux is calculated. At 500° C tungsten has a thermal diffusivity of about $5 \cdot 10^{-5}\,\mathrm{m}^2\,\mathrm{s}^{-1}$ and thermal conductivity of about 100 W/(m · K). The diffusion coefficient can be written as

$$\chi = \frac{\Delta x^2}{\Delta t} \tag{1}$$

described by the diffusion length Δx after time Δt. The diffusion length is thus given by the known thermal diffusivity and the diffusion time between two samples

$$\Delta x = \sqrt{\Delta t \cdot \chi} \approx \sqrt{1.10^{-3}\mathrm{s} \cdot 5 \cdot 10^{-5}\mathrm{m}^2\mathrm{s}^{-1}} \approx 2 \cdot 10^{-4}\mathrm{m} \tag{2}$$

From this we derive the perturbation in the heat flux density δq from a perturbation in the measured temperature δT:

$$\delta q = \kappa \frac{\delta T}{\Delta x} \approx 20\mathrm{kWm}^2\;. \tag{3}$$

Depending on the camera settings like the sample frequency an effective noise of about $\sigma_q = 20$ to $40\,\mathrm{kW\,m}^{-2}$ can be reached in good conditions with the evaluation using THEODOR.

3 Forward Model

3.1 Heat Diffusion

The forward model for the heat transport in the target material is based on the THEODOR code, described in [7], as described in the following. Figure 2 shows the situation: the rectangular cross-section of an isotropic material through which heat is conducted.

The heat transport in the divertor target is described by heat diffusion, with a nonlinear diffusion coefficient κ with respect to the temperature. The corresponding nonlinear second order partial differential equation of the temperature T reads

$$\frac{\partial T}{\partial t}\rho c_p = \nabla\left(\kappa(T)\nabla T\right) . \tag{4}$$

Here ρ and c_p are the mass density and specific heat capacity of the material. In this equation, the temperature is substituted by the heat potential

$$u(\kappa) = \int_0^T \kappa(T')\mathrm{d}T' \tag{5}$$

leading to the semilinear differential equation

$$\frac{\mathrm{d}u}{\mathrm{d}t} = \frac{1}{\rho c_p}\chi(u)\Delta u . \tag{6}$$

This system is solved using the finite difference implicit Euler scheme with operator splitting. The derivative is split into a part along the surface Δ_x and a part into the depth of the tile Δ_y. This leads to two tridiagonal systems, which are solved successively using the Thomas Algorithm [11]. The heat flux is deduced from the heat potential gradient in the first three layers:

$$q = \kappa(T)\nabla T = \nabla u \tag{7}$$

The same numerical scheme is used as a forward model, by setting the heat flux to the surface as boundary condition instead of the temperature. The relevant result of an iteration is then the surface temperature for the given heat flux distribution.

3.2 Measurement System

From the surface temperature modelled as described above the expected count rate of the IR camera [12] is deduced, depending on the calibration and integration time of the camera and the emissivity of the surface. The functional dependence

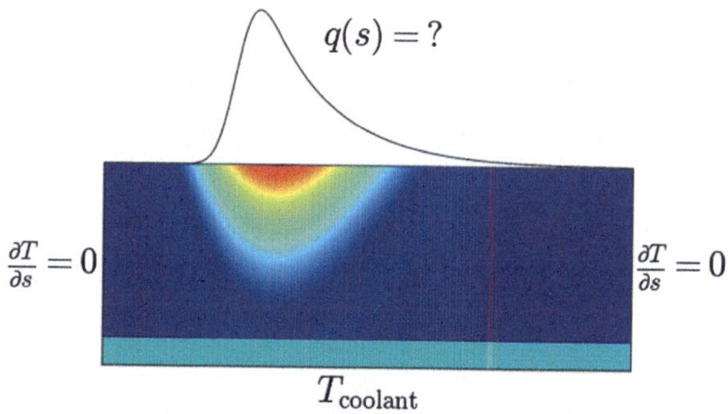

Fig. 2 Sketch of the cross-section of the target material with the temperature encoded in the colour. From measured surface temperatures, the spatially resolved heat flux density $q(s)$ impinging onto the surface has to be deduced. The lateral boundary conditions allow no heat transport while the back side is in contact with a coolant

between surface temperature and count rate is determined by Plancks Law of radiation. It describes the emission of electromagnetic radiation from an ideal black body with finite temperature. For the presented investigation the photon rate—energy rate divided by energy per photon—in a certain wavelength detected by the camera is of relevance. The photon flux emitted by a surface into the half-space is calculated as

$$\Gamma(T) = \int\int \varepsilon \frac{2\pi c}{\lambda^4} \frac{1}{\exp\left(\frac{hc}{\lambda kT}\right) - 1} \,\mathrm{d}A\,\mathrm{d}\lambda \tag{8}$$

ε represents the emissivity of the surface, c the speed of light, k denotes the Boltzmann constant and h the Planck constant. The equation can be approximated for an effective wavelength λ_{eff} and a constant c_0 describing the optical system:

$$\Gamma(T) \simeq c_0 \frac{2\pi c\varepsilon}{\lambda_{eff}^4} \frac{1}{\exp\left(\frac{hc}{\lambda_{eff} kT}\right) - 1} \tag{9}$$

The calibration coefficient c_0 depends on the aperture and sensitivity of the camera and losses in the optical system. Solving this equation for the temperature yields:

$$T = \frac{hc}{\lambda_{eff} k} \frac{1}{\ln\left(\frac{2\pi c\varepsilon}{\lambda_{eff}^4} \cdot \frac{c_0}{\Gamma} + 1\right)} \tag{10}$$

The likelihood is determined using a normal distribution, with the uncertainty deduced from the modelled counts per pixel.

4 Heatflux Model: Adaptive Kernel

To describe the 1D profile, an adaptive resolution approach is used. A description can be found in [4]. Instead of imposing smoothness via a penalty on derivatives of the function f, a structure — furthermore called hidden image h— is smoothed by a kernel function g:

$$f = g \times h \tag{11}$$

The wider and smoother the kernel, the smoother the resulting function. By using not a fixed width, but treating the width of every kernel in a discrete space as hyper parameters the resolution can be adjusted to reflect the information content of the measurement. The value of $f(x_0)$ of N Gaussian kernel at positions x_i with widths b_i is given by

$$f(x_0) = \sum_i \frac{h_i}{\sqrt{2\pi} b_i} \exp\left(-\frac{1}{2}\left(\frac{x_i - x_0}{b_i}\right)^2\right). \tag{12}$$

This approach has been shown to work well for positive additive distributions (PADs) [4] like spectroscopy and depth profiles. The noise level for our application is expected to vary in time and space, as the amount of emitted radiation depends on the temperature of the surface area, which can be strongly peaked. Instead of using a global regularisation term, the adaptive kernel approach allows a self-consistent determination of the *best* resolution.

4.1 Effective Number of Degrees of Freedom (eDOF)

The degrees of freedom is an important quantity for model comparison, as additional DOFs typically improve the likelihood, with not necessarily gaining more information about the system. In vector notation Eq. (12) is written

$$\mathbf{f} = \mathbf{Bh} \tag{13}$$

with **B** the transfer matrix. For the adaptive kernel, no explicit model comparison is necessary, as the complexity of the model is described by the transfer matrix, mapping the hidden image into the data space. In the simplest case, **B** is the unit matrix, corresponding to no smoothing and the hidden image being identical to the function f. On the other hand, **B** has not to be a square matrix, for more or less kernel than cells in the data space for over- respective under-sampling.

In the following, assume a square transfer matrix, mapping N values of the hidden image space onto N measurement locations. In this case, $2 \times N$ parameters are used, to describe N data points. With increasing width of the kernel, the flexibility of the model is reduced. For the limit of all widths larger than the system size $b \to \infty$ the effective degree of freedom (eDOF) is 1, as the resulting function consists only of a constant, with N contributions from the hidden image all being mutually anti-correlated. The other limit $b_i \to 0$ represents independent delta peaks. The function is therefore described by the N hidden image values, a point-wise reconstruction with N degrees of freedom. The eDOF can be determined by the eigenvalues of the squared transfer matrix

$$eDOF = \sum_{i=1}^{N} \sqrt{\text{eigenvalue}_i (B^T \cdot B)} \tag{14}$$

reaching from 1 to N.

5 Exploring the Parameter Space

For production work optimisation routines from NAG [10] and Minuit [2] are used, searching for the mode of the posterior distribution. The C++ code uses the *adept* [8] library to efficiently determine the Jacobian of the posterior with respect to all input parameters.

With 10^4–10^5 iterations per time step and run-times of few milliseconds per time step it takes about 10–100 s per time step and therefore the evaluation of typical 10.000 frames takes a few days. The minimisation time and stability benefit significantly from using the automatic differentiation. For the absolute time effort, though the evaluation of the Jacobian takes about 5 times the time of a regular iteration, drops by about a factor of 4. The precision of the gradient information is higher than with finite difference approaches.

Alternatively, a Marcov Chain Monte Carlo procedure is used to get insight into the marginal distributions of the single parameters. The thermal system is in good approximation linear for the common temperature gradients and though the photon flux scales about quadratic with the temperature it is considered linear as the uncertainty of well below 1 °C is negligible compared to the absolute temperature starting from about 400 K. Therefore, a normal distribution of the parameter is found, leading to a normal distributed posterior.

6 Synthetic Data as Benchmark

The abilities and limitations of the reconstruction are best shown based on the synthetic data, where the reference is known. Therefore, a 1D heat flux profile evolving in time as it is common in measurements is defined. For this case, a sample frequency of 1 kHz and spatial resolution of 1 mm is used, with an integration time of 10 µs. Figure 3 on the top left shows the heat flux density encoded in a colour map with the vertical axis representing the position and the horizontal axis the evolution in time. Using the forward model, the resulting surface temperature is calculated. Given the calibration factors and a typical integration time the signal of a virtual instance of the camera is determined. In the end, a realisation of noise according to the normal distribution is added. Figure 3 shows a comparison between the classic deterministic evaluation and the probabilistic evaluation with the form free adaptive kernel approach for synthetic data. For the deterministic approach, the count rate and integration time used for the probabilistic evaluation is converted to the corresponding temperature, as it is done in the actual measurement.

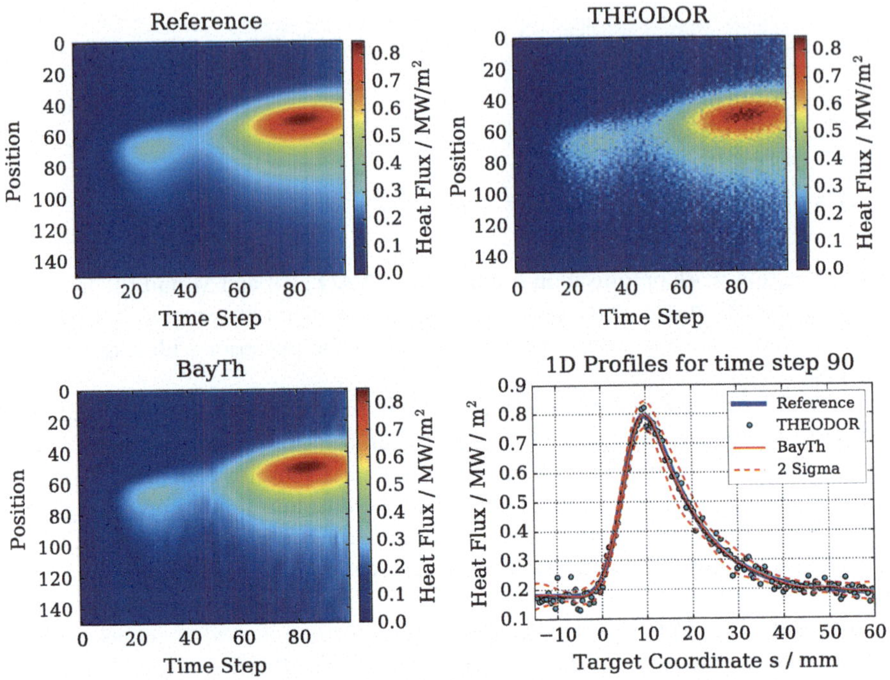

Fig. 3 Comparison between deterministic and probabilistic reconstruction of synthetic data. Top left: synthetic reference signal. Top right: deterministic reconstruction with THEODOR. Bottom left: probabilistic, form free reconstruction with the Bayesian THEODOR *BayTh*. Bottom right: direct comparison of 1D profiles. The dashed line represents the 2 sigma interval derived from the parameter distribution

For the specific case shown, the standard deviation is reduced from $\pm 25\,\mathrm{kW\,m^{-2}}$ for the deterministic evaluation to $\pm 3.9\,\mathrm{kW\,m^{-2}}$ for the probabilistic method. The adaptive kernel are able to describe the slopes of the profile, with a tendency to minor ringing at the ends of the profile. The latter is due to the additive nature of the kernel, which allows only an inward pointing gradient when no contributions from outside of the data range are allowed. This limitation is mitigated by placing an additional kernel element just outside of the reconstruction area.

7 Processing Measured Data

An example for measured data with a rich set of features is used to illustrate the possibilities of the method presented in this contribution. Figure 4 shows a profile from a study with magnetic perturbation [3], resulting in a splitting of the heat flux pattern. Though a reduced resolution of 1 Kernel only every two pixel was used, the fine features around 6 cm target position are well reproduced. Interesting is also the profile at the right half, where new structures are identified, inaccessible otherwise.

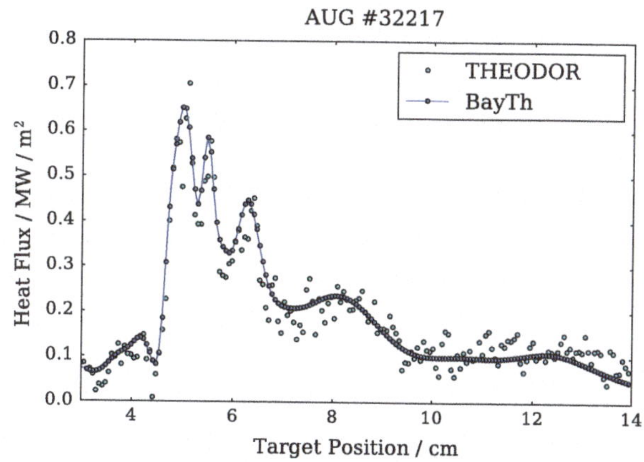

Fig. 4 Comparison of deterministic to probabilistic THEODOR: The adaptive kernel reconstruct fine peaks where necessary without over-fitting

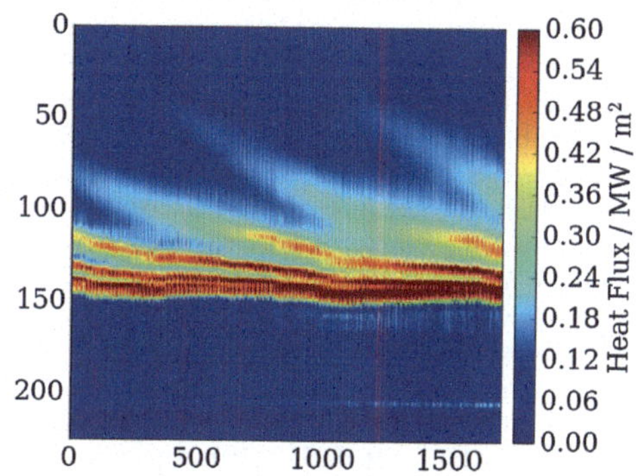

Fig. 5 Time evolution of the lobe structure calculated with BayTh. The heat flux is plotted as a heat map showing the position in pixel versus the frame number. A single profile is shown in Fig. 4

When looking at the time evolution in Fig. 5 the positions of the so called lobes is changing. Beside the possibility to determine the position of the peaks in the profile in a wider range, the hidden image shows the source of the contributions.

The main imperfection of the measurement is represented by hot-spots, as seen in Fig. 5 just below pixel 200. Imperfections of the surface or dust particles sticking to the surface change the emissivity. Foreign particles also lower thermal conduction to the bulk, changing the temperature response. Both changes tend to result in an overestimation of the surface temperature.

8 Conclusions

The probabilistic evaluation of infrared data allows to infer more details and reliable error bars for the reconstructions of heat flux pattern onto a surface.

Compared to the common deterministic approach with THEODOR, the probabilistic evaluation with BayTh has some striking benefits. Comparisons based on the synthetic data show a reduction of the standard deviation to the reference by a factor of 5–10. Evaluating experimental data, work in progress, reveals features predicted by a theory which were formerly not even accessible with post-processing.

Next steps will include improvements regarding systematic measurement errors like hot-spots. A long-term goal is an extension for 2D reconstructions.

Acknowledgements This work has been carried out within the framework of the EUROfusion Consortium and has received funding from the Euratom research and training programme 2014–2018 under grant agreement No 633053. The views and opinions expressed herein do not necessarily reflect those of the European Commission.

References

1. Bazylev, B., Janeschitz, G., Landman, I., Pestchanyi, S., Loarte, A., Federici, G., Merola, M., Linke, J., Zhitlukhin, A., Podkovyrov, V., Klimov, N., Safronov, V.: Iter transient consequences for material damage: modelling versus experiments. Phys. Scr. **2007**(T128), 229 (2007). https://doi.org/10.1088/0031-8949/2007/T128/044
2. Brun, R., Rademakers, F.: Root - an object oriented data analysis framework. In: Proceedings AIHENP'96 Workshop, Lausanne (1996): Nucl. Inst. Methods Phys. Res. A **389**, 81–86 (1997). http://root.cern.ch/
3. Faitsch, M., Sieglin, B., Eich, T., Herrmann, A., Suttrop, W.: Divertor heat load in asdex upgrade l-mode in presence of external magnetic perturbation. Plasma Phys. Control. Fusion (2017). https://doi.org/10.1088/1361-6587/aa75e7
4. Fischer, R., Mayer, M., von der Linden, W., Dose, V.: Enhancement of the energy resolution in ion-beam experiments with the maximum-entropy method. Phys. Rev. **55**, 6667–6673 (1997). https://doi.org/10.1103/PhysRevE.55.6667
5. Goldston, R.: Downstream heat flux profile versus midplane T profile in tokamaks. Phys. Plasmas **17**(1), 012503 (2010). https://doi.org/10.1063/1.3280011
6. Goldston, R.: Heuristic drift-based model of the power scrape-off width in low-gas-puff H-mode tokamaks. Nucl. Fusion **52**(1), 013,009 (2012). https://doi.org/10.1088/0029-5515/52/1/013009
7. Herrmann, A., Junker, W.: G"unther, K., Bosch, S., Kaufmann, M., Neuhauser, J., Pautasso, G., Richter, T., Schneider, R.: Energy flux to the asdex-upgrade diverter plates determined by thermography and calorimetry. Plasma Phys. Control. Fusion **37**(1), 17 (1995). https://doi.org/10.1088/0741-3335/37/1/002
8. Hogan, R.: Fast reverse-mode automatic differentiation using expression templates in C++. ACM Trans. Math. Softw. **40**(4), 26:1–26:24 (2014). https://doi.org/10.1145/2560359
9. Li, M., You, J.H.: Interpretation of the deep cracking phenomenon of tungsten monoblock targets observed in high-heat-flux fatigue tests at 20 mw/m2. Fusion Eng. Des. **101**, 1–8 (2015). https://doi.org/10.1016/j.fusengdes.2015.09.008
10. The NAG C Library, The Numerical Algorithms Group (NAG), Oxford, United Kingdom. www.nag.com. Accessed 15 July 2016
11. Press, W., Teukolsky, S.A., Vetterling, W.T., Flannery, B.P.: Numerical Recipes, 3rd edn. Cambridge University Press, Cambridge (2007)
12. Sieglin, B., Faitsch, M., Herrmann, A., Brucker, B., Eich, T., Kammerloher, L., Martinov, S.: Real time capable infrared thermography for asdex upgrade. Rev. Sci. Instrum. **86**(11), 113502 (2015). https://doi.org/10.1063/1.4935580
13. Stangeby, P., Chankin, A.: Simple models for the radial and poloidal E $\times$ B drifts in the scrape-off layer of a divertor tokamak: Effects on in/out asymmetries. Nucl. Fusion **36**(7), 839 (1996). https://doi.org/10.1088/0029-5515/36/7/I02

Schrödinger's Zebra: Applying Mutual Information Maximization to Graphical Halftoning

Antal Spector-Zabusky and Donald Spector

Abstract The graphical process of halftoning is, fundamentally, a communication process: an image made from a continuous set of possible grays, for example, is to be represented recognizably by elements that are only black or white. With this in mind, we ask what a halftoning algorithm would look like that maximizes the mutual information between images and their halftoned renditions. Here, we find such an algorithm and explore its properties. The algorithm is inherently probabilistic and bears an information theoretic similarity to features of quantum mechanical measurements, so we dub the method *quantum halftoning*. The algorithm provides greater discrimination of medium gray shades, and less so very dark or very light shades, as we show via both the algorithm's mathematical structure and examples of its application. We note, in passing, some generalized applications of this algorithm. Finally, we conclude by showing that our methodology offers a tool to investigate Bayesian priors of the human visual system, and spell out a scheme to use the results of this paper to do so.

Keywords Graphical process · Information theory · Quantum halftoning

1 Introduction

Graphic artists have known since at least 1850 [6] that it is possible to take images made from a continuous set of shades of gray and render them using only black and white elements. Doing so is known as halftoning [7], and it can be accomplished in a variety of ways [8]. Halftoning relies on the visual processes of observers to

A. Spector-Zabusky
Department of Computer and Information Science, University of Pennsylvania, Philadelphia, PA 19104, USA
e-mail: antals@seas.upenn.edu

D. Spector (✉)
Department of Physics, Hobart and William Smith Colleges, Geneva, NY 14456, USA
e-mail: spector@hws.edu

A. Polpo et al. (eds.), *Bayesian Inference and Maximum Entropy Methods in Science and Engineering*, Springer Proceedings in Mathematics & Statistics 239, https://doi.org/10.1007/978-3-319-91143-4_7

interpret the mixture of black and white elements as the original image. One can use halftoning not just on grayscale images, but for simplicity, we focus on that context.

Although halftoning is generally treated as a graphical or visual tool, at its heart, halftoning is an information theoretic communication process. Alice, the artist, has a grayscale image to convey to Bob, the beholder. However, Alice is constrained, and can only send Bob a black and white image. Even though this will eliminate information, Alice still wants to convey the original image as best she can.

This point of view leads us here to introduce a new halftoning algorithm, based on the information theory [5], designed to maximize the mutual information [1] between original images and their halftoned versions. The mathematics of the algorithm bears a similarity to certain problems associated with quantum measurement theory [9], and so we dub the method *quantum halftoning*.

In the next section, we formulate halftoning using the language of information theory, after which we determine the precise halftone mappings that maximize the relevant mutual information. This then allows us to specify the quantum halftoning algorithm, an inherently probabilistic algorithm. We explore the output of the algorithm and the images it produces. Along the way, we highlight the formal connection between this algorithm and quantum mechanics. Finally, we close by showing how our methodology provides a tool for exploring the Bayesian priors of the human visual system.

There is a large literature on halftoning, including deterministic and probabilistic algorithms, with well-known techniques such as dithering that incorporates two-dimensional information [7, 8]. What is new here is the use of an information theoretic measure to create the halftoning algorithm. We note that our method is most sensitive to variations in mid-range grays, and less so in very light or very dark regions, and standard methods appear to produce visually more faithful images. This potential weakness, however, is actually a strength: it means our method provides a tool to determine the Bayesian priors of the human visual system with respect to brightness, as we discuss at the end of the paper.

2 Information Theory and Halftoning

We are interested in halftoning procedures for grayscale images in which every grayscale pixel of the original image is replaced by a square array of N pixels, each of which must be either black or white. We refer to the gray pixels of the original image as the *inpixels* and the black and white pixels of the halftoned image as the *outpixels*; thus, there are N outpixels for each inpixel.

We can represent each shade of gray according to its brightness using the real numbers from 0 to 1, where 0 is black and 1 is white; the smaller the number, the darker the gray. As a viewer is primarily impacted by the proportion of black and white pixels in a given area, we will not be interested in the particular arrangement of black and white outpixels corresponding to a particular inpixel, only the fraction that are black.[1] Thus, to specify a halftoning algorithm in this context is to specify a function $p(x)$ that indicates what fraction of the N outpixels used to represent an inpixel of gray shade x should be black. Because $p(x)$ is intended to produce halftoned images, we impose the condition that $p(x)$ be monotonically nonincreasing so that lighter grays do not get encoded with a larger fraction of black pixels.

The essence of our algorithm is that we choose the function $p(x)$ so as to maximize an appropriate measure of the mutual information between the original and halftoned images. This means that we view $p(x)$ not as a fraction, but as a probability, and thus $p(x)$ is not actually the fraction of outpixels for a given gray inpixel that are black, but rather the probability that each such outpixel is black.

To maximize the mutual information, we must know the distribution that characterizes x; since our goal is an all-purpose algorithm, usable on any initial image, we assume that all values of x are equally likely, and so x is characterized by the uniform distribution $F(x) = 1$. (At the end of the paper, we reconsider this assumption, with some important implications.) We now need to find the $p(x)$ that maximizes the mutual information between the original distribution $F(x)$ and the results of applying $p(x)$ a total of N times. Two features are clear: $p(x)$ will depend on x (indeed, we expect $p(0) = 1$ and $p(1) = 0$, which in fact will be the case), and $p(x)$ will depend on N. Thus, going forward, we will denote the relevant probabilities $p_N(x)$.

Fortuitously, our problem is mathematically equivalent to one considered by Wootters [9]: namely, how Alice can best communicate a real number uniformly distributed on a closed interval to Bob by sending Bob a weighted coin that he can toss exactly N times. Thus, the mutual information we obtain has the same form as that obtained in this coin-toss procedure. Furthermore, Wootters establishes a connection between this information theoretic problem and measurements in quantum mechanics, and so we dub our halftoning method *quantum halftoning*.

We now return to constructing the requisite mutual information. If one has random variables X and Y, the mutual information $I(X : Y)$ is given by $H(X) - H(X|Y)$, where $H(X)$ is the entropy associated with X, and $H(X|Y)$ is the conditional entropy of X given Y. For our problem, the two variables are the shade of gray x, with distribution $F(x) = 1$, and the number of black pixels n_b, arising from N applications of $p_N(x)$. Since x is continuous and n_b is discrete, we use integrals for x, where we have sums for n_b, which means using the differential entropy [1] where necessary. The mutual information here then takes the same form as in Wootters's coin-toss procedure [9], namely,

$$I(n_b : x) = -\sum_{n_b=0}^{N} \overline{P}(n_b) \log \overline{P}(n_b) + \int_0^1 dx \Big[\sum_{n_b=0}^{N} \hat{P}\big(n_b | p_N(x)\big) \log \hat{P}\big(n_b | p_N(x)\big) \Big], \tag{1}$$

where $\overline{P}(n_b)$ is the probability of getting n_b black pixels averaged over all possible values of x, and the conditional probability $\hat{P}$ is the binomial distribution,

---

[1] From a purely information theoretic point of view, one could of course distinguish among different arrays of N pixels that have the same fraction of black pixels, but this information is not generally accessible in a meaningful way to human viewers.

$$\hat{P}(n_b|q) = \frac{N!}{n_b!(N-n_b)!} q^{n_b}(1-q)^{N-n_b} . \quad (2)$$

With these results, we now seek to find the $p_N(x)$, the probability that each of the N outpixels corresponding to an original spot of grayness x will be black, that will maximize the mutual information in Eq. (1).

Naively, one might expect something like $p_N(x) = 1 - x$, i.e., that as the gray gets lighter, the probability of generating black pixels should go down. However, this is not the case. One can get a sense of what is going on by considering the case $N = 1$, where each gray inpixel is turned into either a black or a white outpixel. If the original shade is a darker gray, i.e., $x < 0.5$, then 0 (black) is always closer to x than is 1 (white). As a consequence, the mutual information is maximized in this case, when $p_1(x) = 1$ for $x < 0.5$ and $p_1(x) = 0$ for $x > 0.5$; there is never any advantage to mixing in the alternate result, and so we get a step function. (The precise value of $p_1(0.5)$ is irrelevant, since $x = 0.5$ is a set of measure zero.) Put another way, one pixel can only convey a single bit of information, and so the best that can be done is to divide the interval into two regions of equal size.

Some features of the $p_N(x)$ that maximize the mutual information were found by Wootters [9]. For any $N > 2$, while there are values of x for which it is desirable for $p_N(x)$ to be other than 1 or 0, the stepping nature of the solution persists. The optimal solution takes the interval [0, 1] and divides it into L distinct subintervals (L depends on N), with some widths $w_1, \ldots, w_L$, where $\sum_{j=1}^{L} w_j = 1$; for each of these subintervals, there is a constant q_j (with $j = 1, \ldots, L$) such that $p_N(x) = q_1$ when $0 \le x \le w_1$, and $p_N(x) = q_j$ when $w_1 + \cdots + w_{j-1} < x \le w_1 + \cdots + w_j$ for $j = 2, \ldots, L$. Requiring that $p_N(x)$ be monotonically nonincreasing means, then, that a plot of this function looks like a descending staircase. The value of L needed to maximize the mutual information is bounded from above by $\lfloor N/2 \rfloor + 2$, though numerical work shows the necessary L is generally much smaller than that.

With these results, the mutual information Eq. (1) takes the simplified form [9]

$$I(n_b : x) = -\sum_{n_b=0}^{N} \overline{P}(n_b) \log \overline{P}(n_b) + \sum_{j=1}^{L} w_j \Big[\sum_{n_b=0}^{N} \hat{P}(n_b|q_j) \log \hat{P}(n_b|q_j) \Big], \quad (3)$$

with $\overline{P}(n_b) = \sum_{j=1}^{L} w_j \hat{P}(n_b|q_j)$. When the mutual information is maximized, the staircases are symmetric, with the symmetry given by the relations $w_{L+1-j} = w_j$ and $q_{L+1-j} = 1 - q_j$. In addition, $q_1 = 1$ and $q_L = 0$. Beyond that, the particular values of L, w_j, and q_j depend on N, and we will need to compute these values to implement our algorithm.

Since for $N > 2$, it is not feasible to solve for the optimal $p_N(x)$ analytically, we used Mathematica to obtain the optimal $p_N(x)$ numerically for all $N \le 36$. In order not to distort the aspect ratio, we will always replace each inpixel from the original image with a square block of outpixels, so we present the results for

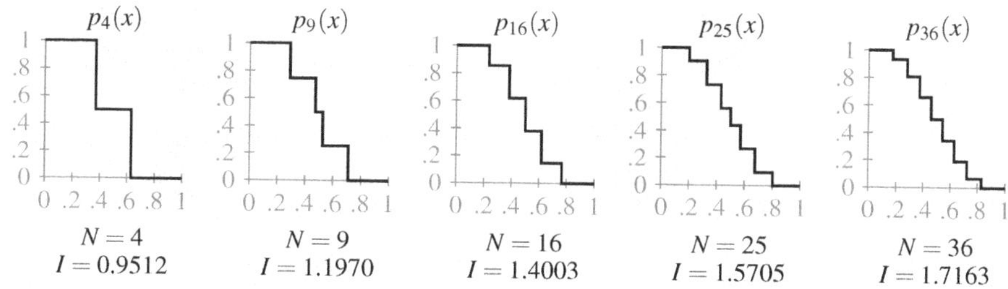

Fig. 1 The probability distributions $p_N(x)$ that maximize the mutual information I for five cases of interest. The mutual information I associated with each optimal $p_N(x)$ is listed underneath each graph. The varying sizes of the steps within each "staircase" make the algorithm more sensitive for medium than light or dark grays

$N = 4, 9, 16, 25$, and 36, displaying these results graphically in Fig. 1 and numerically (along with the $N = 1$ case) in the appendix.[2]

One noteworthy result is that the number of staircase steps is less than one might expect. For example, in the case of $N = 4$ pixels, a deterministic halftoning algorithm could use five possible black pixel ratios: 100, 75, 50, 25, and 0%. Yet in a probabilistic algorithm, dividing up the interval [0, 1] into five regions with $p(x)$ taking on these five distinct values yields a smaller value for the mutual information than the solution presented here with the three regions. Our optimal $p_4(x)$ yields a mutual information of 0.9512. Dividing the interval into five regions of equal width, with $p(x)$ taking the values 1.00, 0.75, 0.50, 0.25, and 0.00 in those regions, yields a mutual information of only 0.7912. Using five regions with those same five "natural" values for $p(x)$, but having the middle three regions centered, respectively, on $x = 0.25, 0.50$, and 0.75, each taking up a quarter of the interval, with $p(x)$ taking the values 1 and 0 on the ends, yields an even smaller mutual information of 0.6426.

3 Quantum Halftoning

We now specify the quantum halftoning algorithm. The crux of the process is replacing each gray pixel, its shade specified by a real number x, with a square of N black and white outpixels. We do this in a way that maximizes the mutual information, using the results from the preceding section.

The inputs to our algorithm are a grayscale image and the parameter N (which must be a perfect square), and the output is a halftoned image, where each inpixel has been mapped to a block of N black and white outpixels. The method for turning each inpixel into N outpixels relies on the probability distribution $p_N(x)$ that maximizes the mutual information in Eq. (1). In particular, given an inpixel with a shade of gray corresponding to the real number x (recall that $x \in [0, 1]$), we set the N outpixels

[2] Incidentally, the limit of these curves as $N \to \infty$ is $p(x) = \cos^2(\pi x/2)$ [9].

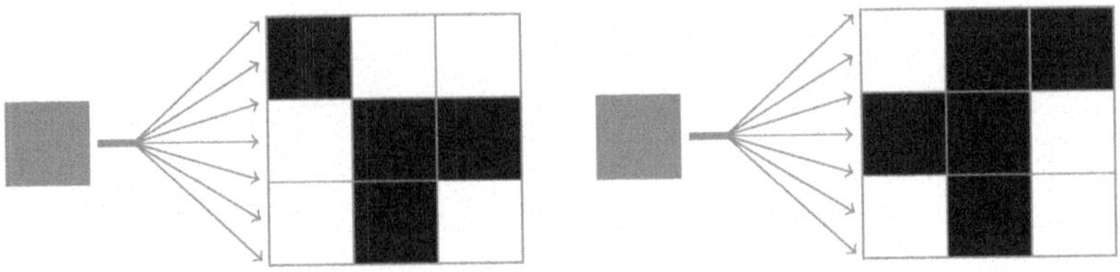

Fig. 2 Two different iterations of a grayscale pixel being expanded into a 3×3 square of black and white pixels

corresponding to this inpixel randomly to black or white, where black is chosen with probability $p_N(x)$ and white with probability $1 - p_N(x)$. An example of how a grayscale pixel might be expanded is shown in Fig. 2. The various N-outpixel blocks are positioned relative to each other just as their associated inpixels were, and this produces the quantum halftoned image.[3]

Note that the probabilistic interpretation of $p_N(x)$ is natural here, since the values of $p_N(x)$ that maximize the mutual information will not generally be multiples of $1/N$. This means $p_N(x)$ is the *expected* value, rather than the *actual* value, of the fraction of black outpixels coming from an inpixel of shade x.

As this algorithm is inherently nondeterministic, the quantum halftoned version of an image will not necessarily be the same each time. The probabilistic nature of the algorithm has advantages and disadvantages. On the one hand, there is a small chance that an inpixel for which $p_N(x) = 0.4$ could be mapped into a square of N outpixels that are, say, 70% black. On the other hand, if the original image has an extended region of a single shade of gray, the stochastic nature of the algorithm allows us to get an average number of black pixels closer to $p_N(x)$ over the extended region than possible over a single block of N pixels or in a deterministic algorithm that treats each block of N pixels independently, without dithering or other such adjustments.

Note that one can store the information needed to reconstruct the quantum halftoned image in a compressed format. For concreteness, consider the case $N = 25$, for which the quantum halftoning algorithm invokes one of eight possible probabilities. With just three bits per inpixel, we can specify the probability that the corresponding outpixels should be black. This information could be stored, and then, at the time of viewing, used to construct a quantum halftoned image. Note that in this scenario, the precise form of the image is not determined till it is generated.

We observe that the connection between our halftoning algorithm and quantum measurements is both qualitative and quantitative. In quantum mechanics, there is a continuous collection of possible states, but when certain properties are measured, there is only a discrete set of possible outcomes. For example, when light encounters a linear polarizer that has its transmission axis oriented vertically, each photon individually winds up either vertically polarized and transmitted, or horizontally polarized and absorbed. The probability of each outcome is fixed by the rules of

[3]Scenarios in which $s \times s$ sets of inpixels are averaged and then mapped to N bits are also possible.

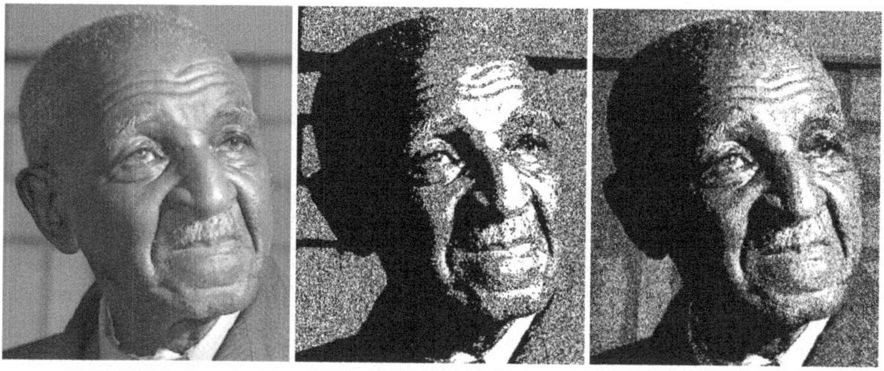

(a) Original image (b) Quantum halftoned, $N = 4$ (c) Quantum halftoned, $N = 9$

Fig. 3 George Washington Carver [3], quantum halftoned at different values of N. Note the extra details, such as the wrinkles in the forehead, that emerge at higher N

quantum mechanics. Thus, a continuous set of possibilities is pigeonholed into one of only two possible outcomes via a probabilistic process, as in our halftoning algorithm. Furthermore, if we restrict to incoming photons that are linearly polarized,[4] the probabilities for the vertical and horizontal outcomes in quantum mechanics are precisely those that maximize, in the large N limit [9], the same mutual information Eq. (1) used in our halftoning algorithm.

4 Implementation and Examples

We implemented the quantum halftoning algorithm in the GHC [10] implementation of the Haskell programming language. The program takes as input the value of $\sqrt{N}$ and a grayscale image in a standard format. It first normalizes the gray shade of each inpixel to lie within [0, 1]. It then takes each inpixel of the original image and represents it by a square of N outpixels in the halftoned image. These outpixels are (pseudo)randomly filled with either black or white according to $p_N(x)$; when x is the inpixel's shade of gray, each corresponding outpixel is black with probability $p_N(x)$ and white otherwise. Of course, this enlarges the pixel count of the image.

To explore the algorithm, we examined how the quantum halftoned versions of some grayscale images varied with N, as well as how they varied under repeated applications of the algorithm at fixed N. We display some examples here.

The dependence on N is exemplified by the images of George Washington Carver in Fig. 3. As one would expect, the halftoned image is more faithful to the original

[4]Or, more generally, if we consider real-vector-space quantum mechanics, with real rather than complex probability amplitudes.

(a) Original image (b) Quantum halftoned (c) Quantum halftoned (d) Quantum halftoned

Fig. 4 Neil Armstrong [4], quantum halftoned multiple times with $N = 9$. There are differences between the images, but only slight ones, such as that the white patches in Armstrong's face have slightly different boundaries from each other

image at larger N, but the halftoned version is already quite recognizable at $N = 4$, even though $p_4(x)$ has only three possible values.

Due to the nondeterministic nature of quantum halftoning, each application of the algorithm will provide a slightly different image, even at fixed N. We find that even for small N, these fluctuations do not affect the integrity of the image. In Fig. 4, we see three separate $N = 9$ quantum halftoned versions of an image of Neil Armstrong. The fluctuations are more evident in regions where the original image is a medium gray, neither very dark nor very light, but the fluctuations average out to produce effectively equivalent images.

As quantum halftoning is probabilistic, one can formulate dynamic implementations of the algorithm, using the algorithm repeatedly to update the output after the image is initially rendered. This does add flicker, but enables the observer to average the outpixels over time as well as space. One can also apply quantum halftoning to color images in the standard way, for example by working in RGB color space and applying the algorithm separately to the red, green, and blue values of the inpixels. Additionally, an alternative way to view our results is not as a halftoning algorithm, but as a lossy deterministic compression algorithm, in which each inpixel replaced by a pixel of the gray shade $1 - p_N(x)$ (so only a small set of grays is possible), which would provide a kind of visualization of mutual information maximization.

5 Obtaining Insights Regarding Human Vision

At the start of the paper, we took the distribution on the brightness x of the input image pixels to be the uniform distribution, $F(x) = 1$, and then constructed the mutual information to maximize. While this might seem reasonable barring knowledge of the original images to be rendered, it is not clear that this is actually the correct choice. Mathematically, the choice of brightness is somewhat arbitrary. After all, one could just as well define an alternative variable $y(x)$, where $y(0) = 0$, $y(1) = 1$, and y

Table 1 Optimal $p_1(x)$
Mutual information 0.6931

Min x	Max x	Prob.
0	0.5000	1.0000
0.5000	1	0.0000

Table 2 Optimal $p_4(x)$
Mutual information 0.9512

Min x	Max x	Prob.
0	0.3700	1.0000
0.3700	0.6300	0.5000
0.6300	1	0.0000

Table 3 Optimal $p_9(x)$
Mutual information 1.1970

Min x	Max x	Prob.
0	0.2887	1.0000
0.2887	0.4735	0.7463
0.4735	0.5265	0.5000
0.5265	0.7113	0.2537
0.7113	1	0.0000

increases monotonically, and apply the logic of quantum halftoning to that variable. Physically, halftoning as a method relies on the properties of human vision, but the human vision has evolved in the context of particular visual stimuli with particular impacts on the ability to survive and propagate one's genes. As a consequence, it is plausible, indeed likely, that the human visual system is not premised on all brightnesses being equally expected.

Fortunately, our method allows for the exploration of the built-in Bayesian priors of human vision. In particular, one can try a variety of potential distributions for $F(x)$ other than the uniform distribution, maximize the mutual information with respect to those new $F(x)$ expressions, and produce sets of images with those new schemes. Then, by surveying people to see which images appear to them to be the most faithful to the original images, we would have a means of measuring the Bayesian priors of the human visual system with respect to brightness. The ability to do this is a distinctive consequence of our halftoning method.

Technically, there is a way to do this that enables us to use the precise mathematical results of this paper, without repeatedly computing and maximizing new mutual information expressions. Rather than directly repeating the above analysis for different $F(x)$, we instead can map x to various possibilities $y(x)$ as suggested above; set $F(y) = 1$, implicitly determining a new $F(x)$; and then construct the quantum halftoned images in y-space rather than x-space. This would allow us to use the same staircase functions, just over y instead, simplifying the process of exploring alterna-

Table 4 Optimal $p_{16}(x)$ Mutual information 1.4003

Min x	Max x	Prob.
0	0.2343	1.0000
0.2343	0.3821	0.8553
0.3821	0.5000	0.6187
0.5000	0.6179	0.3813
0.6179	0.7657	0.1447
0.7657	1	0.0000

Table 5 Optimal $p_{25}(x)$ Mutual information 1.5705

Min x	Max x	Prob.
0	0.1976	1.0000
0.1976	0.3238	0.9047
0.3238	0.4290	0.7332
0.4290	0.5000	0.5618
0.5000	0.5710	0.4382
0.5710	0.6762	0.2668
0.6762	0.8024	0.0953
0.8024	1	0.0000

Table 6 Optimal $p_{36}(x)$ Mutual information 1.7163

Min x	Max x	Prob.
0	0.1706	1.0000
0.1706	0.2792	0.9333
0.2792	0.3720	0.8088
0.3720	0.4580	0.6593
0.4580	0.5420	0.5000
0.5420	0.6280	0.3407
0.6280	0.7208	0.1912
0.7208	0.8294	0.0667
0.8294	1	0.0000

tives to the uniform distribution over brightness, which would, in turn, simplify the execution of such a study.

6 Conclusion

In this paper, we have formulated quantum halftoning, a stochastic method of halftoning grounded in information theory. This method, designed to maximize mutual information rather than directly mirroring brightness level, produces recognizable images, but with some distinctive features, most notably a greater sensitivity to changes in medium gray levels and a lesser sensitivity to very light or dark grays. The algorithm is not only stochastic, but also shares an information theoretic structure that is mathematically analogous to features of quantum mechanics restricted to real probability amplitudes. Perhaps, the most compelling feature of our technique is that it offers a methodology for determining the Bayesian priors of the human visual system with respect to brightness (and, indeed, to see how much these vary across individuals and cultures), which we are currently exploring.

Acknowledgements This work was supported in part by NSF award #1521523 (Antal Spector-Zabusky) and a grant from FQXi (Donald Spector).

Appendix

The Tables 1, 2, 3, 4, 5 and 6 give the probability distributions that maximize the mutual information for the cases of interest to us, where $p_N(x)$ is the probability for producing a black pixel when the grayness is given by x and there are N outpixels per inpixel.

References

1. Cover, T.M., Thomas, J.A.: Elements of Information Theory, 2nd edn. Wiley Interscience, New York (2006)
2. Griffiths, D.J.: Introduction to Quantum Mechanics, 2nd edn. Pearson Prentice Hall, USA (2014)
3. Image of George Washington Carver. https://commons.wikimedia.org/wiki/File:George_Washington_Carver-crop.jpg (2018). Accessed 21 Feb 2018
4. Image of Neil Armstrong. https://www.dfrc.nasa.gov/Gallery/Photo/Pilots/Large/E-3342.jpg (2018). Accessed 21 Feb 2018
5. Shannon, C.E.: A mathematical theory of communication. Bell Syst. Tech. J. **27**, 379 (1948)
6. Stulik, D.C., Kaplan, A.: Halftone. Getty Conservation Institute. https://www.getty.edu/conservation/publications_resources/pdf_publications/pdf/atlas_halftone.pdf (2013). Accessed 28 June 2017
7. Ulichney, R.: Digital Halftoning. The MIT Press, Cambridge (1987)
8. Ulichney, R.: A review of halftoning techniques. SPIE **3963**, 378 (2000)
9. Wootters, W.K.: Communicating through probabilities: does quantum theory optimize the transfer of information? Entropy **15**, 3130 (2013)
10. The Glasgow Haskell Compiler. See information at https://www.haskell.org/ghc/. Accessed 28 June 2017

Regression of Fluctuating System Properties: Baryonic Tully–Fisher Scaling in Disk Galaxies

Geert Verdoolaege

Abstract In various interesting physical systems, important properties or dynamics display a strongly fluctuating behavior that can best be described using probability distributions. Examples are fluid turbulence, plasma instabilities, textured images, porous media and cosmological structure. In order to quantitatively compare such phenomena, a similarity measure between distributions is needed, such as the Rao geodesic distance on the corresponding probabilistic manifold. This can form the basis for validation of theoretical models against experimental data and classification of regimes, but also for regression between fluctuating properties. This is the primary motivation for geodesic least squares (GLS) as a robust regression technique, with general applicability. In this contribution, we further clarify this motivation and we apply GLS to Tully–Fisher scaling of baryonic mass vs. rotation velocity in disk galaxies. We show that GLS is well suited to estimate the coefficients and tightness of the scaling. This is relevant for constraining galaxy formation models and for testing alternatives to the Lambda cold dark matter cosmological model.

Keywords Regression analysis · Information geometry · Rao geodesic distance · Tully-Fisher scaling

1 Introduction

In many parametric regression problems, robustness of the estimates is an essential criterion, sometimes even more important than goodness-of-fit. Here, by "robustness" we mean not only resilience against outliers, but also relative insensitivity to model uncertainty. A multitude of techniques, Bayesian and non-Bayesian, have been developed ensuring robustness in the presence of various departures from the

G. Verdoolaege (✉)
Department of Applied Physics, Ghent University, B-9000 Ghent, Belgium
e-mail: geert.verdoolaege@ugent.be

G. Verdoolaege
Laboratory for Plasma Physics, Royal Military Academy, B-1000 Brussels, Belgium

A. Polpo et al. (eds.), *Bayesian Inference and Maximum Entropy Methods in Science and Engineering*, Springer Proceedings in Mathematics & Statistics 239, https://doi.org/10.1007/978-3-319-91143-4_8

regression model. However, it can be difficult for the nonexpert user to make the right choices of methods and implementation details. This constitutes a major obstacle for adoption of the right techniques by practitioners in various application domains with little tradition in the data sciences.

In this paper, we advocate the use of a simple but powerful robust regression method, called *geodesic least squares* (GLS), that was previously introduced in [1] and [2]. The purpose of the present contribution is, first, to generalize to a certain extent the theoretical underpinnings of the method, and second, to compare the performance of the method in a practical application from astronomy with other methods, including a standard robust Bayesian approach.

The motivation for GLS can be explained using a simple example that essentially describes a very common situation. Imagine the turbulent flow of a fluid through a pipe with a variable cross-section. The regression task consists of finding a relation between the flow speed of the fluid (response variable) and the cross-section of the pipe (predictor variable), based on the reading from flow meters positioned at different locations along the axis of the pipe. Of course, the flow speed measured by any of these meters fluctuates in time. Thus, even when the predictor variable is held fixed, the response variable is not constant. So far, this is a common regression problem, which the vast majority of practitioners from applied fields would solve by calculating time averages and performing least squares regression between these average flow speeds and the measured cross-section of the pipe.[1] A standard maximum likelihood or Bayesian solution would be possible too, basing the likelihood on the distribution of velocity fluctuations (neglecting measurement uncertainty), which we assume to be known from a previous experiment. However, in addition to the turbulent fluctuations there can be other sources of uncertainty. In the present example, this could for instance be due to a variable pumping speed due to a malfunctioning pump. If the flow readings are taken sequentially, this could introduce additional uncertainty not captured by the distributional properties of the intrinsic turbulent fluctuations. This is a case of incorrectly specified uncertainties, which can be handled using various Bayesian approaches depending on the specific problem (see, e.g., [3]). The solution provided by GLS is to consider, on the one hand, the distribution (likelihood) of the flow velocity that would be expected if all assumptions regarding the deterministic and probabilistic components of the regression model were correct. We call this the *modeled distribution*. On the other hand, the distribution of the data is characterized in a generic way using as few assumptions as possible, referred to here as the *observed distribution*. Then, similar to minimization by the least squares method of the sum of squared Euclidean distances between a measurement of the dependent variable and its modeled value, GLS estimates the model parameters by minimizing the sum of squared Rao geodesic distances between the observed and modeled distribution. This introduces extra flexibility ('elasticity') in the analysis, which, in practice, yields excellent robustness properties. Effectively, GLS performs regression between probability distributions on a Riemannian probabilistic manifold. It can also be char-

[1] In fact, any measurement with finite precision is an average over some smaller scale, e.g., the measurement of the cross-section of the pipe.

acterized as a minimum distance method, generalizing likelihood-based techniques, although there are important differences with standard minimum distance estimation (MDE). Another typical example of the application of GLS was treated in [4], relating the properties of a repetitive instability in tokamak plasmas. The distributions of two properties of the instability were determined under stationary plasma conditions and then the regression was carried out between those distributions. Indeed, GLS can take into account uncertainty on all variables (predictor and response). There are many other examples where the GLS approach is natural, e.g., involving signals, images, porous media, cosmological structure, etc., although the method itself is of general applicability.

After explaining the motivation for GLS based on the regression between fluctuating system properties, in this contribution we illustrate the applicability of GLS to common regression problems by estimating a key scaling law in astrophysics: the baryonic Tully–Fisher relation. This is a remarkably tight relation between the total baryonic mass of disk galaxies and their rotational velocity, of great practical and theoretical significance in astrophysics and cosmology.

2 GLS Regression: Principles and Motivation

In parameter estimation problems like regression analysis, the likelihood compares measured quantities with their value predicted by the model, under stationary experimental conditions, determined by fixed, or stationary predictor variables. Hence, the likelihood serves as a distance measure between the measurement and the model. Maximization of the joint likelihood for all measurements is equivalent to the minimization of the Kullback–Leibler divergence (KLD) between the empirical ("observed") distribution and the theoretical ("modeled") distribution of the residuals. In general, MDE techniques can be made more robust against model uncertainty by relying on similarity measures other than the KLD. The Hellinger divergence (closely related to the Bhattacharyya distance) is a common choice [5], first applied to regression in [6].

We follow a somewhat different approach, minimizing the Rao geodesic distance (GD) between the observed and modeled distributions. Consider a parametric multiple regression model involving m predictor variables ξ_j $(j = 1, \ldots, m)$ and a single response variable η, all assumed to be infinitely precise. Suppose that N measurements are acquired for the predictor variables, resulting in measurements ξ_{Ij} $(I = 1, \ldots, N)$. The regression model can be written as follows:

$$\eta_I = f(\xi_{I1}, \ldots, \xi_{Im}, \beta_1, \ldots, \beta_p) \equiv f(\{\xi_{Ij}\}, \{\beta_k\}), \qquad \forall I = 1, \ldots, N. \tag{1}$$

Here, f is the regression model function, in general nonlinear and characterized by p parameters β_k $(k = 1, \ldots, p)$. In regression analysis within the astronomy community, it is customary to add a noise variable to the idealized relation (1). This so-called *intrinsic scatter* serves to model the intrinsic uncertainty on the theoretical relation, i.e., uncertainty not related to the measurement process. We take another route for capturing model uncertainty, however.

In any realistic situation, we have no access to the quantities η_I and ξ_{Ij}. Instead, we assume that at each "measurement site" I a series of n_I measurements $x_{i_I j}$, resp. y_{i_I} is collected for the noisy predictor variables x_j and the response variable y ($i_I = 1, \ldots, n_I$). In this paper, we assume that the measurement model describes fluctuation of the data around a point that lies exactly on the regression function. This need not be the case in reality, which is one of the potential causes of model uncertainty. Nevertheless, if there are multiple measurements at each measurement site, then this can provide useful information on the true distribution of the data under stationary conditions. A common situation is where, at fixed I, the $x_{i_I j}$ and y_{i_I} represent measurements of noisy stationary signals. In the remainder of the paper, we will assume independent Gaussian noise, but this can be generalized to multivariate or non-Gaussian distributions. In the independent Gaussian case, we have

$$\begin{aligned} y_{i_I} &= \eta_I + \varepsilon_{y,i_I}, \qquad \varepsilon_{y,i_I} \sim \mathcal{N}(0, \sigma_{y,I}^2), \\ x_{i_I j} &= \xi_{Ij} + \varepsilon_{x,i_I j}, \quad \varepsilon_{x,i_I j} \sim \mathcal{N}(0, \sigma_{x,Ij}^2). \end{aligned} \tag{2}$$

Notice that ,in general, the standard deviations can be different at each measurement site. For instance, in many real-world situations, such as the one discussed in this paper, there is a constant relative error on the measurements, so the standard deviation can be modeled as being proportional to the measurement itself. Of course, the noise described by the $\sigma_{y,I}$ and $\sigma_{x,Ij}$ need not be the only source of uncertainty contributing to fluctuation of the data around the regression model. This is the case of interest in this paper, where other uncertainty sources such as model uncertainty are present (cf. the intrinsic scatter mentioned before), which could even be more important than the noise at the individual measurement sites and about which little is known. For now we assume that the standard deviations $\sigma_{y,I}$ and $\sigma_{x,Ij}$ were estimated prior to the regression analysis. This may be as simple as calculating the standard deviation of the y_{i_I} and $x_{i_I j}$ at each measurement site. We also include the possibility where $n_I = 1$ for some or all I, in which case the noise variables $\sigma_{y,I}$ and $\sigma_{x,Ij}$ could be given by the error bars obtained from previous experiments or an uncertainty analysis.

In reality, the true model points $(\eta_I, \xi_{I1}, \ldots, \xi_{Im})$ from which the data are assumed to be generated are unknown, but we can estimate them by calculating averages $\bar{y}_I \equiv 1/n_I \sum_{i_I}^{n_I} y_{i_I}$ and $\bar{x}_{Ij} \equiv 1/n_I \sum_{i_I}^{n_I} x_{i_I j}$, which are expected to be distributed according to $\mathcal{N}(0, \sigma_{y,I}^2/n_I)$ and $\mathcal{N}(0, \sigma_{x,Ij}^2/n_I)$, respectively. Now suppose that the model given by (1) and (2) were exact, meaning that $\sigma_{x,Ij}$ and $\sigma_{y,I}$ would characterize the only uncertainty sources, then the joint likelihood of the average data would be given by

$$p(\{\bar{y}_I\}, \{\bar{x}_{Ij}\}|C_\xi) = \prod_{I=1}^{N} \frac{1}{\sqrt{2\pi/n_I}\sigma_{y,I}} \exp\left\{-\frac{1}{2}\frac{\left[\bar{y}_I - f(\{\xi_{Ij}\}, \{\beta_k\})\right]^2}{\sigma_{y,I}^2/n_I}\right\}$$
$$\times \prod_{j=1}^{m} \frac{1}{\sqrt{2\pi/n_I}\sigma_{x,Ij}} \exp\left\{-\frac{1}{2}\frac{\left[\bar{x}_{Ij} - \xi_{Ij}\right]^2}{\sigma_{x,Ij}^2/n_I}\right\}. \quad (3)$$

Here, C_ξ stands for the collection $\{\beta_k\}, \{\xi_{Ij}\}, \{\sigma_{x,Ij}\}, \{\sigma_{y,I}\}$, the notation $\{\bar{x}_{Ij}\}$ referring to the set of $\bar{x}_{Ij}$ for all I and j, and similar for other sets. Also, we use the same indices for summation and for indicating set members, in order not to complicate the notation. As the ξ_{Ij} are not known, they have to be marginalized over. This is usually accomplished by decomposing the line from the measurement to the unknown point on the model in a perpendicular and parallel component w.r.t. the model, and assuming a uniform prior on the coordinates along the model surface [3, 7]. For a linear model, effectively this comes down to inserting the measurement values into the model equation, and propagating the uncertainty on the predictor variables through the model. Treatment of a nonlinear model is more complicated, but can be simplified by a linear approximation of the model in the vicinity of the model point nearest to the data point. Alternatively, one can perform Gaussian error propagation to obtain an approximate normal conditional likelihood for $\{\bar{y}_I\}$:

$$p_{\text{mod}}(\{\bar{y}_I\}|C_x) = \prod_{I=1}^{N} \frac{1}{\sqrt{2\pi}\sigma_{\text{mod},I}} \exp\left\{-\frac{1}{2}\frac{\left[\bar{y}_I - f(\{\bar{x}_{Ij}\}, \{\beta_k\})\right]^2}{\sigma_{\text{mod},I}^2}\right\}. \quad (4)$$

In this expression, C_x stands for the collection $\{\beta_k\}, \{\bar{x}_{Ij}\}, \{\sigma_{x,Ij}\}, \{\sigma_{y,I}\}$. The uncertainty on the predictor variables propagates through the function f and adds to the conditional uncertainty on the response variable, determined by $\sigma_{\text{mod},I}$. For example, referring to $f(\{\bar{x}_{Ij}\}, \{\beta_k\})$ as the modeled mean $\mu_{\text{mod},I}$, for a linear model we have (with relabeled β_k):

$$\mu_{\text{mod},I} \equiv \beta_0 + \beta_1 x_{I1} + \ldots + \beta_m x_{Im},$$
$$\sigma_{\text{mod},I}^2 \equiv \sigma_{y,I}^2 + \beta_1^2\sigma_{x,I1}^2 + \ldots + \beta_m^2\sigma_{x,Im}^2.$$

In the literature, uninformative priors for the model parameters β_k have been derived as well, based on the transformation invariance [8]. We use these priors for comparison of GLS with the standard Bayesian analysis.

Now, suppose for a moment that one would proceed with the maximum likelihood method to estimate the parameters β_k. From (4), one sees that this is equivalent to minimization of the sum of squared Mahalanobis distances between each observed

$\bar{y}_I$ and its corresponding value $f\left(\{\bar{x}_{Ij}\}, \{\beta_k\}\right)$ determined by the model function f.[2] The Mahalanobis distance can be regarded as the distance between two univariate Gaussian clusters of points with centroids given by $\bar{y}_I$ and $f\left(\{\bar{x}_{Ij}\}, \{\beta_k\}\right)$, each with the same standard deviation, in the present case $\sigma_{\text{mod},I}$. Interestingly, it is also a special case of the Rao GD, namely the GD between the corresponding normal distributions with those means and common standard deviation [9]. It is therefore natural to generalize this to the case where not only the means of the distributions, but also the standard deviations are allowed to differ. One could choose to generalize the Mahalanobis distance to the Bhattacharyya distance or the Hellinger divergence, but we prefer the Rao geodesic distance owing to its solid mathematical foundations and intuitive geometric interpretation.

By allowing the standard deviation of the observed and modeled distribution to be different, the method is rendered robust, as the actual distribution of the data is allowed to deviate from the modeled distribution. So, on the one hand, we consider at each measurement site I the modeled distribution $\mathcal{N}\left(f\left(\{\bar{x}_{Ij}\}, \{\beta_k\}\right), \sigma^2_{\text{mod},I}\right)$. On the other hand, we have the observed distribution p_{obs}, which has to rely on as few assumptions as possible regarding the regression model, in an attempt to "let the data speak for themselves." We here only assume that it also is a Gaussian distribution, $p_{\text{obs}} = \mathcal{N}(\bar{y}_I, \sigma^2_{\text{obs},I})$, centered on the actually observed average $\bar{y}_I$, and with an unknown standard deviation $\sigma_{\text{obs},I}$, to be estimated from the data. Although this can all be generalized, the normal distribution offers a computational advantage, as the corresponding expression for the GD has a closed form [10]. In addition, we already mentioned that, in principle, $\sigma_{\text{obs},I}$ can be different at each measurement site, but in practice, it is clear that we will need to introduce some sort of regularization to render the model identifiable. In this paper, we either assume $\sigma_{\text{obs},I}$ a constant s_{obs}, or proportional to the response variable, $\sigma_{\text{obs},I} = r_{\text{obs}}|\bar{y}_I|$. The parameters s_{obs} or r_{obs} have to be estimated from the data. More complicated (parametrized) relations between $\sigma_{\text{obs},I}$ and the response variable or other data would be possible too, but one should be careful not to put too many restrictions on p_{obs}, thereby defeating its purpose.

GLS now proceeds by minimizing the total GD between, on the one hand, the joint observed distribution of the N values $\bar{y}_I$ and, on the other hand, the joint modeled distribution. Owing to the independence assumption in this example, we can write this in terms of products of the corresponding marginal distributions (including all dependencies and with γ_{obs} either s_{obs} or r_{obs}):

$$\left\{\beta_k, \gamma_{\text{obs}}\right\} = \underset{\beta_k, \gamma_{\text{obs}} \in \mathbb{R}}{\operatorname{argmin}} \sum_{I=1}^{N} \text{GD}^2\left[p_{\text{obs}}\left(Y|\bar{y}_I, \gamma_{\text{obs}}\right), p_{\text{mod}}\left(Y|C_x\right)\right]. \quad (5)$$

[2]Under the assumption of symmetry of the likelihood distribution and homoscedasticity, this reduces to minimization of the sum of squared differences (Euclidean distances) between each measured $\bar{y}_I$ and predicted $f\left(\{\bar{x}_{Ij}\}, \{\beta_k\}\right)$.

Here, the variable Y models the site averages. In addition, note that the parameters β_k occur both in the mean and the variance of the modeled distribution. Furthermore, in (5) we have used the property that the squared GD between products of distributions can be written as the sum of squared GDs between the corresponding factors [10]. Hence, the optimization procedure involves, at each measurement site, matching not only $\bar{y}_I$ with $f(\{\bar{x}_{Ij}\}, \{\beta_k\})$, but also $\sigma_{\text{obs},I}$ with $\sigma_{\text{mod},I}$, in a way dictated by the geometry of the likelihood distribution. As will be shown in the experiments, the result is that GLS is relatively insensitive to uncertainties in both the stochastic and deterministic components of the regression model. The same quality renders the method also robust against outliers. In the experiments below, we employed a classic active-set algorithm to carry out the optimization. Furthermore, presently the GLS method does not directly offer confidence (or credible) intervals on the estimated quantities. Future work will address this issue in more detail, but for now error estimates were derived by a bootstrap procedure.

From the conceptual point of view, GLS performs regression between points (distributions) on a Riemannian probabilistic manifold, describing the data corresponding to the response variable at each measurement site as a whole through either the observed or the modeled distribution. It is important to stress that this is quite different from treating the data at each measurement site in a pointwise way, i.e., using each individual y_{i_I}. Our method respects the intrinsic nature of the fluctuating quantity described by the variable y. For instance, if, for fixed I, y_{i_I} is a series of samples from a stationary signal, then comparing the measured signal distribution with the predicted distribution can be seen as more natural than comparing each individual sample with its predicted value. Furthermore, in MDE regression usually the data distribution is characterized using a kernel density estimate. Although this offers great flexibility, the disadvantage is that this estimate could be based on the data from different measurement sites. In addition, our parametric approach can be an advantage if few measurements are available. Finally, the geometrical view on regression analysis can be illustrated by visualizing the probabilistic manifold [2].

3 Application of GLS to Tully–Fisher Scaling

3.1 *The Baryonic Tully–Fisher Relation*

The baryonic Tully–Fisher relation (BTFR) between the total (stellar + gaseous) baryonic mass M_b of disk galaxies and their rotational velocity V_f is of fundamental importance in astrophysics and cosmology [11, 12]. It is a remarkably simple and tight empirical relation of the form

$$M_b = \beta_0 V_f^{\beta_1}. \tag{6}$$

Here, M_b is expressed in solar masses $M_\odot$ and V_f in km s^{-1}. The BTFR not only serves as one of the tools for determining cosmic distances, but also provides constraints on galaxy formation and evolution models. In addition, it serves as a test for the Lambda cold dark matter paradigm (ΛCDM), particularly in evaluating alternatives such as modified Newtonian dynamics (MOND). Indeed, whereas in ΛCDM the BTFR is a consequence of various complex processes and thus should demonstrate significant intrinsic scatter, MOND predicts a relation with zero intrinsic scatter and a well-defined exponent β_1 with a value of exactly 4.

In this scaling problem, we use data from 47 gas-rich galaxies, as detailed in [12]. The advantage of the gas-rich galaxies is that their masses can be more accurately measured than those of star-dominated galaxies, which are traditionally used to define the Tully–Fisher relation. The rotation velocity V_f is measured in the flat part of the galaxy rotation curve, determined from spectral Doppler shifts. The measurements are plotted in Fig. 1a on the logarithmic scale and in Fig. 1b on the original scale.

In this application, clearly $n_I = 1$ for all I, so little information can be obtained regarding the distribution of the data from the single measurement at each site. However, the data in [12] also contain estimates of the observational errors, which we treat here as a single standard deviation. This suggests a measurement error on the response variable proportional to M_b, about 38%, i.e., a constant error bar on the logarithmic scale.

3.2 *Regression Analysis*

Owing to the power law character of most scaling laws, they are often estimated by linear regression on a logarithmic scale. However, it is known that this may lead to unreliable estimates, as the logarithm (heavily) distorts the distribution of the data [2, 13]. This is, in particular, the case if the estimation is carried out using simple OLS or when there are outliers in the data. In contrast, we will show that GLS regression produces consistent results on both the logarithmic and original scales, demonstrating its robustness.

In view of the proportional error on M_b, the observed standard deviation in GLS is modeled here as a constant $\sigma_{\text{obs},I} \equiv s_{\text{obs}}$ on the logarithmic scale and as $\sigma_{\text{obs},I} = r_{\text{obs}} M_b$ on the original scale. Estimation of these parameters is of interest to get an idea of the intrinsic scatter on the BTFR.

We compare the results of GLS regression with OLS and a Bayesian approach. In the latter, uncertainty on the predictor variables was taken into account into the likelihood. In the case of nonlinear power law regression, the likelihood was approximated by a Gaussian, as the full treatment with marginalization over the model points is too computationally intensive to incorporate in an MCMC simulation [3]. Uncertainty in the specified error bars was modeled through a scale factor with a Jeffreys prior [3]. We also tested the GLS algorithm using the KLD as a similarity measure between the observed and modeled distribution, instead of the Rao GD. We will refer to this algorithm as "Kullback–Leibler least squares," or KLS.

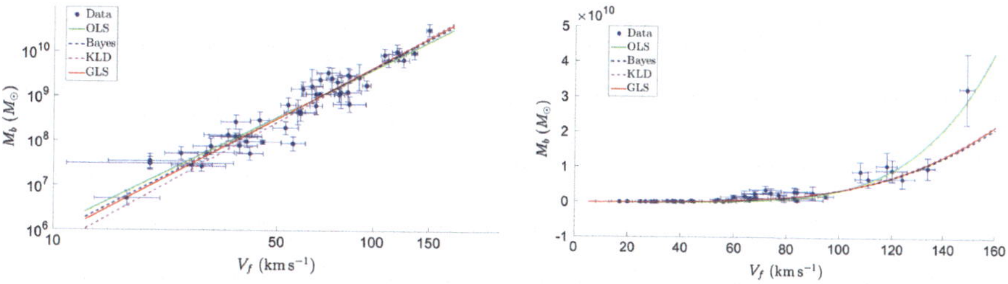

Fig. 1 The BTFR data and estimated regression functions by OLS, a robust Bayesian method, KLS and GLS. **a** Logarithmic scale. **b** Original scale

In order to get a feeling of the uncertainty of the estimates obtained from the optimization routines, 100 bootstrap samples were created from the data, yielding average parameter estimates and their standard deviations on the basis of the results from OLS, KLS, and GLS. Similar estimates were obtained from the MCMC chain in the robust Bayesian approach.

The parameter estimates estimated by the various methods, as well as their standard deviations, are given in Table 1. Figure 1 shows the corresponding regression curves. It is interesting to compare the results obtained by regression on the logarithmic scale, with those derived using nonlinear regression analysis. On the logarithmic scale the data follow a rather clear linear pattern, hence the estimates by the various methods are similar. However, in the nonlinear case, the best fit is somewhat less clear at first sight. Although the Bayesian, KLS, and GLS methods agree relatively well, the OLS parameter estimates are very different from the linear case. Most noticeably, the nonlinear OLS estimate for the exponent β_1 is heavily influenced by the point with the largest value of V_f and $M_b \approx 3 \times 10^{10}\, M_\odot$. The other methods are much less attracted by this point because of the large corresponding error bar on M_b. Thus, part of the danger of the logarithmic transformation is due to its influence on the error bars in the presence of model uncertainty. The differences between the parameter estimates by the other methods are much less pronounced, although the consistency appears to be best in the case of GLS. This is in agreement with the good robustness quality of GLS compared to other methods seen in previous analyses [2, 4].

It is also worth pointing out that the scale factor r_{obs} (observed relative error) was estimated by GLS to amount to roughly 63%. This is considerably larger than the value of 38% predicted by the model (and dominated by σ_{M_b}), possibly indicating that the scatter on the scaling law is not due to measurement error alone.

Table 1 Average regression estimates and their standard deviations for the BTFR obtained with OLS, KLS, and GLS from 100 bootstrap samples. Similar results were derived by MCMC sampling with the robust Bayesian method. The units of the parameters have been left out for simplicity. (a) Logarithmic scale. (b) Original scale

Method	β_0	β_1
OLS	360 ± 220	3.57 ± 0.15
Bayes	220 ± 220	3.72 ± 0.19
KLS	80 ± 80	3.98 ± 0.23
GLS	140 ± 82	3.80 ± 0.16

a

Method	β_0	β_1
OLS	$(1.0 \pm 2.3) \times 10^3$	4.94 ± 1.40
Bayes	88 ± 140	3.81 ± 0.20
KLS	120 ± 100	3.91 ± 0.19
GLS	130 ± 130	3.79 ± 0.21

b

4 Conclusion

We have introduced and motivated geodesic least squares, a versatile and robust regression method based on the regression between probability distributions describing fluctuating or otherwise uncertain system properties. Part of the strength of the method is its simplicity, allowing straightforward application by users in the various application fields, without the need for parameter tuning. We have applied GLS to baryonic Tully–Fisher scaling, thereby demonstrating the robustness of the method and providing an alternative means for testing cosmological models based on the estimated intrinsic scatter.

References

1. Verdoolaege, G., : In: Bayesian Inference and Maximum Entropy Methods in Science and Engineering, AIP Conference Proceedings, vol. 1636. Canberra, Australia (2014)
2. Verdoolaege, G.: Entropy **17**(7), 4602 (2015)
3. von der Linden, W., Dose, V., von Toussaint, U.: Bayesian Probability Theory. Applications in the Physical Sciences. Cambridge University Press, Cambridge (2014)
4. Verdoolaege, G., Shabbir, A.: JET Contributors, In: Bayesian Inference and Maximum Entropy Methods in Science and Engineering, AIP Conference Proceedings, vol. 1853, art. no. 100002 (8 pp.) Ghent, Belgium (2017)
5. Beran, R.: Ann. Stat. **5**(3), 445 (1977)
6. Pak, R.: Stat. Probab. Lett. **26**(3), 263 (1996)
7. Werman, M., Keren, D.: IEEE Trans. Pattern Anal. Mach. Intell. **23**(5), 528 (2001)
8. Preuss, R., Dose, V., : In: Bayesian Inference and Maximum Entropy Methods in Science and Engineering, AIP Conference Proceedings, vol. 803. Melville, NY (2005)

9. Rao, C.: Differential Geometry in Statistical Inference. Institute of Mathematical Statistics, Hayward, CA (1987)
10. Burbea, J., Rao, C.: J. Multivariate Anal. **12**(4), 575 (1982)
11. Tully, R., Fisher, J.: Astron. Astrophys. **54**(3), 661 (1977)
12. McGaugh S.: Astron. J. **143**(2), 40 (15 pp.) (2012)
13. Xiao, X., et al.: Ecology **92**(10), 1887 (2011)

Bayesian Portfolio Optimization for Electricity Generation Planning

Hellinton H. Takada, Julio M. Stern, Oswaldo L. V. Costa and Celma de O. Ribeiro

Abstract Nowadays, there are several electricity generation technologies based on the different sources, such as wind, biomass, gas, coal, and so on. Considering the uncertainties associated with the future costs of such technologies is crucial for planning purposes. In the literature, the allocation of resources in the available technologies have been solved as a mean-variance optimization problem using the expected costs and the correspondent covariance matrix. However, in practice, the expected values and the covariance matrix of interest are not exactly known parameters. Consequently, the optimal allocations obtained from the mean-variance optimization are not robust to possible errors in the estimation of such parameters. Additionally, there are specialists in the electricity generation technologies participating in the planning process and, obviously, the consideration of useful prior information based on their previous experience is of utmost importance. The Bayesian models consider not only the uncertainty in the parameters, but also the prior information from the specialists. In this paper, we introduce the Bayesian mean-variance optimization to solve the electricity generation planning problem using both improper and proper prior distributions for the parameters. In order to illustrate our approach, we present an application comparing the Bayesian with the naive mean-variance optimal portfolios.

Keywords Statistics · Inference methods · Energy analysis · Policy issues

---

H. H. Takada (✉)
Quantitative Research, Itaú Asset Management, São Paulo, Brazil
e-mail: hellinton.takada@itau-unibanco.com.br

J. M. Stern
Institute of Mathematics and Statistics, University of São Paulo, São Paulo, Brazil
e-mail: jstern@ime.usp.br

O. L. V. Costa · C. de O. Ribeiro
Polytechnic School, University of São Paulo, São Paulo, Brazil
e-mail: oswaldo@lac.usp.br

C. de O. Ribeiro
e-mail: celma@usp.br

A. Polpo et al. (eds.), *Bayesian Inference and Maximum Entropy Methods in Science and Engineering*, Springer Proceedings in Mathematics & Statistics 239, https://doi.org/10.1007/978-3-319-91143-4_9

1 Introduction

In the early and middle years of the nineteenth century, the fundamental principles of electricity generation were discovered by scientists such as Alessandro Volta, André Ampère, Benjamin Franklin and Michael Faraday [5]. Since then, already in the last years of the nineteenth century, the electricity generation plants started to be built together with the transmission networks [9]. During the time, the mankind has developed several electricity generation technologies based on the different sources, such as wind, biomass, gas, coal, nuclear, and so on. Evidently, each technology has associated costs, sustainability, and security of supply characteristics, efficiency, and environmental concerns.

The worldwide demand for energy has been increasing over the last decades and it will continue to grow [10]. Consequently, for both countries and companies, the long-term planning of the electricity generation infrastructure is of utmost importance. Actually, it should be part of the central objectives of any energy policy. The achievement of an optimally designed electricity generation infrastructure bends toward a more balanced portfolio allocation among the different available technologies. In addition, it is also important to distinguish in the planning process the already existing electricity producing plants with maintenance costs from the ones desired to be built. Obviously, drastic changes of the electricity investment allocations is not feasible.

The U.S. Energy Information Administration has not only historical data on the average annual operation, maintenance, and fuel costs for existing power plants by major fuel or energy source types, but also projections for electricity generation costs [18]. However, even so, the costs have a significant uncertainty. For instance, future control on CO_2 emission and the corresponding mechanisms will surely impact the electricity generation costs. Precisely, the future price of an emitted ton of CO_2 is uncertain and this uncertainty should be taken into account in the planning process. Consequently, electricity generation policies solely relying on the evolution of historical average costs of electricity generation technologies are unsatisfactory.

Considering the costs as random variables, in the literature, the allocation of resources in the available electricity generation technologies has been solved as a mean-variance optimization problem using the expected values and covariance matrix of the technology costs in megawatt hours (see, for instance, [1, 2, 14, 15]). The mean-variance optimization, introduced by Markowitz [13], was the first mathematical formalization of investment diversification and it is part of the modern portfolio theory (MPT). The mean-variance optimized portfolios compose the called efficient frontier, a set of portfolios that dominate all other feasible portfolios in terms of their mean and variance tradeoff. Obviously, in the MPT the random variables of interest are the returns of the risky assets instead of the costs of the technologies.

In practice, the expected values and the covariance matrix of the electricity generation technology costs for a future time horizon are not exactly known. Noticeably, the usefulness of the allocations obtained from the mean-variance optimization depends on the preciseness of such parameters. For instance, in the MPT context, it was shown in [3] that small changes in the expected returns can produce large changes in asset allocation decisions. Consequently, several robust versions of the mean-variance optimization were proposed in the MPT literature to consider uncertainties on the expected returns and covariance matrix (see, for instance, [4, 8, 11]). Particularly in [6], for the first time in the electricity planning context, it was presented a robust portfolio optimization approach to deal with uncertainties in the input parameters.

In the electricity planning processes, it is usual to have the participation of specialists in the electricity generation technologies of interest. Undoubtedly, a natural way of conducting a comprehensive planning process is to take into account the available data together with the prior experience of the participant specialists. Bayesian approaches treat probability distributions as uncertain and subject to updates as new information becomes available. Consequently, the Bayesian approach has been successfully applied in the MPT context to take into account not only the beliefs of the investors but also the uncertainties in the expected returns and the correspondent covariance matrix (see, for instance, [3, 16, 17]). The Bayesian mean-variance portfolio optimizations could take into account both the estimation uncertainty and the specialist prior information.

In this paper, our objective is the introduction of the Bayesian approach to electricity generation planning. First, we give a brief review of the classical mean-variance optimization with the basic notation and fundamental concepts. Then, the Bayesian approach is presented using both improper and proper priors. For illustration purposes, an application comparing the Bayesian with the naive mean-variance optimal portfolios is given. Finally, some final comments are presented.

2 Classical Approach

Traditionally, the classical or naive mean-variance optimization presumes that cost and risk, the last one measured as the portfolio volatility, are known when making portfolio-selection decisions. Therefore, a rational planner would prefer a portfolio with a lower expected cost for a given level of risk. Alternatively, a preferred portfolio is one that minimizes risk for a given expected cost level. The set of portfolios that are optimal is called the efficient frontier. No rational planner would select a portfolio lying above the efficient frontier, since that would mean accepting a higher cost for the same amount of risk as an efficient portfolio. Equivalently, it would mean accepting greater risk for the same expected cost as an efficient portfolio.

As already mentioned and following [6, 12], it is important to distinguish in the planning process an already existing electricity producing plant using technology i, with random cost C_i^e in USD/MWh, from a prospective idea of using i, with random cost C_i^p in USD/MWh. The random vectors of costs for existing and prospective technologies when there are N different technologies are given by

$$\mathbf{C}^e \equiv \left(C_1^e \; C_2^e \; \ldots \; C_N^e\right)' \quad \text{and} \quad \mathbf{C}^p \equiv \left(C_1^p \; C_2^p \; \ldots \; C_N^p\right)', \tag{1}$$

respectively. It is also usual to assume that the random costs are multivariate normal

$$\mathbf{C}^e|\boldsymbol{\mu}^e, \Sigma^e \sim \mathrm{N}\left(\boldsymbol{\mu}^e, \Sigma^e\right) \quad \text{and} \quad \mathbf{C}^p|\boldsymbol{\mu}^p, \Sigma^p \sim \mathrm{N}\left(\boldsymbol{\mu}^p, \Sigma^p\right), \tag{2}$$

where $\boldsymbol{\mu}^e = (\mu_i^e)_{N\times 1}$ and $\boldsymbol{\mu}^p = (\mu_i^p)_{N\times 1}$ are mean vectors and Σ^e and Σ^p are $N \times N$ covariance matrices. The means μ_i^e and μ_i^p are different, because maintenance costs are different from the costs of building a new plant. Additionally, the risk of maintenance σ_i^e is also different from the risk of building a new plant σ_i^p. However, since the technology is the same, the correlation between C_i^e and C_i^p is equal to $\rho_{C_i^e, C_i^p} = 1$. Thus, we can write almost surely (with probability 1) that (see Proposition 1.1.2 from [7])

$$C_i^e = \frac{\sigma_i^e}{\sigma_i^p}\left(C_i^p - \mu_i^p\right) + \mu_i^e. \tag{3}$$

Essentially, the Eq. 3 says that the source of uncertainty for both C_i^e and C_i^p is the same. Additionally, $\Sigma^e = \mathrm{diag}\,(\boldsymbol{\sigma}^e)\,\mathbf{R}\,\mathrm{diag}\,(\boldsymbol{\sigma}^e)$ and $\Sigma^p = \mathrm{diag}\,(\boldsymbol{\sigma}^p)\,\mathbf{R}\,\mathrm{diag}\,(\boldsymbol{\sigma}^p)$, where the correlation matrix $\mathbf{R}$ is the same for both the existing and the prospective costs, $\boldsymbol{\sigma}^e = \left(\sigma_i^e\right)_{N\times 1}$ and $\boldsymbol{\sigma}^p = \left(\sigma_i^p\right)_{N\times 1}$.

Defining $\mathbf{C} = \left(\mathbf{C}^{e\prime}\ \mathbf{C}^{p\prime}\right)'$, it follows that

$$\mathbf{C}|\boldsymbol{\mu}, \Sigma \sim \mathrm{N}\left(\boldsymbol{\mu}, \Sigma\right), \tag{4}$$

where

$$\boldsymbol{\mu} = \left(\boldsymbol{\mu}^{e\prime}\ \boldsymbol{\mu}^{p\prime}\right)' \quad \text{and} \quad \Sigma = \begin{pmatrix} \Sigma^e & \mathrm{diag}\,(\boldsymbol{\sigma}^e)\,\mathbf{R}\,\mathrm{diag}\,(\boldsymbol{\sigma}^p) \\ \mathrm{diag}\,(\boldsymbol{\sigma}^e)\,\mathbf{R}\,\mathrm{diag}\,(\boldsymbol{\sigma}^p) & \Sigma^p \end{pmatrix}. \tag{5}$$

The portfolio weights are the proportions of the total budget allocated in each technology. The allocation vectors in the existent and prospective technologies are denoted by $\boldsymbol{\omega}^e = (\omega_i^e)_{N\times 1}$ and $\boldsymbol{\omega}^p = (\omega_i^p)_{N\times 1}$, respectively. Naturally, $0 \le \omega_i^e \le 1$, $\forall i = 1, 2, \ldots, N$; $0 \le \omega_i^p \le 1$, $\forall i = 1, 2, \ldots, N$; and

$$\sum_{i=1}^{N}\left(\omega_i^e + \omega_i^p\right) = 1. \tag{6}$$

Defining $\boldsymbol{\omega} = \left(\boldsymbol{\omega}^{e\prime}\ \boldsymbol{\omega}^{p\prime}\right)'$, we denote by Ω the set of admissible electricity generation mix so that we must have $\boldsymbol{\omega} \in \Omega$. The set Ω will represent constraints like Eq. 6, $\boldsymbol{\omega}'\mathbf{1}_{2N} = 1$ ($\mathbf{1}_{2N}$ is a $2N \times 1$ vector of ones), and minimum and/or maximum values for the allocations ($\boldsymbol{\omega}_{\min} \le \boldsymbol{\omega}$ and/or $\boldsymbol{\omega} \le \boldsymbol{\omega}_{\max}$). Using the $\boldsymbol{\omega}$ definition, the total cost of the portfolio is given by

$$\mathscr{C} = \boldsymbol{\omega}'\mathbf{C}. \tag{7}$$

Using the previous Eq. 7, the expected cost of the portfolio is given by

$$\mathrm{E}\left[\mathscr{C}\right] = \boldsymbol{\omega}'\boldsymbol{\mu} \tag{8}$$

and the variance of the portfolio is given by

$$\mathrm{Var}\left[\mathscr{C}\right] = \boldsymbol{\omega}'\boldsymbol{\Sigma}\boldsymbol{\omega}. \tag{9}$$

For the case in which the vector of expected costs $\boldsymbol{\mu}$ and the covariance matrix $\boldsymbol{\Sigma}$ are known, three kinds of mean-variance problems are usually considered in the literature. The first approach minimizes the variance of the costs conditional on a target maximum expected cost c. The target maximum expected cost $c \in \Re_+$ is provided by the electricity energy policy planner which represents the maximum allowable expected energy cost. Formally, the problem is written as follows

$$\min_{\boldsymbol{\omega}} \boldsymbol{\omega}'\boldsymbol{\Sigma}\boldsymbol{\omega} \tag{10}$$

$$\text{s. t. } \boldsymbol{\omega}'\boldsymbol{\mu} \leq c, \ \ \boldsymbol{\omega} \in \Omega. \tag{11}$$

The second approach, a dual form of the first approach, minimizes the expected cost conditional on a maximum value s^2 for the variance of the costs. The value $s^2 \in \Re_+$, provided by the policy planner, represents the maximum value that the variance of the cost could achieve. Formally, the problem is written as follows

$$\min_{\boldsymbol{\omega}} \boldsymbol{\omega}'\boldsymbol{\mu} \tag{12}$$

$$\text{s. t. } \boldsymbol{\omega}'\boldsymbol{\Sigma}\boldsymbol{\omega} \leq s^2, \ \ \boldsymbol{\omega} \in \Omega. \tag{13}$$

The third approach minimizes a combination of the expectation and variance of the costs, weighted by a risk aversion parameter $\lambda > 0$. Higher value of λ indicates a greater risk aversion. Formally, the problem is written as follows

$$\min_{\boldsymbol{\omega}} \boldsymbol{\omega}'\boldsymbol{\mu} + \lambda\boldsymbol{\omega}'\boldsymbol{\Sigma}\boldsymbol{\omega} \tag{14}$$

$$\text{s. t. } \boldsymbol{\omega} \in \Omega. \tag{15}$$

Trivially, using quadratic programming solvers, the previous three problems can be solved for the case in which the vector of expected costs $\boldsymbol{\mu}$ and covariance matrix $\boldsymbol{\Sigma}$ are assumed to be known.

3 Bayesian Approach

In terms of modeling, the Bayesian approaches address estimation risk from a conceptually different perspective. Instead of treating the unknown parameters as constants, they are considered random. Additionally, the belief or prior knowledge of the specialist about the input parameters is combined with the observed data. The Bayesian models yield an entire distribution of predicted costs which explicitly takes into account the estimation and predictive uncertainty.

The predictive, posterior, or updated distribution of the unknown parameters $\boldsymbol{\mu}$ and Σ, according to the Bayes' theorem, is given by

$$\mathrm{p}\left(\boldsymbol{\mu}, \Sigma | \mathbf{c}_1, \ldots, \mathbf{c}_T\right) \propto \mathrm{L}\left(\boldsymbol{\mu}, \Sigma | \mathbf{c}_1, \ldots, \mathbf{c}_T\right) \pi\left(\boldsymbol{\mu}, \Sigma\right), \tag{16}$$

where $\mathbf{c}_1, \ldots, \mathbf{c}_T$ are recorded observations; $\pi(\cdot)$ is the prior distribution; and $\mathrm{L}(\cdot|\cdot)$ is the likelihood function. Particularly, the likelihood function is given by

$$\mathrm{L}\left(\boldsymbol{\mu}, \Sigma | \mathbf{c}_1, \ldots, \mathbf{c}_T\right) \propto |\Sigma|^{-\frac{T}{2}} \exp\left[-\frac{1}{2}\sum_{i=1}^{T}\left(\mathbf{c}_i - \boldsymbol{\mu}\right)' \Sigma^{-1}\left(\mathbf{c}_i - \boldsymbol{\mu}\right)\right]. \tag{17}$$

In the following subsections, we present the predictive distributions using improper and proper priors for the unknown parameters $\boldsymbol{\mu}$ and Σ.

3.1 Improper Prior Case

In many cases, our prior beliefs are vague and thus difficult to translate into an informative prior. Therefore, we want to reflect our uncertainty about the model parameters without substantially influencing the predictive parameter inference. The so-called noninformative priors, also called vague or diffuse priors, are employed to that end. We consider the case when the investor is uncertain about the distribution of both parameters, $\boldsymbol{\mu}$ and Σ, and has no particular prior knowledge of them. This uncertainty can be represented by a improper or diffuse prior, which is typically taken to be the Jeffreys' prior,

$$\pi\left(\boldsymbol{\mu}, \Sigma\right) \propto |\Sigma|^{-\frac{(2N+1)}{2}}, \tag{18}$$

where $\boldsymbol{\mu}$ and Σ are considered independent in the prior, and $\boldsymbol{\mu}$ is not restricted. The prior is noninformative in the sense that only changes in the data exert an influence on the predictive distribution of the parameters.

When the sample mean, $\hat{\boldsymbol{\mu}}$, and sample covariance matrix, $\hat{\Sigma}$, are given, it is straightforward to verify that the predictive distribution of the costs is a multivariate Student's t-distribution

$$\mathbf{C}|\tilde{\boldsymbol{\mu}}, \tilde{\Sigma} \sim \mathrm{t}_{T-2N}\left(\tilde{\boldsymbol{\mu}}, \tilde{\Sigma}\right), T - 2N \geq 2, \tag{19}$$

where the predictive mean and covariance matrix are, respectively,

$$\tilde{\boldsymbol{\mu}} = \hat{\boldsymbol{\mu}} \text{ and } \tilde{\Sigma} = \frac{\left(1 + T^{-1}\right)(T-1)}{T - 2N - 2}\hat{\Sigma}. \tag{20}$$

The predictive covariance here represents the sample covariance scaled up by a factor, reflecting estimation risk. For a given number of technologies N, the uncertainty $\tilde{\Sigma}$ decreases as more historical data become available. Actually, when N is fixed and $T \to \infty$, we have $\tilde{\Sigma} \to \hat{\Sigma}$. On the other hand, with a fixed number of historical observations T, increasing the number of technologies N respecting the constraint $T - 2N - 2 > 0$, leads to higher uncertainty and estimation risk, since the relative amount of available data declines. In practice, there is relevant information coming from specialists on energy costs. Consequently, in the next subsection, we present a study with proper priors.

3.2 Proper Prior Case

The specialists have informative beliefs about the mean and covariance of technology costs. In this subsection, we adopt conjugate priors because it is an algebraic convenience producing a closed expression for the posterior. The conjugate prior for the unknown covariance matrix of the normal distribution is the inverse Wishart distribution while the conjugate prior for the mean vector of the normal distribution (conditional on Σ) is the multivariate normal:

$$\boldsymbol{\mu}|\Sigma \sim \mathrm{N}\left(\boldsymbol{\eta}, \frac{1}{\tau}\Sigma\right), \Sigma \sim \mathrm{W}^{-1}(\Psi, \nu), \tag{21}$$

where $\boldsymbol{\eta}$ is the vector of expected costs based on the specialist experience, $\tau \in \Re_+$ represents the strength of the confidence the specialist places on the value of $\boldsymbol{\eta}$, Ψ is the covariance matrix based on the specialist experience, $\nu \in \Re$ represents the degrees of freedom of the inverse Wishart distribution reflecting the confidence about Ψ. Lower values of τ and ν indicates higher uncertainty about $\boldsymbol{\eta}$ and Ψ, respectively.

As in the improper prior case, the predictive distribution of the costs is a multivariate Student's t-distribution

$$\mathbf{C}|\breve{\boldsymbol{\mu}}, \breve{\Sigma} \sim \mathrm{t}_{T-2N}\left(\breve{\boldsymbol{\mu}}, \breve{\Sigma}\right), T - 2N \geq 2, \tag{22}$$

where the predictive mean and covariance matrix are, respectively,

$$\breve{\boldsymbol{\mu}} = \frac{\tau}{T+\tau}\boldsymbol{\eta} + \frac{T}{T+\tau}\hat{\boldsymbol{\mu}} \tag{23}$$

and

$$\breve{\Sigma} = \frac{T+1}{T(\nu+2N-1)}\left[\Psi + (T-1)\hat{\Sigma} + \frac{T\tau}{T+\tau}(\eta-\hat{\mu})(\eta-\hat{\mu})'\right]. \tag{24}$$

We notice that the predictive mean $\breve{\mu}$ is a weighted average of the prior mean, η, and the sample mean, $\hat{\mu}$. In other words, the sample mean is shrunk toward the prior mean. Actually, the predictive mean and predictive covariance matrix are not proportional to the sample estimates. The improper prior case is appropriate to employ when we do not suspect that the sample mean or sample covariance matrix contains substantial estimation errors. Otherwise, the proper prior case is better when the planner believes that in the future the expectation and covariance matrix of the costs will differ substantially from the historical ones.

4 Results

In this section, we present an application to illustrate the robust Bayesian approaches. In [12], the vector of expected costs and standard deviations are given for 8 different technologies (differentiating between existent and prospective cases). Additionally, the correlation matrix of the technologies is also given. For the purpose of our application, we consider the data from [12] as the sample estimates of the parameters $\hat{\mu}$ and $\hat{\Sigma}$. The naive mean-variance efficient frontier obtained using $\hat{\mu}$ and $\hat{\Sigma}$ is presented in Figs. 1 and 2 (repeated in the two graphics). It is important to notice that the portfolios above the efficient frontier are inefficient and the portfolios below the efficient frontier are unrealizable.

In the improper prior case, illustrated in Fig. 1, the efficient frontier changes depending on the value of T. As already mentioned, the predictive covariance of the improper case is the sample covariance scaled up by a factor that approaches to one when T increases. Obviously, we do not have here T representing the actual size of the sample used in the estimation. Actually, for us, T is not only a proxy to the size of the sample used in the estimation but also the degree of confidence the planner has on the estimations based only on the historical data. Consequently, decreasing the value of T shifts the efficient frontier to the right. The same shift to the right was observed in [6] using the robust mean-variance optimization when decreasing the degree of confidence the planner has on the estimations. However, the robust mean-variance optimization is computationally more expensive than our approach because the first requires more optimizations.

In the proper prior case, the hyperparameters η and Ψ represent the prior information of the specialist about the expected value and covariance matrix of the technology costs, respectively. Since we do not have such parameters for the situation described in [12], we assume, for illustration purposes, that η and Ψ are obtained increasing in 10% the parameters $\hat{\mu}$, $\hat{\Sigma}^e$ and $\hat{\Sigma}^p$. In Fig. 2, we present the obtained efficient frontiers for different values of τ with $T = 50$ and $\nu = 34$. Noticeably, the

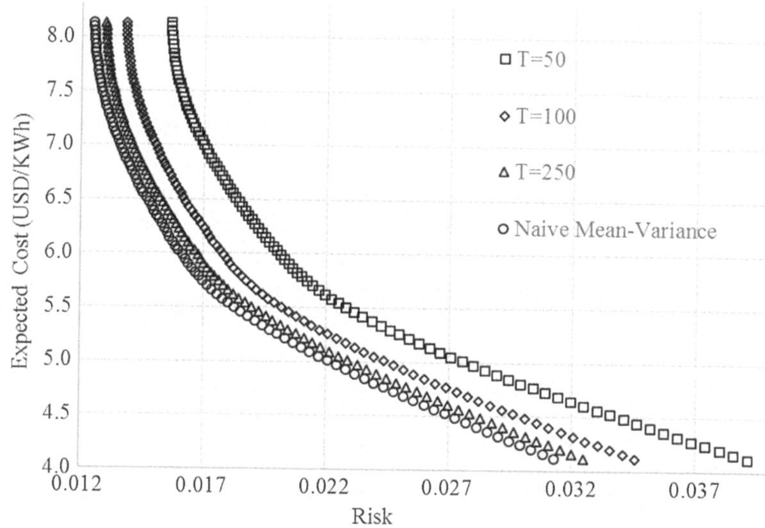

Fig. 1 Efficient frontiers using naive and Bayesian approaches for the improper prior case for some values of T

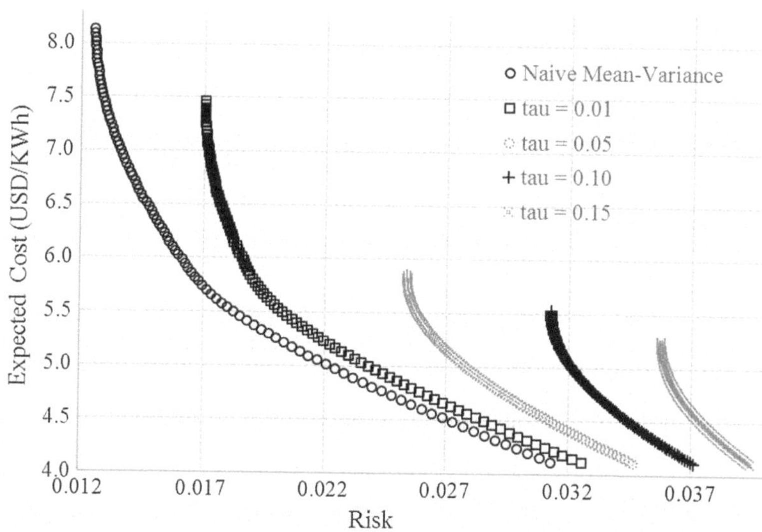

Fig. 2 Efficient frontiers using naive and Bayesian approaches for the proper prior case for some values of τ

resulting efficient frontiers are not simple shifts of the naive mean-variance frontier. Consequently, as already mentioned, the informative proper prior case is better than the improper prior when the planner believes that in the future the costs will differ substantially from the historical ones.

5 Final Remarks

In this paper, we introduce the use of the Bayesian mean-variance optimization in the electricity generation planning. We illustrate the application of the approach using improper and proper priors. Comparing with the existent robust approach to electricity portfolio selection, the Bayesian approach has the advantage of not only dealing with the estimation uncertainty, but also considering the prior information of the specialists in the planning process. Particularly, in the proper prior case, we have assumed that the covariance matrix of the expected value of the costs are proportional to the covariance matrix of the costs. In practice, the assumption is not necessarily valid. For future research, we suggest the investigation of changing the proper priors to give more flexibility to the electricity generation planner.

Acknowledgements The authors are grateful for the support of IME-USP, the Institute of Mathematics and Statistics of the University of São Paulo; FAPESP - the State of São Paulo Research Foundation (grants CEPID 2013/07375-0 and 2014/50279-4); and CNPq - the Brazilian National Counsel of Technological and Scientific Development (grant PQ301206/2011-2).

References

1. Awerbuch, S.: Portfolio-based electricity generation planning: policy implications for renewable and energy security. Mitig. Adapt. Strateg. Glob. Chang. **11**, 693–710 (2006)
2. Awerbuch, S., Berger, M.: Applying Portfolio Theory to EU Electricity Planning and Policy-Making. International Energy Agency/Emerging Energy Technologies, Paris (2003)
3. Black, F., Litterman, R.: Asset allocation: combining investor views with market equilibrium. J. Fixed Income **1**, 7–18 (1991)
4. Bertsimas, D., Brown, D.B., Caramanis, C.: Theory and applications of robust optimization. SIAM Rev. **53**(3), 464–501 (2011)
5. Breeze, P.: Power Gener. Technol. Elsevier, Oxford (2014)
6. Costa, O.L.V., Ribeiro, C.O., Rego, E.E., Stern, J.M., Parente, V., Kileber, S.: Robust portfolio optimization for electricity planning and policy-making. Submitted for publication
7. Davis, M.H.A., Vinter, R.B.: Stochastic Modelling and Control. Chapman and Hall, New York (1985)
8. Fabozzi, F.J., Kolm, P.N., Pachamanova, D.A., Focardi, S.M.: Robust Portfolio Optimization and Management. Wiley, Hoboken (2007)
9. Hughes, T.P.: Networks of Power: Electrification in Western Society, 1880–1930. The John Hopkins University Press, Baltimore (1983)
10. International Energy Agency, World Energy Outlook (2016)
11. Kim, J.H., Kim, W.C., Fabozzi, F.J.: Recent developments in robust portfolios with a worst-case approach. J. Optim. Theory Appl. **161**(1), 103–121 (2014)
12. Losekann, L., Marrero, G.A., Ramos-Real, F.J., Almeida, E.L.F.: Efficient power generating portfolio in brazil: conciliating cost, emissions and risk. Energy Policy **62**, 301–314 (2013)
13. Markowitz, H.: Portfolio Selection: Efficient Diversification of Investments. Wiley, New York (1959)
14. Marrero, G.A., Puch, L.A., Ramos-Real, F.J.: Mean-variance portfolio methods for energy policy risk management. Int. Rev. Econ. Financ. **40**, 246–264 (2015)
15. Marrero, G.A., Ramos-Real, F.J.: Electricity generation cost in isolated system: the complementarities of natural gas and renewables in the canary islands. Renew. Sustain. Energy Rev. **14**, 2808–2818 (2010)

16. Meucci, A.: Risk and Asset Allocation. Springer, New York (2005)
17. Rachev, S.T., Hsu, J.S.J., Bagasheva, B.S., Fabozzi, F.J.: Bayesian Methods in Finance. Wiley, Hoboken (2008)
18. U.S. Energy Information Administration, Annual Energy Outlook (2016)

Bayesian Variable Selection Methods for Log-Gaussian Cox Processes

Jony Arrais Pinto Junior and Patrícia Viana da Silva

Abstract Point patterns are very common in present days of many researchers. The desire to understand the spatial distribution and investigate connections between point patterns and p covariates, that is possibly associated with the event of interest, arises naturally. Generally, not all of the p covariates are useful. Therefore it would be handy to identify the covariate which is, and just use those. Variable selection is an important step when setting a parsimonious model and still occupies the minds of many statisticians. In this work, we investigated Bayesian variable selection methods in the context of point pattern. This work concentrated on the following methods: Kuo and Mallick, Gibbs Variable Selection, and Stochastics Search Variable Selection for log-Gaussian Cox processes. The methods were evaluated in several scenarios: with a different number of covariates that should be included in the model, absence, and presence of multicollinearity and fixed and random effect model. Our results suggest that the three methods, specially Stochastics Search Variable Selection, can work very well with the absence of multicollinearity. We implemented the methods in BUGS.

Keywords Point pattern · Log-Gaussian Cox process · Bayesian variable selection

1 Introduction

Geo-referenced data is very common nowadays. Scientific studies use these types of data in a lot of research areas, such as Ecology, Geography, Seismology, and Epidemiology. Sometimes, the event of interest is known, for instance, infection of

J. A. Pinto Junior (✉)
Instituto de Matemática e Estatística (UFF), Rua Professor Marcos Waldemar de Freitas Reis, s/n, Bloco H - Campus do Gragoatá - São Domingos, Niterói - RJ, Brazil
e-mail: jarrais@id.uff.br

P. Viana da Silva
Faculdade de Matematica (UFU), Av Joao Naves de Avila, 2121, Uberlândia, Brazil
e-mail: patriciaviana@ufu.br

A. Polpo et al. (eds.), *Bayesian Inference and Maximum Entropy Methods in Science and Engineering*, Springer Proceedings in Mathematics & Statistics 239, https://doi.org/10.1007/978-3-319-91143-4_10

trees by a plague, deaths from stroke and vehicle theft, but the locations of occurrences of the event are unknown. Point pattern is the set of these locations and it is usually the result of a dynamic process that occurs both in space and in time.

Models for these types of data are usually built with point pattern processes, for more details see [5] and [7]. In the initial approach, exploratory methods based on the distances were used. This approach had an issue because they did not specify likelihoods, which it made hard to compare different alternatives as Likelihood-based methods including homogeneous and non-homogeneous Poisson processes, Cox processes, and log-Gaussian Cox processes. The last method will be used in this work. An important class of hierarchical models along these lines, with the introduction of spatial covariate effects, were suggested by [2].

To select the subset of covariates should be included into the model, several methods have been proposed. Classical methods delete or add covariates by means of mean-squared error or modified mean squared error criteria. (backward, forward, and stepwise). Also, Bayesian methods have been proposed: Bayesian information criteria — BIC [16], asymptotic information criterion—AIC [1] among others. All of these methods have problems to handle the number of possible submodels (2^p) being considered for p covariates. A more automatic data-driven tool is needed to identify a parsimonious model.

More recently, several Bayesian variable selection methods use an indicator to select variables and a second auxiliary variable to quantify the effect of a covariate in a regression problem. Reference [14] shows a review of the Bayesian variable selection methods and investigated how the different methods perform in practice in the context of generalized linear models. With this approach is possible to estimate the posterior probability that a subset of covariates is “in” the model, i.e., the posterior inclusion probability by the occurrences of a particular model in the MCMC simulations.

Few studies to select variables to log-Gaussian Cox processes has been proposed so far, thus, the purpose of this work is to compare the performance of methods based on the indicator model selection for it. The idea is to investigate, in the background of point pattern data, how the Kuo and Mallick (KM), Gibbs Variable Selection (GVS) e Stochastic Search Variable Selection (SSVS) methods perform in several scenarios: with different number of covariates that should be included into the model absence, and presence of multicollinearity and fixed and random effect model.

In Sect. 2, we will define a spacial point pattern process and describes a Log-Gaussian Cox Process. In Sect. 3, we will define the methods of Bayesian variable selection that will be considered in this work. Section 4 describes the simulation study and presents its results. Section 5 concludes about the three methods of variable selection.

2 Spatial Point Pattern Process

Consider the spatial point pattern process $X = \{X(s) : s \in S\}$, which S is a set of indexes. $X(s) = 1$, if the event of interest occurred in s and $X(s) = 0$ otherwise.

They are useful models for the statistical analysis of geo-referenced observed point patterns. This stochastic process is responsible to control both the location and the occurrence number of the event in the region of space.

The most common point pattern process is the (non-homogeneous) Poisson process with intensity function denoted by $\Lambda(\cdot) = \{\Lambda(s) : s \in S\}$. The notation used in this work will be $X \sim PP(\Lambda(\cdot))$.

We assume $S \subseteq \Re^d$, $d > 0$. When $d = 2$, X is a spatial process on the plane. In practice, only a finite set of point locations contained in a region $B \subset S$ will be observed. A realization of X can be unequivocally identified with an occurrence set $\{s_i\}_{i=1}^n = \{s_1, s_2, \ldots, s_n\}$, $s_i \in S$, $s_i \in S, \forall i$ and $n \geq 1$, where all observed events take place.

Two important features of a spatial point process are related to the moments of the process. The first is the stationarity of X, which guarantees that its distribution is invariant to translations in $\Re^d$ and the second is isotropy, i.e., X will be isotropic if its distribution is invariant over rotations in the origin of $\Re^d$. An example of a point process that has these two properties is defined below.

Definition 1 (*Isotropic Gaussian Process*) A process $T(\cdot)$, defined in S, is said to be isotropic Gaussian if $\forall n > 1$ and any set of locations $\{s_1, \ldots, s_n\} \in S$,

$$(T(s_1), \ldots, T(s_n))^T \sim N_n(\mu \mathbf{1}, \tau^{-1} R_\gamma),$$

in which $\mu \in \Re$ is mean, $\mathbf{1} = (1, \ldots, 1)^T$, $\tau \in \Re_+$ is a precision parameter and R_γ is a correlation matrix with elements $R_{i,l} = \rho_\gamma(\|s_i - s_l\|)$ defined through a correlation function ρ_γ, $\gamma \in \Re_+$, depending on s_i and s_l only through their distance, for $i, l = 1, \ldots, n$, denoted by $T(\cdot) \sim GP(\mu, \tau, \rho_\gamma)$.

2.1 Log-Gaussian Cox Process

The Cox process is a doubly stochastic Poisson point process, i.e., a non-homogeneous Poisson process with stochastic intensity parameter $\Lambda(\cdot)$. This parameter is assigned as some location-dependent function in the same underlying space which the Poisson process is defined. If the logarithm of the intensity of the Cox process is a Gaussian process, then this point process is a Cox log-Gaussian process [13].

The full model formulation for log-Gaussian Cox process in continuous space is given by

$$X \sim PP(\Lambda(\cdot)), \tag{1}$$

a Poisson point process non-homogeneous. The intensity will assume a multiplicative decomposition of the intensity function with

$$\Lambda(s) = r(s)\lambda(s), \forall s \in S, \tag{2}$$

where $r(s)$ representing a known offset, usually required for standardization and $\lambda(s)$ is a stochastic component. Furthermore, a log-linear model is proposed for $\lambda(s)$,

$$\log \lambda(s) = z(s)^T \boldsymbol{\theta} + w(s), \tag{3}$$

where $z(s) = (z_1(s), z_2(s), \ldots, z_p(s))^T$ is a vector of p covariates related to locations, $\boldsymbol{\theta} \in \Re^p$ is the vector of effects of the spatial covariates and

$$w(\cdot) \sim GP(\mu, \tau_w, \rho_\gamma), \tag{4}$$

is an isotropic Gaussian process with spatial autocorrelation function, ρ_γ. w is responsible to capture the spatial heterogeneity.

To complete the model, consider the following priors distributions $\boldsymbol{\theta} \sim N(\boldsymbol{m}, G)$, $\mu \sim N(a, b)$, $\tau_w \sim Gamma(c, d)$, and $\gamma \sim Gamma(e, f)$.

In order to perform inference on the difficult-to-treat likelihood (defined in a continuous space), the space discretization approach is commonly used even in point processes through time [8]. Here, it is extended to the spatial domain discretizing the continuous space in N subareas, assuming the model defined by the expressions (1) to (4) and that the spatial auto-correlation is exponential function of the distances

$$X \sim PP(\Lambda_k), \tag{5}$$

$$\Lambda_k = r_k \lambda_k, \tag{6}$$

$$\log \lambda_k = z_k^T \boldsymbol{\theta} + w_k, \tag{7}$$

$$\boldsymbol{w} \sim N_N(\boldsymbol{0}, \tau_w^{-1} R_\gamma), \tag{8}$$

for $k = 1, 2, \ldots, N$, with $R_\gamma = [R_{k_1,k_2}]_{k_1,k_2=1,2,\ldots,N}$, $R_{k_1,k_2} = \exp\left\{-\frac{d_{k_1,k_2}}{\gamma}\right\}$ and d_{k_1,k_2} is the distance between k_1 and k_2.

The spatial discretization of the intensity above can also be found in several works [3, 11, 15]. In fact, [17] showed that the posterior distributions of the intensities are well approximated and converge to the posterior distribution of the exact, continuously varying intensity when the volumes of the subareas tend to 0.

3 Bayesian Variable Selection

In the context to distinguish between large and small effects, we could use a prior distribution that expresses the belief that there are coefficients close to zero and larger coefficients. We can easily construct that priors as a mixture of two distributions, one with a "spike" at zero and the other with mass spread over a wide range of plausible values. They are called spike and slab priors, respectively, and very useful for variable selection purposes because they allow classifying the regression coefficients into two groups: the first one that consists of large, important regressors and the second one with small, negligible effects.

To present the methods that will be studied in this work, we will need to define two auxiliary variables. The first one is an indicator variable I_j (where $I_j = 1$ indicates presence, and $I_j = 0$ absence of covariate j in the model) that will denote whether the variable is in the slab or spike part of the prior. The second one is the effect size β_j, $j = 1, \ldots, p$, where $\beta_j = \theta_j$, if $I_j = 1$, for instance, if we define $\theta_j = I_j\beta_j$. When $I_j = 0$, the variable β_j can be defined in several ways.

Suppose the $\beta_j|I_j = 1$ is drawn from $N(0, \nu^2)$, where ν is constant, we will refer this as fixed effect models and when ν is an unknown parameter to be estimated, as a random effect model. We used the terminology of classical statistics.

The different ways of manage θ_j, β_j, and I_j define the following variable selection methods.

3.1 *Kuo and Mallick (KM)*

The first approach proposed by Kuo and Mallick [10] takes $\theta_j = I_j\beta_j$ into account. Also, the indicators and effects are independent a prior so $P(I_j, \beta_j) = P(I_j)P(\beta_j)$. Therefore, the β_j values are sampled from full conditional distribution, but if $I_j = 0$ it is the prior distribution.

3.2 *Gibbs Variable Selection (GVS)*

Gibbs Variable Selection (GVS) was proposed by [6] and also uses $\theta_j = I_j\beta_j$ but the dependence is assumed, i.e., $P(I_j, \beta_j) = P(\beta_j|I_j)P(I_j)$. Unlike Sect. 3.1, this method use a mixture of distributions for β_j, $p(\beta_j|I_j) = (1 - I_j)N(m, s) + I_jN(0, \nu^2)$. As we can see, the $N(m, s)$ is the spike distribution whereas $N(0, \nu)$ is the slab. m and s need to be chosen so that good values of β_j are proposed when $I_j = 0$. According to [14], the data will determine which values are good but without directly influencing the posterior, e.g., m and s could be mean and variance values of posterior for the full model, respectively.

3.3 *Stochastic Search Variable Selection (SSVS)*

Stochastic Search Variable Selection (SSVS) was introduced by [9]. In this method, $\theta_j = \beta_j$ and the indicators affect the prior distribution of β_j, i.e., $P(I_j, \beta_j) = P(\beta_j|I_j)P(I_j)$. The regressor coefficient is modeled as coming from a mixture of two normal distributions with different variances, i.e., $p(\beta_j|I_j) = (1 - I_j)N(0, \nu^2) + I_jN(0, c\nu^2)$. The spike has density concentrated around zero and the slab has density spread out over large plausible values. Even though a small ν^2 should be taken, in the slab distribution the constant c compensates the variance making a sparse distribution. Unlike GVS method, the posterior distribution is influenced by values of the prior parameters when $I_j = 0$.

3.4 Comparing the Methods

Here, we highlight the main similarities and differences among the methods. The KM and GVS methods define $\theta_j = \beta_j I_j$, unlike SSVS that defines $\theta_j = \beta_j$. The KM is the only method that sampling β_j from the same distribution, when $I_j = 0$ and $I_j = 1$, i.e., consider I_j and β_j independent. The other methods, define the prior distribution from β_j using the slab and spike distributions. But they define in different ways. The GVS method uses the pseudo-prior to sampling β_j, when $I_j = 0$. The SSVS method uses a concentrated around 0 normal distribution for sampling β_j when $I_j = 0$, and a flat normal distribution for sampling β_j when $I_j = 1$, the responsible for the spread of the density is c.

4 Simulation Study

To investigate the performance of the methods, we simulated several scenarios to see how well they can identify the covariates that should be included in the model. For the simulation study, 100 replicated datasets were created. We considered a space S divided into $N = 100$ subareas and the values of the 10 covariates were simulated independently from a standard normal distribution, $N(0, 1)$. We also considered, the values of the offset were simulated from a beta distribution, $Beta(b_1, b_2)$, where b_1 and b_2 were chosen to allowed the total number of cases in the space S were similar in all scenarios studied.

Known values of $\mu = 0$, $\tau = 1$, and $\gamma = 4.24$ were used to simulated w. The value of γ was chosen to guarantee a small correlation (0.05) between the two most distant subareas.

How we considered different numbers of covariates in the model, we use different known values of $\boldsymbol{\theta}$ specified in Table 1.

We also would like to investigate the performance of the methods considering dependence among covariates, i.e., the presence of multicollinearity. We only considered the scenario with three covariates and set a $corr(z_i, z_l) = 0.5^{|i-l|}$, $i, l = 1, \ldots, 10$, and the other characteristics we maintain as previously.

Table 1 Known values of parameters used in simulation

Number of covariates	θ_1	θ_2	θ_3	θ_4	θ_5	θ_6	θ_7	θ_8	θ_9	θ_{10}
3	0.5	1	1.5	0	0	0	0	0	0	0
5	0.5	1	1.5	−1.1	−0.6	0	0	0	0	0
8	0.5	1	1.5	−1.1	−0.6	1.2	−0.75	0.7	0	0

4.1 Prior Distributions

We considered $I_j \sim Bernoulli(0.5)$, suggested by [9], making all models equiprobable and $\tau_w \sim Gamma(1, 0.01)$, relatively vague prior. For all three methods, for the random models, we considered $\beta_j|I_j = 1 \sim N(0, \tau_\beta^{-1})$, $j = 1, \ldots, 10$, with $\tau_\beta \sim Gamma(1, 0.01)$ and for the fixed models we assumed $\tau_\beta = 0.01$.

For KM, $\beta_j|I_j = 0$ has the same distribution than $\beta_j|I_j = 1$. For GVS, we assumed $\beta_j|(I_j = 0) \sim N(0, 0.25)$, $j = 1, \ldots, 10$. We also evaluated the impact of different values of the variance of this distribution, 0.15, 0.20, 0.30, 0.35, and 0.4. The values were chosen so that good values of β_j are proposed when $I_j = 0$. While for SSVS, the prior was constructed so that $P(|\beta_j| < c) < 0.01$, by setting it to be three standard deviations away from the mean, i.e., we assumed $\beta_j|I_j = 0 \sim N(0, (3 \times c)^2)$, $j = 1, \ldots, 10$ and $c = 0.04$. The choices of the hyperparameters from the priors was made following [14].

Note that we are considering $\mu = 0$ and $\gamma = 4.24$ known values. We decided do not estimate γ, because that is a difficulty parameter to be estimated. This well-known difficulty of spatial models was reported in many papers, including [11]. How the goal of this work investigates the performance of the variable selection methods, first, we decided to investigate without this confounding factor. Set $\mu = 0$, without loss of generality.

4.2 Landscapes Definitions

In this work, there are six landscapes to be compared or two basic landscapes to be investigated: fixed effects versus random effects and a different number of covariate should be in the model: three, five and eight covariates. You can see the definition of each landscape below: (a) Fixed effect with three covariates in the model, (b) Random effect with three covariates in the model, (c) Fixed effect with five covariates in the model, (d) Random effect with five covariates in the model, (e) Fixed effect with eight covariates in the model and (f) Random effect with eight covariates in the model.

Table 2 Posterior probability that the correct model was selected for landscape (a) with multicollinearity

Method	Median	1st quartile	3rd quartile
KM	0.769	0.580	0.856
GVS	0.775	0.580	0.840
SSVS	0.763	0.588	0.834

4.3 Results

Results were obtained via MCMC methods by Metropolis–Hastings algorithm and they were implemented in BUGS [12]. Convergence was ascertained by Gelman–Rubin–Brooks criterion [4] using two chains with different starting values. Correlation between successive chain draws was alleviated by thinning at every 10 iterations, after a burn-in period of 2,000 draws. The resulting sample consisted of 2,000 draws.

As an overall qualitative assessment of computational speed and parameter estimation, we can say about speediness that SSVS was the fastest method. Most times, SSVS was more than three times faster than the others two methods. All three methods show good accuracy into estimate the true values of the effects of the covariates.

To compare the three methods about their efficacy to select the correct model in the scenario with no multicollinearity, we used the posterior probability that the correct model was selected. As we can see in Fig. 1, all three methods showed good probabilities to select the correct model. Here, the probability will be the proportions of times that the MCMC chain visited the model that contains the subset of covariates used to simulated the dataset. If we analyze the median of these probabilities in Fig. 1, the worst result for all landescapes is 80%, i.e., in the landscape with the poorest identification of the correct model, 50% of fits select the correct model 80% of times.

Furthermore, in Fig. 1, we have a greater dispersion for the probability of identifying the correct model in the landscapes with a greater number of covariates, for both fixed and random model. For landscapes a-d, the KM showed the worst result to random model when compared to the fixed model, but GVS and SSVS improved their outcomes. These results weren't observed to landscape e-f. The KM method showed some instability for different initial values for I_j. In some cases, convergence wasn't obtained. In addition, for landscapes b and d, SSVS and GVS showed significantly best outcomes than KM.

As we can see in Table 2, even when we consider the presence of multicollinearity the results still promising. Obviously, the outcomes are poorer than the scenario without multicollinearity.

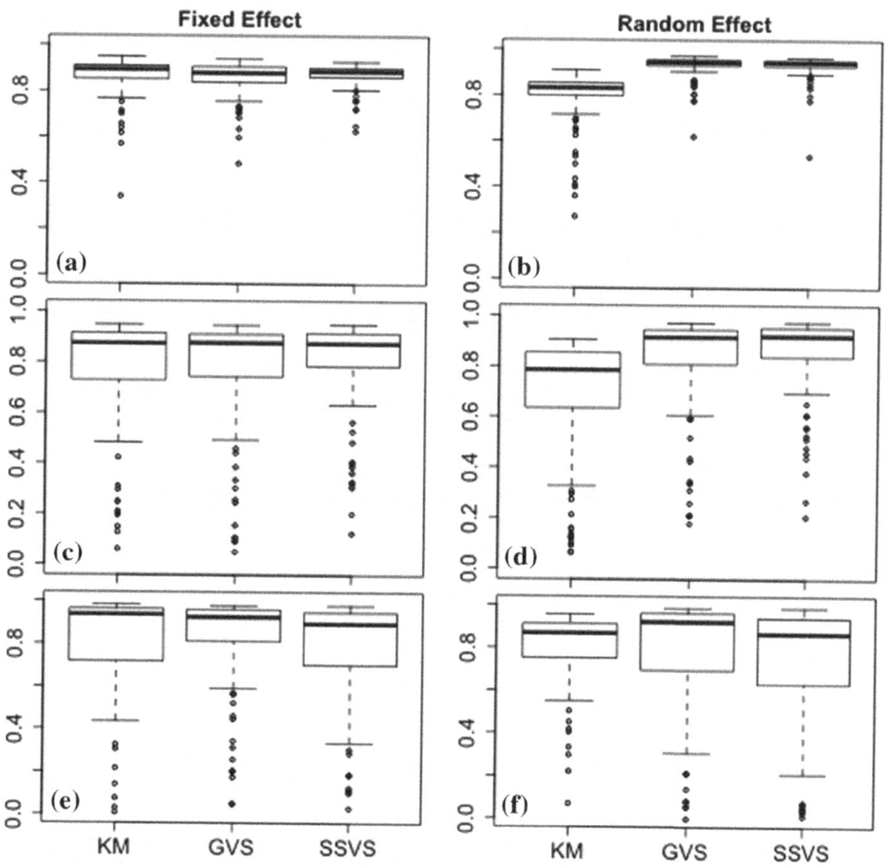

Fig. 1 Boxplot of the medians of the posterior probability that the correct model was selected for all landscapes investigated with no multicollinearity

5 Conclusion

This work investigated the performance of KM, GVS, and SSVS methods for variable selection to log-Gaussian Cox processes models. Useful models to analyses point pattern data. Few studies to select covariates in this field has been proposed so far.

The results showed that three methods have good accuracy to identify the correct model in the scenario with no multicollinearity. They showed the high posterior probability that the correct model was selected. A higher number of covariates showed poorer outcomes when compared with scenarios with fewer covariates, but even in the worst scenario, the median of the desired probability still equal or greater than 80%. In general, KM showed the poorest and SSVS the best performance, including the fact that this method was the fastest of all of them.

Even in scenarios with the presence of multicollinearity, the results seem promising, but a deeper investigation is necessary.

Besides, the methods allow calculating a simple measure to compare them with an easy interpretation. Unlike the usual measures (AIC, BIC, among others).

References

1. Akaike, H.: A new look at the statistical model identification. IEEE Trans. Autom. Control. **19**, 716–23 (1974)
2. Beněs, V., Boldak, K., Møller, J., Waagepetersen, R.: A case study on point process modelling in disease mapping. Image Anal. Stereol. **24**, 159–168 (2005)
3. Brix, A., Møller, J.: Space-time multi type log gaussian cox processes with a view to modelling weeds. Scand. J. Stat. **28**(3), 471–488 (2001)
4. Brooks, S., Gelman, A.: General methods for monitoring convergence of iterative simulations. J. Comput. Graph. Stat. **7**(4), 434–455 (1998). https://doi.org/10.1080/10618600.1998.10474787
5. Cox, D., Isham, V.: Point Processes. Chapman & Hall, London (1980)
6. Dellaportas, P., Forster, J.J., Ntzoufras, I.: Bayesian model and variable selection using MCMC. Technical report, Department of Statistics, Athens University of Economics and Business, Athens Greece (1997)
7. Diggle, P.: Statistical Analysis of Spatial Point Patterns, 2nd edn. Arnold, London (2003)
8. Gamerman, D.: A dynamic approach to the statistical analysis of point process. Biometrika **79**, 39–50 (1992)
9. George, E.I., McCulloch, R.E.: Variable selection via Gibbs sampling. J. Am. Stat. Assoc. **85**, 398–409 (1993)
10. Kuo, L., Mallick, B.: Variable selection for regression models. Sankhyá Indian J. Stat. Ser. B **60**, 65–81 (1998)
11. Liang, S., Carlin, B., Gelfand, A.: Analysis of Minnesota colon and rectum cancer point patterns with spatial and nonspatial covariate information. Ann. Appl. Stat. **3**, 943–962 (2008)
12. Lunn, D.J., Thomas, A., Best, N., Spiegelhalter, D.: Winbugs - a Bayesian modelling framework: concepts, structure, and extensibility. Stat. Comput. **10**(4), 325–337 (2000)
13. Møller, J., Syversveen, A., Waagepetersen, R.: Log Gaussian Cox process. Scand. J. Stat. **25**, 451–482 (1998)
14. O'Hara, R.B., Sillanpaa, M.J.: A review of Bayesian variable selection methods: what, how and which. Bayesian Anal. **4**(1), 85–117 (2009). 10.1214/09-BA403, http://projecteuclid.org/euclid.ba/1340370391
15. Pinto Junior, J.A., Gamerman, D., Paez, M.S., Fonseca Alves, R.H.: Point pattern analysis with spatially varying covariate effects, applied to the study of cerebrovascular deaths. Stat. Med. **34**, 1214–1226 (2014). https://doi.org/10.1002/sim.6389
16. Schwarz, G.: Estimating the Dimension of a Model. Ann. Stat. **6**(2), 461–464 (1978)
17. Waagepetersen, R.: A dynamic approach to the statistical analysis of point process. Convergence of posterioris for discretized log Gaussian Cox process. Stat. Probab. Lett. **66**(3), 229–235 (2004)

Effect of Hindered Diffusion on the Parameter Sensitivity of Magnetic Resonance Spectra

Keith A. Earle, Troy Broderick and Oleks Kazakov

Abstract Magnetic Resonance spectroscopy is a powerful tool for elucidating the details of molecular dynamics. In many important applications, a model of hindered diffusion is useful for summarizing the complex dynamics of ordered media, such as a liquid crystalline environment, as well as the dynamics of proteins in solution or confined to a membrane. In previous work, we have shown how the sensitivity of a magnetic resonance spectrum to the details of molecular dynamics depends on the symmetries of the magnetic tensors for the relevant interactions, e.g., Zeeman, hyperfine, or quadrupolar interactions. If the hindered diffusion is modeled as arising from an orienting potential, then the parameter sensitivity of the magnetic resonance spectrum may be studied by generalizations of methods we have introduced in previous work. In particular, we will show how lineshape calculations using eigenfunction expansions of solutions of the diffusion equation, can be used as inputs to an information-geometric approach to parameter sensitivity estimation. We illustrate our methods using model systems drawn from Nuclear Magnetic Resonance, Electron Spin Resonance, and Nuclear Quadrupole Resonance.

Keywords Hindered diffusion · Lineshape analysis · Magnetic resonance spectroscopy

1 Introduction

Spectral line shape calculations are an important tool for inferring details of mechanism and function whenever molecular dynamics controls the response of the system

K. A. Earle (✉) · T. Broderick · O. Kazakov
University at Albany, 1400 Washington Ave., Albany, NY 12222, USA
e-mail: kearle@albany.edu

T. Broderick
e-mail: tbroderick@albany.edu

O. Kazakov
e-mail: okazkov@albany.edu

A. Polpo et al. (eds.), *Bayesian Inference and Maximum Entropy Methods in Science and Engineering*, Springer Proceedings in Mathematics & Statistics 239, https://doi.org/10.1007/978-3-319-91143-4_11

under study to probing perturbations. In order to provide some context for the work reported on here, we begin with a review of some key concepts in the computation of magnetic resonance line shapes in Sect. 2. We then discuss the problem of hindered diffusion in Sect. 3 and survey some important considerations arising in the context of coordinate transformations in Sect. 3.2 and sketch the derivation of an efficient method for the computation of the elements of the starting vector introduced in Sect. 2. For standard line shape calculations, computation of the elements of the starting vector when hindered diffusion is the relaxation mechanism is the most time-intensive task. When the hindered diffusion parameters are varied this requires the starting vector to be recomputed. The algorithm introduced here converts a multidimensional numerical integration to the solution of a system of homogeneous algebraic equations leading to significant improvements in efficiency and stability. In Sect. 4 we discuss and motivate a simple model that nevertheless captures the essential features of the hindered diffusion problem relevant for this work. Finally, in Sect. 5, we present a qualitative discussion of the results of our numerical studies and connect it to previous work in the parameter inference problem.

2 Magnetic Resonance

The equation of motion for the coupled spin and orientational degrees of freedom which defines the spectral line shape is the Stochastic Liouville Equation (SLE). It incorporates a torque-like term arising from the action of the spin Hamiltonian, and a relaxation term governing the establishment of equilibrium. The SLE may be written as follows

$$\frac{\partial \rho(\Omega, t)}{\partial t} = -\frac{i}{\hbar}\left[\mathscr{H}(\Omega, t), \rho(\Omega, t)\right] - \Gamma(\rho(\Omega, t) - \rho_0(\Omega, t)), \tag{1}$$

where Ω is a set of parameters characterizing the orientational dependence of the relaxation process, assumed here to correspond to rotational diffusion, We discuss relevant parameterizations as well as important reference frames for the magnetic resonance line shape problem in Sect. 3.2. Furthermore, $\mathscr{H}(\Omega, t)$ is the relevant spin Hamiltonian which may contain various magnetic interactions. Note that $\mathscr{H}$ is required to be Hermitian. By an appropriate choice of basis states, matrix elements of $\mathscr{H}$ can always be chosen to be real. This is not a requirement but there are computational advantages to this choice, as we discuss below. The formalism is general enough to incorporate a variety of possible interactions, such as the Zeeman interaction, hyperfine interactions, zero field splittings, quadrupolar interactions, and so on. We can account for time-dependent, transition-inducing interactions by allowing the applied magnetic field **H** to be an oscillatory function of time in the Zeeman interaction, and this is necessary for describing nonlinear spin responses [1]. For this work, we will focus on the linear response regime, but the goal is to keep the notation as general as possible until we consider a specific model system in Sect. 4.

The quantity $\boldsymbol{\rho}(\Omega, t)$ is the spin density matrix, which we expand in a complete set of spin multipole operators [2]. The quantity $\boldsymbol{\rho}_0(\Omega, t)$ is the equilibrium density matrix when the time-dependence of the Zeeman interaction may be neglected. When the time-dependence of the Zeeman interaction may not be neglected, $\boldsymbol{\rho}_0(\Omega, t)$ can be interpreted as the steady-state density matrix [1]. Note that $\Gamma \boldsymbol{\rho}_0(\Omega, t) \equiv 0$. Given that the torque term and the relaxation term appear in an additive fashion in the Stochastic Liouville Equation 1, it is useful to expand both $\boldsymbol{\rho}(\Omega, t)$ and $\boldsymbol{\rho}_0(\Omega, t)$ in a product state of spin multipoles and eigenfunctions of the Γ operator. Some useful properties of eigenfunctions of Γ are discussed in Sect. 3. Note that in order for Eq. 1 to have solutions that remain bounded as $t \to \infty$ the eigenvalues of Γ must have positive real part. For the class of problems considered here Γ is a purely dissipative process, and so the eigenvalues are purely real and positive.

It is useful to think of the Stochastic Liouville Equation 1 as a generalization of the phenomenological Bloch equations. The Bloch equations may be recovered from Eq. 1 in the limit that the relaxation operator Γ is approximated as a simple exponential decay and the spin system is assumed to be an ensemble of weakly interacting spins 1/2. In this limit, the classical and quantum mechanical descriptions coincide and one recovers the familiar form

$$\frac{d\mathbf{M}(t)}{dt} = \gamma(\mathbf{M} \times \mathbf{H}) - \mathbf{R}(\mathbf{M} - \mathbf{M}_0),$$

where $\mathbf{M}$ is the net magnetization, $\mathbf{M}_0$ is the equilibrium magnetization and $\mathbf{R}$ is a diagonal relaxation tensor with characteristic relaxation times T_2 corresponding to decay of the transverse (nonequilibrium) magnetization and T_1 corresponding to the characteristic timescale over which equilibrium is established. Finally, γ is the gyromagnetic ratio.

Given that the frequency–domain spectrum is usually the quantity that is most readily compared with experiment, it is useful to take the Fourier transform of Eq. 1. When this is done, the initial conditions corresponding to $\boldsymbol{\rho}(\Omega, 0)$ appear explicitly. The usual assumption is that a coherence corresponding to a nonvanishing transverse magnetization is generated either by a hard pulse of short duration, or gentle, continuous wave excitation. In either case, one recovers the linear response. Solving Eq. 1 for all times $t > 0$ allows one to compute the spin–spin correlation function from which the Fourier transform allows one to compute the frequency–domain spectrum. The references may be consulted for the practical details of how this is done [1, 3, 4]. For the purposes of this work, it is sufficient to note that the spectral line shape calculation may be put in the form

$$I(\Delta\omega) = \langle v | C^{-1}(\Delta\omega) | v \rangle, \tag{2}$$

where $| v \rangle$ is the vector of initial conditions corresponding to the product state of a nonequilibrium coherence characterized by a given spin multipole with the relevant eigenstates of the diffusion operator. In addition, $\Delta\omega$ is the offset from resonance. In a basis, where $\mathscr{H}$ and Γ have real matrix elements C is a complex symmetric

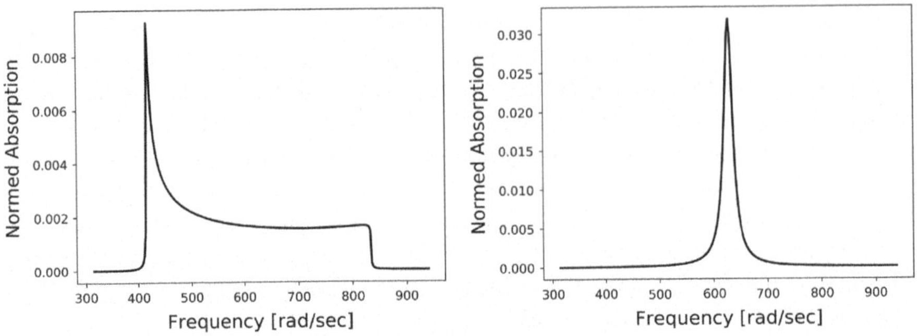

Fig. 1 Magnetic resonance spectra corresponding to axial symmetry. The spectrum on the left corresponds to the rigid limit in which there is little to no residual motion. The spectrum on the right corresponds to the motional narrowing regime. The magnetic tensor anisotropy is the same in both cases. The difference is in the rotational diffusion rate

matrix. When C is complex symmetric there are efficient computational algorithms for tridiagonalizing C [1, 4]. The spectral function can be readily evaluated when C is put into tridiagonal form as discussed elsewhere [1, 4]. A product basis of irreducible spherical tensor operators [4] in spin and orientation space is a convenient choice for this class of problems. In the case of hindered diffusion, treated in Sect. 3, a further symmetrization process is needed to ensure that the complex symmetric form is maintained. For the case of the axial Zeeman interaction treated in Sect. 4, one obtains spectra of the form shown in Fig. 1. The motionally narrowed spectrum, corresponding to the spectrum on the right in Fig. 1, is also characteristic of hindered diffusion, as we will discuss in Sect. 5.

In order to quantify parameter sensitivity, we may take derivatives of the spectrum with respect to the line shape parameters. For the simple model discussed in Sect. 4, the relevant parameters are the rotational diffusion rate R, there taken to be isotropic, the spectral extent, parameterized by the anisotropy of the Zeeman interaction Δ_g, and the strength of the orienting potential λ. Starting from the identity $CC^{-1} = 1$, one may verify that

$$\frac{dI(\Delta\omega)}{d\theta} = \langle dv/d\theta | C^{-1} | v \rangle - \langle v | C^{-1}(dC/d\theta)C^{-1} | v \rangle + \langle v | C^{-1} | dv/d\theta \rangle, \tag{3}$$

where $\theta \in \{R, \Delta_g, \lambda\}$. Equation 3 may be symmetrized by defining a new vector $| w \rangle = | v \rangle + | dv/d\theta \rangle$. Written in terms of $| v \rangle$, $| dv/d\theta \rangle$, and $| w \rangle$ one finds

$$\begin{aligned} \frac{dI(\Delta\omega)}{d\theta} &= \\ &\langle w | C^{-1}(\Delta\omega) | w \rangle - \langle v | C^{-1}(\Delta\omega) | v \rangle - \langle dv/d\theta | C^{-1}(\Delta\omega) | dv/d\theta \rangle \\ &- \langle v | C^{-1}(\Delta\omega)(dC(\Delta\omega)/d\theta)C^{-1} | v \rangle . \end{aligned} \tag{4}$$

As discussed in Sect. 3, the only parameters that affect the starting vector $| \, v \, \rangle$, at least in the linear response regime, arise from the ordering potential $U(\Omega)$. The diffusion constant R and the magnetic anisotropy Δ_g occur linearly in $C(\Delta\omega)$ which leads to significant simplifications of Eq. 4. Note further that the vector $C^{-1}(\Delta\omega) \, | \, v \, \rangle$ arises naturally in the computation of the lineshape and so there is no extra computational cost associated with its evaluation. Due to the form of the Γ operator discussed in Sect. 3, the potential coefficient λ appears bilinearly and is also involved with products of R. One outcome of this dependence is that mixed derivatives of the form $\partial^2 I(\Delta\omega)/\partial R \partial\lambda$ and $\partial^2 I(\Delta\omega)/\partial \Delta_g \partial\lambda$ will be nonvanishing leading to significant correlations of the model parameters.

3 Hindered Diffusion

In this section, we take a closer look at the relaxation operator Γ. For isotropic and hindered diffusion, we define an equilibrium orientational distribution P_0 by the following relation $\Gamma P_0 = 0$, where Γ is an operator that quantifies the diffusion process. Note that this relation may rewritten as

$$P_0^{1/2} P_0^{-1/2} \Gamma P_0^{1/2} P_0^{1/2} = 0$$

This allows us to define an operator $\tilde{\Gamma} \equiv P_0^{-1/2} \Gamma P_0^{1/2}$ so that the equilibrium probability satisfies the symmetrized equation $P_0^{1/2} \tilde{\Gamma} P_0^{1/2} = 0$. This choice, while not a requirement, allows the matrix elements of Eq. 1, or equivalently Eq. 2 to be expressed in complex symmetric form with the advantages of computational efficiency as noted above. In many cases, it is useful to express Γ as a differential operator in terms of the generators of infinitesimal rotations of a diffusing body $\mathbf{L}$ referred to a set of body-fixed axes instantaneously co-moving with the diffusing particle. The rotational diffusion operator may then be expressed as follows

$$\Gamma = \mathbf{L} \cdot \mathbf{R} \cdot \mathbf{L},$$

for isotropic media, where $\mathbf{R}$ can be represented as a diagonal matrix in a symmetry frame of the diffusor. For isotropic media and ordered media where hindered diffusion is operative, it is useful to expand the equilibrium distribution P_0 in terms of eigenfunctions of the diffusion operator. A convenient choice, which also corresponds to the eigenfunctions of the quantum mechanical symmetric top, is the complex conjugate of the Wigner rotation matrix elements $(\mathscr{D}^L_{M\,K}(\Omega))^*$, where Ω parameterizes the relevant infinitesimal rotation. For a thorough discussion of why it is the complex conjugate of the Wigner rotation matrix elements that are the relevant eigenfunctions, see the references [5, 6]. The action of $\mathbf{L}$ on the $\left(\mathscr{D}^L_{M\,K}(\Omega)\right)^*$ may be quantified as follows [5]

$$\mathbf{L}^2\left(\mathscr{D}^L_{M\,K}(\Omega)\right)^* = L(L+1)\left(\mathscr{D}^L_{M\,K}(\Omega)\right)^* \tag{5}$$

$$L_z\left(\mathscr{D}^L_{M\,K}(\Omega)\right)^* = K\left(\mathscr{D}^L_{M\,K}(\Omega)\right)^* \tag{6}$$

$$(L_x \pm iL_y)\left(\mathscr{D}^L_{M\,K}(\Omega)\right)^* = \sqrt{(L\pm K)(L\mp K+1)}\left(\mathscr{D}^L_{M\,K\mp 1}(\Omega)\right)^*. \tag{7}$$

For many systems of interest, the diffusing particle can be characterized as approximately axially symmetric. In that case, there is a fundamental symmetry between states of $\pm K$ related by a rotation by π radians through the body-fixed 'y' axis [6]. It is therefore useful to take symmetric and antisymmetric combinations of $(\mathscr{D}^L_{M\,K}(\Omega))^* \pm (\mathscr{D}^L_{M\,-K}(\Omega))^*$ symmetric top eigenstates. For those cases, where the orienting potential characterizing hindered diffusion also has this symmetry, which is common in practice, it suffices to work with the symmetric combination $(\mathscr{D}^L_{M\,K}(\Omega))^* + (\mathscr{D}^L_{M\,-K}(\Omega))^*$. When only the symmetric contributions are retained, there is a significant reduction in the size of the matrix representation of the equations of motion and an increase in computational efficiency. Detailed examination of the diffusion operator for both isotropic and hindered diffusion [3, 4, 7] confirms that the diffusion operator also has this fundamental K symmetry. For all these reasons, the K-symmetrized basis is preferred for studying magnetic resonance line shapes arising from relaxation due to rotational diffusion. For completeness, the $\tilde{\Gamma}$ operator takes the following form

$$\tilde{\Gamma}(\Omega) = R_{i,i}L_iL_i + \frac{1}{2k_BT}(R_{i,i}L_iL_iU(\Omega)) - \frac{1}{(2k_BT)^2}R_{i,i}(L_iU(\Omega))(L_iU(\Omega)). \tag{8}$$

The (unnormalized) solution for a given potential of the form

$$U(\Omega) = -k_BT\sum_{L,M,K}\lambda^L_{MK}\mathscr{D}^L_{M\,K}(\Omega)$$

is

$$P_0(\Omega) = \exp(-U(\Omega)/k_BT). \tag{9}$$

Note that the potential parameters $\{\lambda^L_{MK}\}$ are dimensionless for consistency with the form of Eq. 9.

As discussed in Sect. 2, the Fourier–Laplace transform of the Stochastic Liouville Equation, Eq. 1, introduces the starting vector of initial conditions. Under the symmetrization operation the elements of the starting vector become proportional to integrals of the following form

$$\int d\Omega\ (P_0(\Omega))^{1/2}\,\mathscr{D}^L_{M\,K}(\Omega).$$

In the usual case of an M-independent potential, the only possible nonvanishing components of the starting vector are dependent on L and K. Furthermore, for those potentials which are symmetric combinations of even values of K, one finds that the only nonvanishing contributions to the starting vector are also symmetric combinations of even K values. The justification for these qualitative considerations may be found from exploring the symmetries of the equilibrium distribution function.

For slow-motional spectra where the required basis sets are large, the integrand is highly oscillatory for large L and K values which makes heavy demands on the stability and accuracy of numerical integration routines. In this work, we propose an alternative method of computation based on the numerical solution of a recurrence relation that is more computationally stable. The derivation relies on well-known properties of the Wigner rotation matrix elements [8] $\mathscr{D}^L_{M\,K}(\Omega)$, and the results can be applied to a variety of standard problems, e.g., diffusion in a cone [9]. Evaluation of the recurrence can be accomplished by singular value decomposition of the matrix representation of the equivalent system of homogeneous linear equations.

3.1 Recurrence

The starting point is the following identity satisfied by the $\mathscr{D}^L_{M\,K}(\Omega)$

$$\sin\beta\frac{\partial}{\partial\beta}\mathscr{D}^L_{M\,K}(\Omega) = -\frac{L(L+1)\sqrt{(L^2-M^2)(L^2-K^2)}}{L(2L+1)}\mathscr{D}^{L-1}_{M\,K}(\Omega) - \frac{MK}{L(L+1)}\mathscr{D}^L_{M\,K}(\Omega) + \frac{L\sqrt{((L+1)^2-M^2)((L+1)^2-K^2)}}{(L+1)(2L+1)}\mathscr{D}^{L+1}_{M\,K}(\Omega), \tag{10}$$

where the meaning of the parameter β is discussed in Sect. 3.2. Multiplying from the left by $\exp(-U(\Omega)/2k_BT)$ and integrating over all Ω leads to simple expressions on the right-hand side of Eq. 10. In order to simplify the left-hand side, the integral may be evaluated by integration by parts. Note that the constant term in the integration is proportional to $\sin^2(\beta)$ evaluated at the endpoints $\beta = 0$ and $\beta = \pi$ which vanishes. The derivatives yield terms proportional to $\mathscr{D}^1_{00}(\Omega)\mathscr{D}^L_{M\,K}(\Omega)$, where $\mathscr{D}^1_{00}(\Omega) = \cos(\beta)$. There are also terms of the form $\mathscr{D}^{L_1}_{M_1\,K_1}(\Omega)\mathscr{D}^{L_2}_{M_2\,K_2}(\Omega)$ that arise from terms bilinear in derivatives of the potential and the original $\mathscr{D}^L_{M\,K}(\Omega)$. These bilinear products of $\mathscr{D}^L_{M\,K}(\Omega)$ may be reduced by a further identity satisfied by the Wigner rotation matrix elements, viz.

$$\mathscr{D}^{L_1}_{M_1\,K_1}(\Omega)\mathscr{D}^{L_2}_{M_2\,K_2}(\Omega) = \sum_{L_3}(2L_3+1)\begin{pmatrix} L_1 & L_2 & L_3 \\ M_1 & M_2 & M_3 \end{pmatrix}\begin{pmatrix} L_1 & L_2 & L_3 \\ K_1 & K_2 & K_3 \end{pmatrix}\left(\mathscr{D}^{L_3}_{M_3\,K_3}(\Omega)\right)^* \tag{11}$$

The quantities in parentheses in Eq. 11 are Wigner 3J symbols. The references should be consulted for more information about them [5, 6, 10]. Three more identities are useful for manipulating Eq. 10. In particular, $\mathscr{D}^2_{00}(\Omega) = (3\cos^2(\beta) - 1)/2$, $\mathscr{D}^0_{00}(\Omega) = 1$, and $(\mathscr{D}^L_{M\,K}(\Omega))^* = (-1)^{M-K}\mathscr{D}^L_{-M\,-K}(\Omega)$. Once these identities have been applied, all terms in the integrated form of Eq. 10 are in the form of the desired elements of the starting vector. A detailed derivation of the recurrence relation will be published elsewhere [11]. For the model system of Sect. 4, the matrix representation

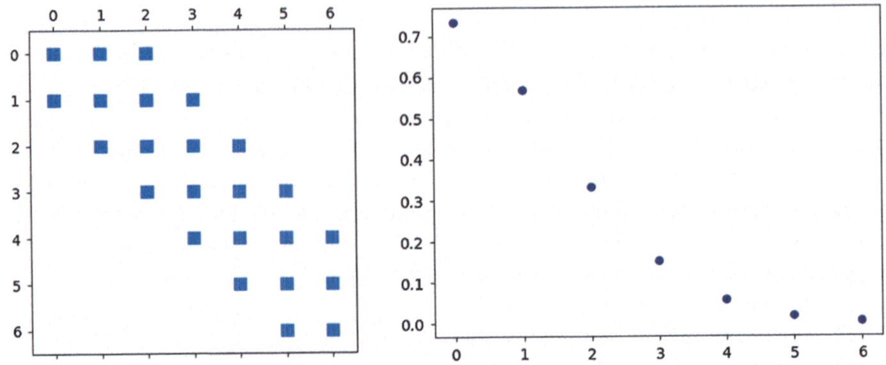

Fig. 2 Schematic presentation of the system of linear equations (left) and their solution (right) corresponding to the recurrence relation for elements of the starting vector presented in this work

of the recurrence relation as well as the unnormalized values of the starting vector elements are shown in Fig. 2 Note that for the model system of Sect. 4, there are only contributions when L is even and $M = K = 0$ and even L. From Fig. 2 it is clear that regardless of the rotational diffusion rate large L values, up to $L = 10$ need to be included in the starting vector. In contrast, only the $L = 0$ term needs to be retained when there is no orienting potential.

3.2 Coordinate Systems

There are a number of coordinate systems that are relevant for characterizing the line shape problem. The spectral lineshape is recorded in the laboratory or 'L' frame, which is defined by the applied magnetic field. For liquid crystalline media, there may be a director or 'D' frame which can be tilted with respect to the laboratory frame. The diffusing particle may have a symmetry axis which defines the diffusion or 'R' frame. Finally, the magnetic interactions may take a simple, diagonal form in a principal axis system or 'M' frame which differs from the 'R' frame symmetry axis. It is convenient to use Wigner rotation matrix elements for quantifying transformations among these coordinate systems. Rotations may be parameterized in various ways. We will use the symbol $\Omega_{F_1 \to F_2}$ to represent the set of parameters that transforms the coordinates from frame F_1 to F_2. The relevant transformations are $\{\Omega_{L\to D}, \Omega_{D\to R}, \Omega_{R\to M}\}$. For the purposes of this work, Euler angles are an adequate scheme, and the form of the coordinate transformations can be chosen to be purely real, which simplifies the evaluation of the matrix elements appearing in the Stochastic Liouville Equation. In particular, the transformation from the 'L' frame to the 'D' frame is typically specified by a rotation by an angle ψ about the laboratory 'y' axis. Such a scheme is shown in Fig. 3. A similar transformation is often used for $\Omega_{R\to M}$. The transformation $\Omega_{D\to R}$

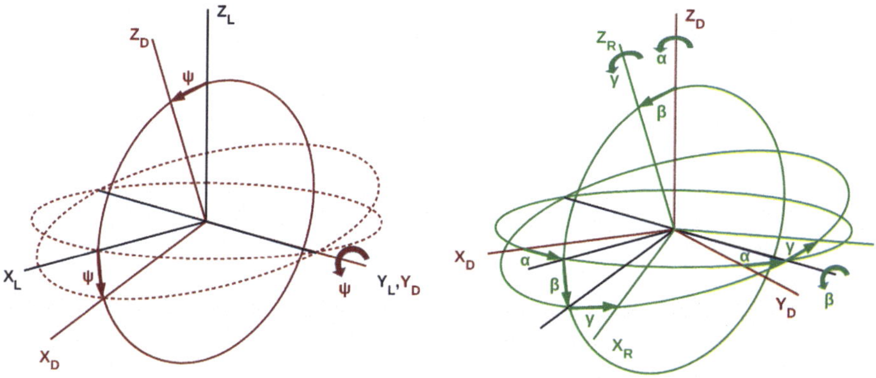

Fig. 3 (Left): Representation of the transformation from the laboratory 'L' frame to the director 'D' frame via a rotation by ψ about the laboratory Y_L axis, or line of nodes; (Right) Representation of the transformation from the director 'D' frame to the rotational diffusion 'R' frame via a rotation by α about the director Z_D axis, followed by a rotation by β about the line of nodes, and finally, a rotation by γ about the new 'z' axis, also known as the figure axis, Z_R

is a stochastic function of time involving the full parameterization of the rotation group. The relevant parameters are also shown in Fig. 3.

In order to compute a spectrum, it is necessary to perform an ensemble average over all members of the ensemble. By the ergodic hypothesis, this is equivalent to an average over all orientations. We have included this detailed look at the relevant coordinate transformations as they are often a source of confusion and errors. A thorough discussion of the relevant issues is available elsewhere [4].

4 Simple Model

In order to apply some of the concepts developed in this work, we consider a simple model of an *approximately* isotropic rapidly tumbling diffusor characterized by a Zeeman interaction with axial symmetry in an orienting potential that has cylindrical symmetry. An example of such a system might be an ^{15}N nucleus on a flexible tether at the end of a slowly diffusing acyl chain in a lipid bilayer. For simplicity, we consider the case of no director tilt, that is $\psi = 0$ in Fig. 3. The case of no orienting potential has been treated separately [12]. In that case, the matrix representation of Eq. 1 is complex symmetric and tridiagonal [12]. For such systems, stable numerical computation of the spectral line shape is a standard problem. In the presence of a simple orienting potential of the form $U(\Omega) = -k_B T \lambda^2_{00} \mathscr{D}^2_{00}(\Omega)$ the matrix representation of the Eq. 1 becomes quintdiagonal. Due to the symmetrization of the Γ operator, the matrix representation of Eq. 1 remains complex symmetric. Efficient computation of the line shape involves tridiagonalization followed by application of standard numerical techniques. A detailed report will be published elsewhere [11]. We discuss some important qualitative features in Sect. 5.

5 Discussion

The effect of an orienting potential on the line shape depends on several key parameters. In the context of the simple model discussed in Sect. 4 these are: the magnitude of the magnetic anisotropy $\frac{2}{3}(g_\perp - g_\|)$, the average value of the Zeeman interaction $(2g_\perp + g_\|)/3$, the rotational diffusion rate, R, and the strength of the potential λ^2_{00}. Consider the rapid tumbling case shown in the spectrum on the right in Fig. 1. In the absence of a potential, all orientations of the spin probe are equally likely, and the spectrum is centered on the trace of the Zeeman tensor, an invariant quantity. For a potential with a positive λ^2_{00}, probe orientations parallel to the laboratory 'z' axis are preferred and the averaging process is incomplete. In this case, the spectrum will shift upfield toward the $g_\| = g_z$ end of the spectrum. This corresponds to the case of nematic ordering. If a director tilt were allowed, the upfield shift is modified consistent with the relative orientation of the director with respect to the applied field. For the case of discotic ordering, where $\lambda^2_{00} < 0$, the rapid tumbling spectrum shifts downfield toward the $g_\perp = g_x = g_y$ end of the spectrum. The larger the potential, the more pronounced the spectral shift. If the average value of the Zeeman interaction is known, then this can be a sensitive means for inferring the degree of ordering in the system. The general expressions for the spectral derivatives given in Eq. 4 can then be used as inputs to the methods described elsewhere to infer quantitatively the degree of parameter sensitivity [13]. A more detailed presentation of these results will be given in a forthcoming publication [11]. For the slow tumbling case, corresponding to the left-hand spectrum in Fig. 1, similar effects from variation of the orienting potential also arise. In this range of rotational tumbling, however, the spectrum still appears to be motionally narrowed, corresponding to the right-hand spectrum of Fig. 1. The nematic versus discotic case gives rise to spectra with very different amplitudes and more pronounced upfield versus downfield shifts, respectively. The analytical basis for these qualitative observations arises from the nonvanishing of the mixed parameter derivatives introduced in Sect. 2.

References

1. Schneider, D.J., Freed, J.H.: Spin relaxation and motional dynamics. In: Hirschfelder, J.O., Wyatt, R.E., Coalson, R.D. (eds.) Lasers, Molecules and Methods, vol. 73, pp. 387–527. Wiley, New York (1989). Chap. 10
2. Blum, K.: Density Matrix Theory and Applications, 2nd edn. Plenum (1996)
3. Meirovitch, Eva, Igner, Dan, Eva, Igner, Moro, Giorgio, Freed, Jack H.: Electron-spin relaxation and ordering in supercooled nematic liquid crystals. J. Chem. Phys. **77**, 3915–3938 (1982)

4. Schneider, D.J., Freed, J.H.: Calculating slow motional magnetic resonance spectra: a user's guide. In: Berliner, L., Reuben, J. (eds.) Biological Magnetic Resonsnce, vol. 8. Plenum (1989). Chap. 1
5. Biedenharn, L.C., Louck, J.D.: Angular Momentum in Quantum Physics. Cambridge (1985)
6. Zare, R.N.: Angular Momentum: Understanding Spatial Aspects in Chemistry and Physics. Wiley, New York (1988)
7. Lee, S., Budil, D.E., Freed, J.H.: Theory of two-dimensional Fourier transform electron spin resonance for ordered and viscous fluids. J. Chem. Phys. **101**, 5529–5558 (1994). There are some typos in the expression for asymmetric diffusion
8. Varshalovich, D.A., Moskalev, A.N., Khersonskii, V.K.: Quantum Theory of Angular Momentum. World Scientific (1988)
9. Earle, K.A., Smirnov, A.I.: High field ESR: applications to protein structure and dynamics. In: Grinberg, O.Y., Berliner, L. (eds.) Biological Magnetic Resonance, vol. 22, pp. 95–143. Kluwer (2004). Chap. 4
10. Brink, D.M., Satchler, G.R.: Angular Momentum. Clarendon Press, Oxford (1968)
11. Earle, K.A.: Efficient computation of starting vector elements in magnetic Resonance line shape problems. In: Applied Magnetic Resonance (2017)
12. Moro, G., Segre, U.: Ultraslow motions and asymptotic lineshapes in ESR. J. Mag. Reson. **83**(1), 65–78 (1989)
13. Earle, K.A., Mainali, L., Sahu, I.D., Schneider, D.J.: Magnetic resonance spectra and statistical geometry. Appl. Mag. Reson. **37**, 865–880 (2010)

The Random Bernstein Polynomial Smoothing Via ABC Method

Leandro A. Ferreira and Victor Fossaluza

Abstract In recent years, many statistical inference problems have been solved by using Markov Chain Monte Carlo (MCMC) techniques. However, it is necessary to derivate the analytical form for the likelihood function. Although the level of computing has increased steadily, there is a limitation caused by the difficulty or the misunderstanding of how computing the likelihood function. The Approximate Bayesian Computation (ABC) method dispenses the use of the likelihood function by simulating candidates of posterior distributions and using an algorithm to accept or reject the proposed candidates. This work presents an alternative nonparametric estimation method of smoothing empirical distributions with random Bernstein polynomials via ABC method. The Bernstein prior is obtained by rewriting the Bernstein polynomial in terms of k mixtures / m mixtures of beta densities and mixing weights. A study of simulation and a real example are presented to illustrate the method proposed.

Keywords Approximation Bayesian Computation · Bayesian estimation Bernstein polynomials · Nonparametric inference

1 Intoduction

Bernstein polynomial smoothing is a nonparametric estimation method widely used as an alternative tool in statistical inference problems [4] in contrast to parametric estimation tools that assumes some family of parameters. A natural issue is how to define a Bayesian approach.

Petrone (1999) [7] proposed a version of Bernstein polynomial smoothing through *Markov Chain Monte Carlo* (*MCMC*) techniques, however, there were many difficult of implementation in the algorithm [6–8]. The MCMC method usually presents some difficulties due to the need of knowing the likelihood function, which is not always

L. A. Ferreira (✉)
Departament of Statistics, University of São Paulo, Rua do Matão,1010, São Paulo, Brazil
e-mail: ferreira.laf@gmail.com

V. Fossaluza
e-mail: victor.ime@gmail.com

A. Polpo et al. (eds.), *Bayesian Inference and Maximum Entropy Methods in Science and Engineering*, Springer Proceedings in Mathematics & Statistics 239, https://doi.org/10.1007/978-3-319-91143-4_12

the case. In this aspect, the *Approximate Bayesian Computation* (*ABC*) method could be more interesting [2, 5].

In practice, the ABC method dispenses the use of the likelihood function. Therefore, the presentation of a Bayesian version of Bernstein polynomial smoothing it will be feasible in terms of this approach.

The paper is organized as follows: In Sect. 2, the ABC method is presented; In Sect. 3, the Random Bernstein polynomial smoothing is showed; Sect. 4 presents the results; Sect. 5 the summary and conclusions.

2 Approximate Bayesian Computation

In this section, the ABC method will be briefly presented allowing the better understanding of the next steps of the nonparametric method of density estimation [5]. According to Marin et al. [5], the likelihood may be unavailable for mathematical reasons (it is not available in closed form as a function of θ) or for computational reasons (it is too expensive too calculate).

To illustrate the importance of this, consider $\theta \in \mathbb{R}_+$, $x \in \mathbb{R}_+$ and $f(x|\theta) \propto \sum \alpha_i f_i(x|\theta))$, $\alpha_i = \frac{1}{2}$, for $i = 1, 2$. Assume that $f_1(x|\theta) \sim Exp(\theta)$ and $f_2(x|\theta) \sim \Gamma(\theta, \theta^2)$, therefore, the likelihood function is given by

$$L(\theta) \propto \frac{1}{2\Gamma(\theta)} \prod \left(\theta e^{-\theta x} + \theta^{2\theta} x^{\theta-1} e^{-\theta^2 x} \right)$$

The *maximum likelihood estimation* is not feasible, indicating the use of another method.

The ABC method dispenses the use of a likelihood function, however, we should be able to replicate the observed experiment, that is, generating a sample of a posterior "candidate" to compare with the observed sample.

The main idea of ABC can be illustrated using the algorithm described as follows

for $i = 1$ *to* N **do**
  repeat
    Generate θ' from the prior distribution $\pi(.)$
    Generate y from the likelihood $f(.|\theta)$
  until $\Delta\,(T(y), T(x)) \leq \varepsilon$
  set $\theta_i = \theta'$
end

Algorithm 1: ABC - Approximated Bayesian Computation

where the parameters of the algorithm are T, a function defining a statistic (which most often is not sufficient); Δ, a distance on T and ε, a tolerance level, $\varepsilon > 0$.

In next section, this method will be applied on Bernstein polynomial considering the "entire sample" as statistic sufficient T; The *Kolmogorov–Smirnov distance*, (*K–S hereby*) as Δ on T.

In the Sect. 4, to show the "goodness of fit," the symmetrized *Kullback–Leibler*, (K–L hereby), *Euclidian* and *K–S* distances will be used, that can be calculated to vectors s and q in expressions (1) for the Euclidian, (2) for the symmetrized *K–L* and (3) for the *K–S*.

$$D(s,q) = \left\{ \sum_{i=1}^{N} (q_i - s_i)^2 \right\}^{\frac{1}{2}}; \tag{1}$$

$$D(s,q) = \frac{1}{2} \left\{ \sum_{i=1}^{N} q_i \log \frac{q_i}{s_i} + \sum_{i=1}^{N} s_i \log \frac{s_i}{q_i} \right\}; \tag{2}$$

$$D(s,q) = sup|F_n(q) - F_n(s)|, \tag{3}$$

where $F_n(\cdot)$ represents the empirical distribution function.

3 Random Bernstein Polynomial

Let $H : [0,1]^k \to \mathbb{R}$ be a continuous function. The Bernstein polynomial of degree m for the function H is given by

$$B_H^m(x_1, \ldots x_k) = \sum_{j_1=0}^{m} \cdots \sum_{j_k=0}^{m} H\left(\frac{j_1}{m}, \ldots, \frac{j_k}{m}\right) \prod_{i=1}^{k} \binom{m}{j_i} x_i^{j_i} (1 - x_i)^{m-j_i}.$$

The univariate case of Bernstein polynomial is given by

$$B_F(x) = \sum_{j=0}^{m} F\left(\frac{j}{m}\right) \binom{m}{j} x^j (1-x)^{m-j},$$

Petrone (1999a) [7, 8] showed that the polynomial may be rewritten in terms of mixtures of beta densities, as follows

$$\sum_{j=1}^{m} w_{j,m} \beta(x; j, m-j+1),$$

where $\beta(\cdot; a, b)$ denotes a beta density with parameters (a, b) and $w_{j,m} = F\left(\frac{j}{m}\right) - F\left(\frac{j-1}{m}\right)$, weights of the mixture.

Assuming that

$$b(x|m, w_m) = \sum_{j=1}^{m} w_{j,m}\beta(x; j, m - j + 1), \tag{4}$$

$b(\cdot; m, w_m)$ is defined as the *Bernstein density* with parameters m and $w_m = (w_{1,m}, \ldots, w_{m,m})$.

To use the expression (4), the priority may be taken as

$$m \sim Poisson(\lambda)$$
$$w_w|m \sim Dirichlet(\alpha).$$

Using the expression (4), given previously, a sample was generated and compared with the observed sample using a chosen the *K–S* distance.

In the Sect. 4, the results of method will be presented.

4 Results and Discussion

The aim of this section is to show the performance of the *Random Bernstein smoothing via ABC method*. The results are expressed by graphs and in terms *K–L*, *Euclidian* and *K–S* distances, see Sect. (2) to know how to calculate it.

All algorithms were performed using R Software version 3.4.1. [3]

Babu et al. [1] suggests, based on the empirical results, the use of the degree as $\lambda = n/\log(n)$.

Example 1 Consider samples of size 50, 100, 200, and 400 from $\beta(\cdot; 2, 5)$. Hierarchically, w_m is sampled from a Dirichlet distribution with parameter $\alpha = (\alpha_1, \ldots, \alpha_m)$. Here, F is considerer as the empirical distribution function.

As observed in Fig. 1, the estimation via Random Bernstein (RB) fits well the observed sample. The Tables 1, 2, 3 and 4 shows the distances among them.

According to Tables 1, 2, 3 and 4 the estimate given by Random Bernstein method is better than theoretical density under all metrics used, except for Euclidian distance.

Example 2 Consider a MELD Score [10], a scale of the severity of a disease, of 482 patients on waiting list for liver transplant attended at Clinics Hospital, São Paulo Medical School (HC-FM-USP) between January 2012 and December 2013 [9].

Table 1 Table of comparison between observed Sample (Sample), Theoretical (Theo) Density, and Random Bernstein (RB) Smoothing for $n = 50$

$n = 50$	Sample x RB	Theo x Sample	Theo x RB
Kullback–Leibler	90.69	85.56	63.78
K–S	0.62	0.5	0.2
Euclidian	0.11	0.13	0.04

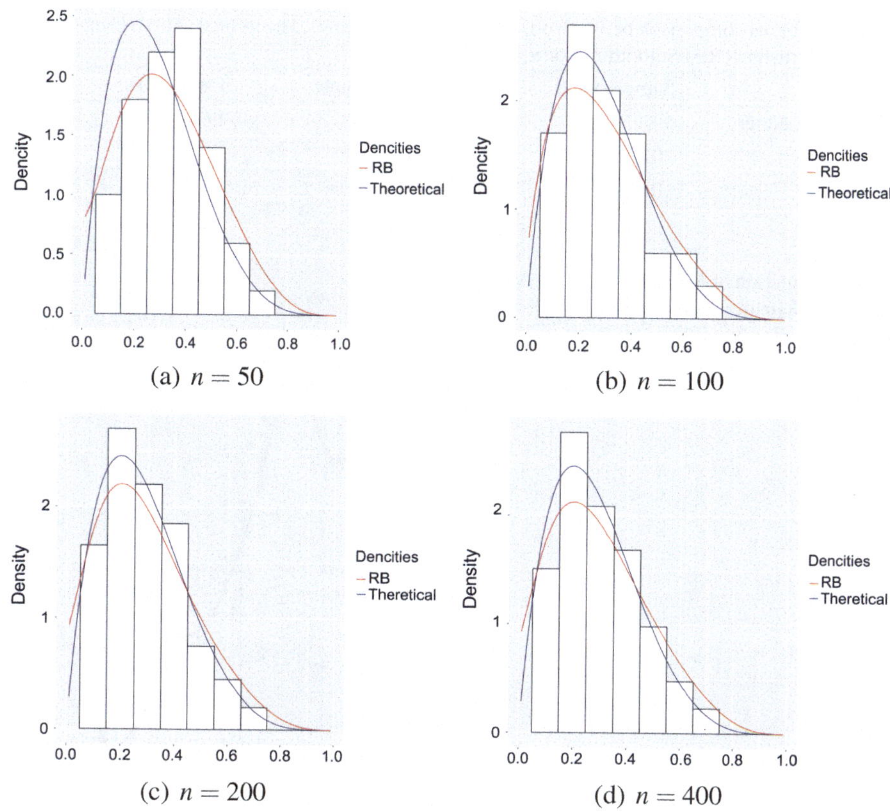

Fig. 1 Estimation of beta densities using samples size of 50, 100, 200, and 400

Table 2 Table of comparison between observed Sample (Sample), Theoretical (Theo) Density, and Random Bernstein (RB) Smoothing for $n = 100$

$n = 100$	Sample x RB	Theo x Sample	Theo x RB
Kullback–Leibler	60.42	52.21	41.80
K–S	0.54	0.51	0.09
Euclidian	0.09	0.21	0.20

Table 3 Table of comparison between observed Sample (Sample), Theoretical (Theo) Density, and Random Bernstein (RB) Smoothing for $n = 200$

$n = 200$	Sample x RB	Theo x Sample	Theo x RB
Kullback–Leibler	61.50	50.02	47.63
K–S	0.57	0.52	0.12
Euclidian	0.07	0.07	0.01

Table 4 Table of comparison between observed Sample (Sample), Theoretical (Theo) Density, and Random Bernstein (RB) Smoothing for $n = 400$

$n = 400$	Sample x RB	Theo x Sample	Theo x RB
Kullback–Leibler	59.17	47.87	49.99
K–S	0.56	0.50	0.14
Euclidian	0.05	0.06	0.01

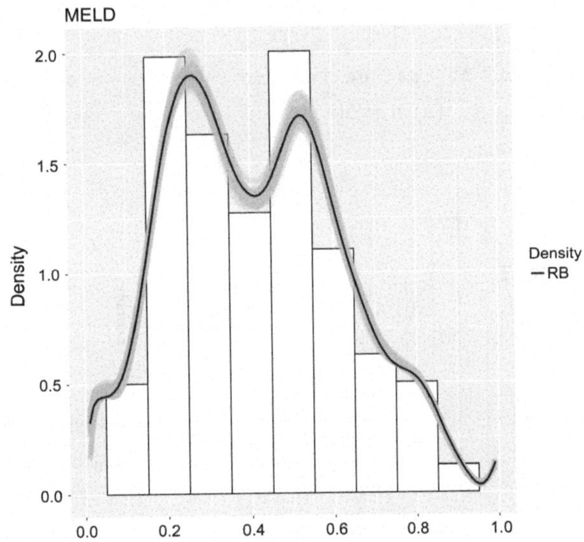

Fig. 2 Estimation of MELD Score sample using Random Bernstein (RB)

Table 5 Table of comparison between MELD Score (MELD) and Random Bernstein (RB) Smoth

	MELD x RB
K–L	49.50
K–S	0.34
Euclidian	0.02

Notice that, the support of Bernstein distribution is [0, 1], therefore to use the method it is necessary to rescale the data into [0, 1]. To rescale the sample, consider the linear transformation $T(\cdot) = \frac{(\cdot)-min(\cdot)}{max(\cdot)-min(\cdot)}$.

As observed in Fig. 2, the estimation via Random Bernstein (RB), in red, fits well the observed MELD Score sample. The estimate of density is given with a *confidence band*, in gray.

To plot the *confidence band*, it was considered 200 estimates, it was taken 95% of them. The Table 5 shows the distances between them.

5 Conclusions

This work presented a technique for nonparametric density estimation using ABC method. Two examples were given to illustrate the method.

The first example, a study of simulation, show the impact of sample size, and the second one, the capacity of estimation of real samples and a measure of uncertainty.

The Random Bernstein smoothing via ABC method may be an important tool for statistical inference; its use may be interesting because it has easier computational implementation when compared to other methods.

Acknowledgements The authors are partially supported by CAPES grants.

References

1. Babu, G.J., Canty, A.J., Chaubey, Y.P.: Application of Bernstein polynomials for smooth estimation of a distribution and density function. J. Stat. Plan. Inference **105**, 377–392 (2002)
2. Blum, K., Gaggiotti, M., François, O.: Approximate Bayesian Computation (ABC) in practice. Trends Ecol. Evol. **25**(7), 410–418 (2010)
3. Core Team, R.: A language and environment for statistical computing. R Foundation for Statistical Computing, Vienna. http://www.R-project.org/ (2016)
4. Fossaluza, V.: Estimation of discretes distributions via Bersntein Copulas. In Port, Estimação de distribuições discretas via Cópulas de Bernstein), IME-USP (2012)
5. Marin, J.-M., Pudlo, P., Robert, C.P., Ryder, R.: Approximate Bayesian Computational methods. Stat. Comput. **21**(2), 289–291 (2011)
6. Petrone, S., Wasserman, L.: Consistency of Bernstein polynomial posteriors. J. R. Stat. Soc. Sen B Stat. Methodol. 64 79–100, (2002)
7. Petrone, S.: Bayesian density estimation using Bernstein polynomials. Can. J. Stat. **27**, 105–126 (1999a)
8. Petrone, S.: Random Bernstein polynomials. Scand. J. Stat. **26**, 373–393 (1999b)
9. Turri, J.A., Descimoni, T., Ferreira, L.A., Diniz, M.A., Haddad, L.B.P., Campolina, A.G.: Higher MELD Score increases the overall cost on the waiting list for liver transplantation. Gastroenterology archives(Online) (2017)
10. Wiesner, et al.: Model for end-stage liver disease (MELD) and allocation of donor livers. Gastroenterology 124(1), 91–6 (2003)

Mean Field Studies of a Society of Interacting Agents

Lucas Silva Simões and Nestor Caticha

Abstract We model a society of agents that interact in pairs by exchanging for/against opinions about issues using an algorithm obtained with methods of Bayesian inference and maximum entropy. The agents gauge the incoming information with respect to the mistrust attributed to the other agents. There is no underlying lattice and all agents interact among themselves. The interaction pair can be described as a dynamics along the gradient of the logarithm of the evidence. By using a symmetric version of the two-body interactions we introduce a Hamiltonian for the whole society. Knowledge of the expected value of the Hamiltonian is relevant information for the state of the society. In the case of uniform mistrust, independent of the pair of agents, the phase diagram of the society in a mean field approximation shows a phase transition that separates an ordered phase where opinions are to a large extent shared by the agents and a disordered phase of dissension of opinions.

Keywords Entropic dynamics · Social systems · Agent models · Mean field

1 Introduction

The realization that Statistical Mechanics (SM) is a theory of information processing, due to Jaynes in the fifties [10, 11], permits applying its methods to other systems which can be considered outside the realm of physics. This would not be a surprise to Maxwell and Boltzmann who expressed hopes about the applicability of the general methods of SM to the mathematical description of human societies. There have been considerable efforts to quantitatively understand human behavior, specially modeling human actions within a society and their large-scale consequences by identifying order parameters that can present general replicable behavior. In spite of being in its

L. S. Simões (✉) · N. Caticha (✉)
Instituto de Física, Universidade de São Paulo, São Paulo, SP CP 66318, Brazil
e-mail: lsimoes@if.usp.br

N. Caticha
e-mail: nestor@if.usp.br

A. Polpo et al. (eds.), *Bayesian Inference and Maximum Entropy Methods in Science and Engineering*, Springer Proceedings in Mathematics & Statistics 239, https://doi.org/10.1007/978-3-319-91143-4_13

infancy, this field has seen the introduction of models that accommodate to a certain extent the regularities in some examples where empirical data is available [3, 5, 6].

An attractive feature of this area is the large abundance of questions we do not yet fully understand. In this paper, we consider a society of agents that exchange discrete for/against opinions about issues. This type of model has been applied to model data obtained by the Moral Foundation Theory group [7–9] by [4]. This paper contribution is the analytical study of the global properties of a model of opinions in which the agents have a definite level of mistrust toward other agents, which was introduced by [1, 4]. Our aim here is not to present the latest model but to show that standard methods of Statistical Mechanics, such as the mean field approximation, which itself has an origin in maximum entropy ideas, can be applied to obtain some analytical understanding of collective properties of large assemblies of interacting agents.

In a set of studies, over the last years [4, 14] the present line of work has been applied to study cultural diversity, making use of sociophysical models that connect moral psychology and neuroscience with statistical mechanics tools. In this work, we expand that understanding with a mean field analysis of an agent-based model. First, in Sect. 2, we describe the opinion exchange dynamics between mistrusting agents, which arises from a generic framework based on Bayesian and maximum entropy methods. This framework and a more general analysis will be presented elsewhere [2]. This leads to a stochastic dynamics along the gradient of a Bayesian evidence (Sect. 3). The pairwise dynamics is not symmetric, making a difference between the emitter and the receiver agents. In Sect. 4, we consider the case of a pairwise interaction in which both agents act as the receiver and emitter, hence permitting to consider a symmetric interaction. This allows to introduce a global Hamiltonian. Finally, we explore a mean field (MF) approximation to this model. We show that this approximation recovers the results obtained in previous works using Monte Carlo style numerical simulations (Sect. 5).

2 The Model

The general framework for the model we use will be presented elsewhere. For the particular model we use, one can check a detailed description of the dynamics in [5]. In this society model each agent is represented by a neural network. An issue is a public input vector $\mathbf{x}$ and the opinion of agent j is σ_j, which takes values 1 or -1. We start considering the exchange of information in which an emitter agent j sends the input vector and σ_j to a receiver agent i. The state of agent i, which we denote by $\hat{\mathbf{w}}_i$, changes in order to accommodate the received information. At a given point, we want to describe the change in the state of incomplete information we have about agent i, when it receives information about the classification of an issue by agent j. Before receiving the information our representation of the state of agent i is given by a prior distribution $Q(\mathbf{w}|\hat{\mathbf{w}}_n\mathbf{C}_n)$. The posterior distribution is given by Bayes theorem

$$P(\mathbf{w}|\sigma_j, \mathbf{x}, \hat{\mathbf{w}}_n \mathbf{C}_n) = \frac{Q(\mathbf{w}|\hat{\mathbf{w}}_n \mathbf{C}_n) P_L(\sigma_i = \sigma_j|\mathbf{wx})}{Z_{n+1}} \tag{1}$$

where the likelihood $P_L(\sigma_i = \sigma_j|\mathbf{wx})$ is the probability that the answer σ_i to question $\mathbf{x}$ is σ_j. The normalizing factor Z_{n+1} is the Bayesian evidence. Since in general the posterior is not in the gaussian family a new posterior $Q(\mathbf{w}|\hat{\mathbf{w}}_{n+1}\mathbf{C}_{n+1})$ is chosen by maximizing the relative entropy between the old and new gaussians, subject to the constraints imposed by the Bayesian posterior P. The evidence will be central to the discussion that follows since the gradient of the energy like function $\mathscr{E}_{n+1} = -\log Z_{n+1}$ determines the change in hyperparameters that implement the learning process. This framework permits reobtaining results that go back to [12] using functional variational methods, to [13] using Bayesian ideas and to [5], where a deduction of Eqs. 2 and 3 below appear. The learning dynamics describes the update of the mean vector and covariance matrix of the Bayesian posterior, or alternatively describe the update of the weights of a neural network that represents an agent, when receiving information about another agent's opinion:

$$\hat{\mathbf{w}}_{n+1} = \hat{\mathbf{w}}_n - \mathbf{C}_n \cdot \nabla_{\hat{\mathbf{w}}_n} \mathscr{E}_{n+1} \tag{2}$$

$$\mathbf{C}_{n+1} = \mathbf{C}_n - \mathbf{C}_n \cdot \left(\mathbf{H}_{\hat{\mathbf{w}}_n} \mathscr{E}_{n+1}\right) \cdot \mathbf{C}_n \tag{3}$$

where $\mathbf{H}_{\hat{\mathbf{w}}_n} \mathscr{E}_{n+1}$ is the Hessian matrix of second derivatives of $\mathscr{E}_{n+1}$ with respect to the elements of $\hat{\mathbf{w}}_n$.

The important element of this dynamics is a "free energy" cost function $\mathscr{E}_{n+1}$, or minus the logarithm of the evidence in the Bayesian learning step. The available information on the agent's architecture enters here. Explicitly it depends on the likelihood, i.e., the probability that a certain opinion is emitted conditioned on the state of the emitting agent. This is better explained in Sect. 3.

3 Moral Agent Learning Model

We consider a specific agent model $\mathscr{M}$ which tries to infer the best value of $\mathbf{w} \in \mathbb{R}^K$ that makes good predictions σ about $\mathbf{x}$ matching its' peers predictions (i.e., conformity-seekers). We choose for the neural network architecture the single-layer perceptron, for the reason that the results are rich, interesting and sufficiently complicated to deserve attention. Furthermore, there are several examples where the empirical data shows that even the performance of humans on certain tasks can be modeled by linear classifiers.

A primal ingredient of the modeling scenario is the parameter ε, which represents the level of mistrust that the receiver agent attributes to the emitter agent. This can be understood as a mistrust about the emitter agent or about the communication channel. In the model ε is the probability that the correct opinion had its signal flipped during the communication.

The model is given by:

$$\mathscr{M} : \begin{cases} \textit{issue/opinion:} & \mathbf{x} \in \mathbb{R}^K; \quad \sigma \in \{-1, +1\} \\ \textit{architecture:} & \sigma' = \text{sign}(\mathbf{x} \cdot \mathbf{w}) \\ & \sigma = \begin{cases} -\sigma' & \text{with probability } \varepsilon \\ \sigma' & \text{with probability } 1 - \varepsilon \end{cases} \\ \textit{inference constraints:} & \mathbb{E}[w^i] = \hat{w}^i, \ \mathbb{E}[w^i w^j] = C_{ij} + \hat{w}^i \hat{w}^j \end{cases} \quad (4)$$

Alternatively, one could interpret the learning situation described by $\mathscr{M}$ as follows: consider a pair of vectors (perceptrons) $\hat{\mathbf{w}}, \mathbf{w} \in \mathbb{R}^K$ which we call receiver and emitter of information, respectively. The receiver $\hat{\mathbf{w}}$ learns from the emitter $\mathbf{w}$ through the presentation of examples $(\mathbf{x}_\mu, \sigma_\mu)$, where we call $\mathbf{x}_\mu \in \mathbb{R}^K$ an issue and $\sigma'_\mu = \text{sign}(\mathbf{w} \cdot \mathbf{x}_\mu)$ the emitter's opinion on the issue. The emitter's opinion is corrupted with a multiplicative noise ε when communicated to the receiver. The reason for this corruption is not analyzed here, but it can be noise in the channel or the possibility of concealed cheating. The likelihood distribution is obtained by marginalizing the joint distribution of the received answer σ while the correct answer should have been σ'

$$P_L(\sigma|\mathbf{x}, \mathbf{w}, \varepsilon) = \sum_{\sigma'} P(\sigma|\sigma'\mathbf{x}, \mathbf{w}, \varepsilon) P(\sigma'|\mathbf{x}, \mathbf{w}, \varepsilon) \quad (5)$$

The flipping probability is $P(\sigma|\sigma'\mathbf{x}, \mathbf{w}, \varepsilon) = \varepsilon$ if $\sigma = -\sigma'$ and $1 - \varepsilon$ if $\sigma = \sigma'$. We also use that $P(\sigma'|\mathbf{x}, \mathbf{w}, \varepsilon)$ is 1 if $\sigma'\mathbf{x}.\mathbf{w} > 0$, to obtain

$$P_L(\sigma|\mathbf{x}, \mathbf{w}, \varepsilon) = \varepsilon\Theta(-\sigma\mathbf{x} \cdot \mathbf{w}) + (1 - \varepsilon)\Theta(\sigma\mathbf{x} \cdot \mathbf{w}) = \varepsilon + (1 - 2\varepsilon)\Theta(\sigma\mathbf{x} \cdot \mathbf{w}) \quad (6)$$

where Θ is the Heaviside step function. The Bayesian evidence after the presentation of the $(n + 1)$th example —which we call $\mathbf{x}$ for simplicity without a time label—is:

$$Z_{n+1} = \int d\mathbf{w} \ P_L(\sigma|\mathbf{x}, \mathbf{w}, \varepsilon) \ Q(\mathbf{w}|\hat{\mathbf{w}}_n \mathbf{C}_n)$$

which, after doing the integral over the $\mathbf{w}$ variables and taking the logarithm, becomes

$$\mathscr{E}_{n+1} = -\gamma_n^2 \log \left[\varepsilon + (1 - 2\varepsilon)\, \Phi\left(\frac{\sigma h_n}{\gamma_n}\right)\right] \quad (7)$$

where we defined the projections $h_n = \frac{1}{\sqrt{K}} \mathbf{x} \cdot \hat{\mathbf{w}}_n$, $\gamma_n^2 = \frac{1}{K}\mathbf{x}^\mathsf{T}\mathbf{C}_n\mathbf{x}$ and Φ is the cumulative distribution function (cdf) of the standard normal distribution. Note that γ_n is directly linked to the covariance of the distribution of the weights. If the uncertainty about the weights is large, so is γ_n and vice-versa.

4 Mean Field Analysis

Equation 7 is an energy like a term that describes how the receiver agent is affected by the emitter agent. If both agents take both roles on a given encounter we can suppose that a symmetric energy term would describe the interactions. This permits going from a dynamical description of the interactions (micro) to a global description of the state of the society (macro) using Statistical Mechanics tools. Still, the calculations are intractable and analytical progress comes at the expense of approximations. In this section, we develop a Mean Field approach to an specific canonical ensemble of social agents in a noisy society.

Let us consider, a society of agents $\{\hat{\mathbf{w}}_i\}$ which we suppose can be described mostly by one specific Hamiltonian $\mathscr{H}$, which is a sum of the energy like terms, suggested from the previous analysis, for all pairs $\langle i, j\rangle$ of agents:

$$\mathscr{H} = -\gamma^2 \sum_{\langle i,j\rangle} \log \left[\varepsilon + (1-2\varepsilon)\,\Phi\left(\frac{1}{\gamma}\left(\sigma_i h_j + \sigma_j h_i\right)\right)\right] = \sum_{\langle i,j\rangle} V_{ij} \tag{8}$$

Societies are dynamic and their global states change in time. However, on a certain time scale, intermediate between the fast update of the individual states of the agents and the very slow change of a society, we might attempt to characterize the state of a society by stipulating that the value of some global function is approximately conserved. The information, or rather the assumption, that a certain quantity is conserved allows to describe the Boltzmann–Gibbs state of the society. We suppose that the mean value $\langle \mathscr{H} \rangle$ is conserved throughout the configuration evolution of the society, that is, $\mathscr{H}$ remains close to some fixed value E. The probability distribution describing this society with this information is then $P_B(\{\hat{\mathbf{w}}_i\}) = \frac{1}{Z_B}\exp\left(-\beta\mathscr{H}\left(\{\hat{\mathbf{w}}_i\}\right)\right)$. Unfortunately we cannot obtain the equilibrium properties of this Hamiltonian and have to rely on further approximations. In order to simplify our model, a mean field approximation is performed, which finds a solution as close as possible to P_B inside a parametric family of probability distributions $P_0 = \Pi_i P_i(\hat{\mathbf{w}}_i)$, which is a product over all agents.

In that case, we do not wish to choose a product distribution indiscriminately; we want to pick the best P_0 approximating P_B given the constraints we have assigned to it. That is, a calculation we can do maximizing the entropy S, as follows:

$$S[P_0||P_B] = -\left\langle \log Z_B + \beta\mathscr{H} + \sum_i \log P_i \right\rangle_{P_0} \tag{9}$$

Usual calculus of variations arguments lead to:

$$\delta S = \int d\hat{\mathbf{w}}_k\, \delta P_k \left[\log Z_B + \log P_k + 1 + \beta \sum_{i\in\partial k} \int d\hat{\mathbf{w}}_i\, P_i V_{ik} \right] \tag{10}$$

where we introduced the notation ∂k meaning *neighbors of* k. This result can only be possible for any variation δP_k if the term in brackets $[\cdots]$ is identically zero itself, so that we get the following result:

$$P_k(\hat{\mathbf{w}}_k) = \frac{1}{Z_k} \exp\left(-\beta \sum_{i\in\partial k} \int d\hat{\mathbf{w}}_i P_i V_{ik}\right) \tag{11}$$

Still, due to the rather complex form of the potential V_{ij}, the equation above is intractable as it is. In that case, we are going to choose (instead of selecting the best one) a mean field probability distribution family similar to the one in Eq. 11.

We proceed comparing the mean field projection done above and the Hamiltonian in (8) to a distribution that take into account only an effective number of neighbors $\nu_{(j)}$ of agent j perceiving effective interactions $m_{(i)} = \langle h_i \rangle$ and $r_{(i)} = \langle \sigma_i \rangle$, in principle depending on agent i. This is inspired by the oldest version of the mean field method and goes back to Curie and Weiss and the idea of the self-consistent method:

$$P_j(\hat{\mathbf{w}}_j) = \frac{1}{Z_j} \left[\varepsilon + (1-2\varepsilon)\, \Phi\left(\frac{1}{\gamma}\left(r_{(i)} h_j + \sigma_j m_{(i)}\right)\right)\right]^{\beta \nu_{(j)} \gamma^2} \tag{12}$$

In fact, one can think of m and r as parameters that describe the overall behavior of the society, such that one agent i receives signals from its neighbors independently of the label i. This does not mean that all the agents are identical, but that their moral vector is drawn from the same probability distribution. We can represent this self-consistently with the following set of equations:

$$m = \int d\hat{\mathbf{w}}\; h(\hat{\mathbf{w}}) P_{\mathrm{MF}}(\hat{\mathbf{w}}) \qquad r = \int d\hat{\mathbf{w}}\; \sigma(\hat{\mathbf{w}}) P_{\mathrm{MF}}(\hat{\mathbf{w}}) \tag{13}$$

Thus, setting ν as constant throughout the society, the mean field probability distribution becomes:

$$P_{\mathrm{MF}}(\hat{\mathbf{w}}) = \frac{1}{\zeta} \left[\varepsilon + (1-2\varepsilon)\, \Phi\left(\frac{r}{\gamma} h(\hat{\mathbf{w}}) + \frac{m}{\gamma}\sigma(\hat{\mathbf{w}})\right)\right]^{\beta \nu \gamma^2} \tag{14}$$

where we recall that $\sigma(\hat{\mathbf{w}}) = \operatorname{sign} h$ and $h(\hat{\mathbf{w}}) = \frac{1}{\sqrt{k}}\mathbf{x} \cdot \hat{\mathbf{w}}$. Note that the effective inverse temperature is $\beta\nu$, where ν can be interpreted as the mean number of neighbors of the agents.

Since we can always rotate the coordinate system to a given orientation, we are going to choose one in which the issue $\mathbf{x}$ aligns with the $\hat{e}_5$ axis (i.e., $\mathbf{x} = \hat{e}_5$). This greatly simplifies our calculation because now $h = \cos\theta$, where θ is the angle between $\mathbf{x}$ and $\hat{\mathbf{w}}$, and other angle integrals are trivial. To see this just recall that, in spherical coordinates, $d\hat{\mathbf{w}} = \sin^3\theta \sin^2\theta_1 \sin\theta_2\, d\theta_1 d\theta_2 d\theta_3 d\theta$.

$$
\begin{aligned}
m &= \frac{1}{\zeta}\int_0^{\pi} d\theta \sin^3\theta \cos\theta \; B(\theta|\varepsilon,\gamma,m,r,\beta,\nu) \\
r &= \frac{1}{\zeta}\int_0^{\pi} d\theta \sin^3\theta \;\mathrm{sign}\,(\cos\theta)\; B(\theta|\varepsilon,\gamma,m,r,\beta,\nu) \\
\zeta &= \int_0^{\pi} d\theta \sin^3\theta \; B(\theta|\varepsilon,\gamma,m,r,\beta,\nu)
\end{aligned}
\tag{15}
$$

where $B(\theta|\varepsilon,\gamma,m,r,\beta,\nu) := \left[\varepsilon + (1-2\varepsilon)\,\Phi\left(\frac{1}{\gamma}\,(r\cos\theta + \mathrm{sign}\,(\cos\theta)\,m)\right)\right]^{\beta\nu\gamma^2}$.

5 Mean Field Results

Equation 15 can only be solved numerically, and we do it by an iterative process. Let $u = (m, r, \zeta)$, then the set of equations is of the type $u = F(u)$. An initial value u_0 is chosen and the map

$$
u_t = (1-\alpha)u_{t-1} + \alpha F(u_{t-1}) \tag{16}
$$

is iterated. This is a fairly standard procedure for solving mean field self-consistent equations. Since it is quite easy to converge no attempt at optimizing α was made. The result can be checked by choosing different starting points u_0. The results are shown in Fig. 1.

It can be seen that there is a phase transition depending on the parameters $\beta\nu$ and γ. We investigate further this transition by looking at the phase diagram in Fig. 2.

We can also change variables in our mean field probability distribution. This is useful because the inner representation $\hat{\mathbf{w}}_i$ is not readily accessible to the experimentalist, whereas the opinion field h in some applications might be:

$$
\begin{aligned}
P(h) &= \int d\mu(\hat{\mathbf{w}})\delta\left(\frac{1}{\sqrt{K}}\hat{\mathbf{w}}\cdot\mathbf{x} - h\right) P_{\mathrm{MF}}(\hat{\mathbf{w}}) \\
&= \frac{1}{C}\,(1-h^2)\left[\varepsilon + (1-2\varepsilon)\,\Phi\left(\frac{r}{\gamma}h + \frac{m}{\gamma}\mathrm{sign}h\right)\right]^{\beta\nu\gamma^2}
\end{aligned}
\tag{17}
$$

Now we can compute other interesting order parameters, such as the variance $v_m = \langle h^2\rangle - \langle h\rangle^2$. The computational results we found are presented in Fig. 3.

As we vary ε to values greater than 0.1 the only change found was that the critical line appeared at larger values of $\beta\nu$ and γ, therefore being more difficult to polarize ($m \neq 0$) the mean field society.

Figure 2 shows that, for fixed β, the phase border can be crossed by increasing the value of γ. This seems paradoxical, since larger γ is associated to a larger norm of the covariance matrix. The explanation of this comes from the fact that the gradient of the evidence, which determines the dynamics in Eq. 2, increases in magnitude with γ when both agents concur. That is, high γ agents rely not only on the novelty

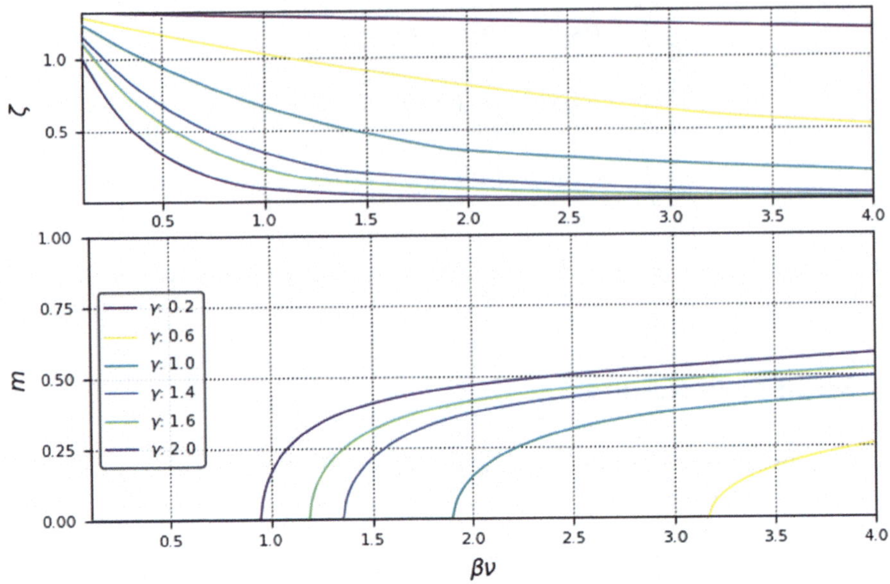

Fig. 1 Solutions of equation 15. Top: The normalization of the MF distribution ζ as a function of the social pressure and number of neighbors ($\beta\nu$). Bottom: The magnetization m. The other order parameter r has a similar behavior to m

brought by disagreement but also learn from corroborating examples. For low γ, agents learn primordially from the novelty of disagreement. Therefore high γ agents will keep on learning even after there is agreement on an issue, resulting in a more ordered society. This same behavior was found in previous works when performing Monte Carlo simulations on non-simplified (that is, without the MF approximation) versions of this model. See for example [4, 5].

6 Discussion

We studied an agent-based model of a society of interacting information processing machines. The agents in this model exchange their opinions on moral issues and learn following a dynamics based upon maximum entropy methods. Depending on the choice of global parameters, such as social pressure, cognitive style, and mistrust level one finds a phase transition from a disordered phase to an ordered one, that is, a phase in which the agents polarize near a common moral belief. This result was to be expected since a similar phase transition appears in [4] and in [5]. As usual in Statistical Mechanics, the phase transition is associated to a collective emergent property of the model. Once the approximations that led to the Hamiltonian form of the theory are taken, the order–disorder transition can be expected from general

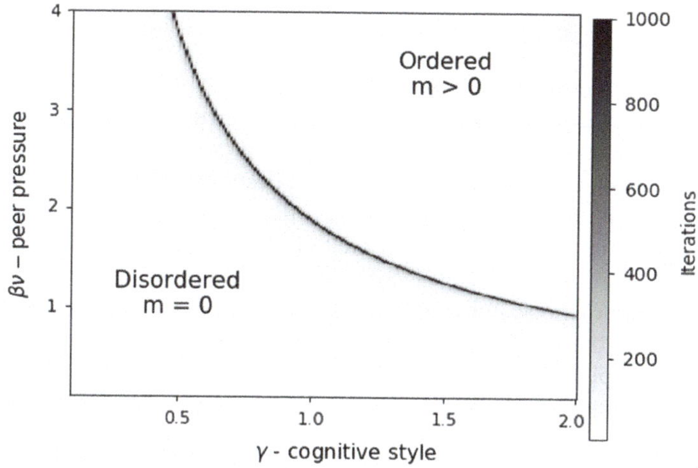

Fig. 2 Phase diagram in the space $\gamma \times \beta\nu$. A phase transition separates an ordered from a disordered phase as signaled by the value of the order parameter m. Here, the value of ε was 0.1 and as it grows toward 0.5 the ordered phase decreases

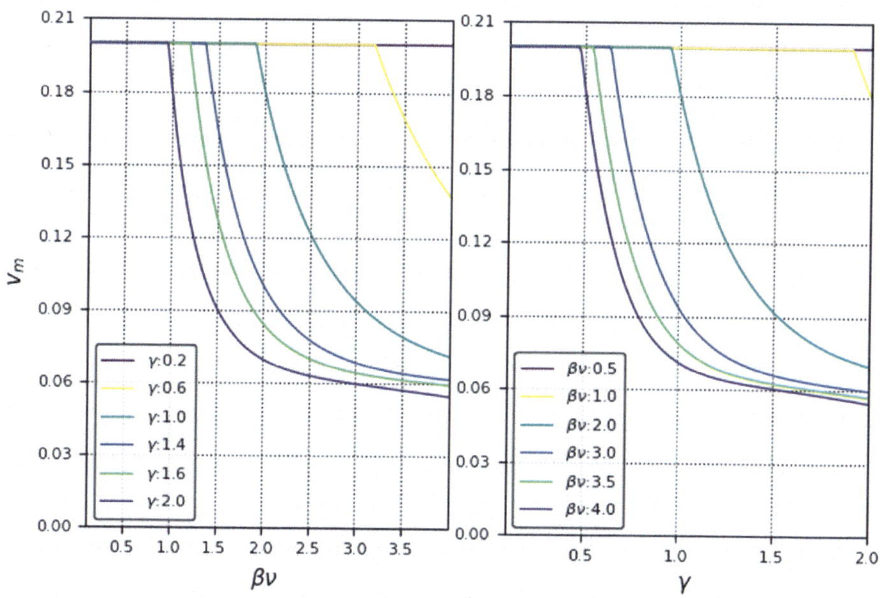

Fig. 3 Variance of the field h with respect to selected values of $\beta\nu$ and γ

arguments. When the empirical data of the Moral Foundation group [7] is considered, the states of societies is found to be in the ordered phase. The disordered state can be seen as a region of parameters where there is no moral uniformity, hence agents don't share common moral values. It is tempting to associate a situation as that to what has been discussed by several authors, specially, Durkheim in the nineteenth century and which he called a state of *anomie*. Whether, this association has any merit deserves further investigation.

Cognitive processes in humans are certainly richer than the simple models we study. We are not trying to describe the emergent properties of societies with any other purpose than being general and a broad description should not attempt to achieve precise numerical validations. The fact that we study mean field versions of the model should not be a very strict limitation. We are making an approximation to an already quite simple model. Despite these shortcomings, this work is to be considered in a line of research that holds the promise of attacking some problems not only of actual, but also perennial interest in the study of societies. Several simple extensions and applications to datasets are under current consideration. The authors will explore extensions of the model in further work.

Acknowledgements LS has a FAPESP fellowship grant n^o2016/15860-3 and thanks CNPq fellowship grant n^o134812/2016-6. Work supported by CNAIPS, the Center for Natural and Artificial Information Processing Systems of the University of São Paulo.

References

1. Alves, F., Caticha, N.: Sympatric multiculturalism in opinion models. In: Giffin, A., Knuth, K.H. (eds.) Bayesian Inference and Maximum Entropy Methods in Science and Engineering: 35th International Workshop on Bayesian Inference and Maximum Entropy Methods in Science and Engineering, July 2015. AIP Conference Proceedings vol. 1757, p. 060005 (2016)
2. Alves, F., Caticha, N.: Entropic Dynamics of Distrust and Opinions of Interacting Agents (In preparation) (2018)
3. Castellano, C., Fortunato, S., Loreto, V.: Statistical physics of social dynamics. Rev. Mod. Phys. **81**(2), 591–646 (2009)
4. Caticha, N., Vicente, R.: Agent-based Social Psychology: From Neurocognitive processes to Social data. Adv. Complex Syst. **14**(5), 711–731 (2011)
5. Caticha, N., Cesar, J., Vicente, R.: For whom will the Bayesian agents vote? Front. Phys. **3**(25), 1–14 (2015)
6. Galam, S.: Sociophysics: A Physicist's Modeling of Psycho-political Phenomena. Springer, New York (2012)
7. Haidt, J.: The emotional dog and its rational tail: a social intuitionist approach to moral judgment. Psychol. Rev. **108**(4), 814–834 (2001)
8. Haidt, J.: The new synthesis in moral psychology. Science **316**(5827), 998–1002 (2007)
9. Haidt, J., Kesebir, S.: Morality. Handbook of Social Psychology, vol. 3:III:22, pp. 797–832 (2010)
10. Jaynes, E.T.: Information theory and statistical mechanics. Phys. Rev. **106**(4), 620–630 (1957)
11. Jaynes, E.T.: Information Theory and Statistical Mechanics. II. Phys. Rev. **108**(2), 171–190 (1957)
12. Kinouchi, O., Caticha, N.: Optimal generalization in perceptions. J. Phys. A **25**(23), 6243–6250 (1992)
13. Opper, M., Winther, O.: A Bayesian approach to on-line learning. In: Saad, D. (ed.) On-Line Learning in Neural Networks, pp. 363–378. Publications of the Newton Institute. Cambridge University Press, Cambridge (1998)
14. Vicente, R., Susemihl, A., Jericó, J.P., Caticha, N.: Moral foundations in an interacting neural networks society: A statistical mechanics analysis. Physica A **400**(c), 124–138 (2014)

The Beginnings of Axiomatic Subjective Probability

Marcio A. Diniz and Sandro Gallo

Abstract We study the origins of the axiomatization of subjective probabilities. Starting with the problem of how to measure subjective probabilities, our main goal was to search for the first *explicit* uses of the definition of subjective probability using betting odds or ratios, i.e., using the Dutch book argument, as it is presently known. We have found two authors prior to Ramsey (The foundations of mathematics and other logical essays. Routledge & Kegan Paul, 1931, [43]) and de Finetti (Fund Math 17:298–329, 1931, [20]) that used the mentioned definition: Émile Borel, in an article of 1924, and Jean-Baptiste Estienne, a French army officer, in a series of four articles published in 1903 and 1904. We tried to identify, in the references given by Borel and Estienne, inspirations common to Ramsey and de Finetti in order to determine, in the literature on the probability of the beginning of the last century, at least some elements that point to specific events that lead to the referred axiomatization. To the best of our knowledge, the genesis of the axiomatic approach in the subjective school was not traced yet, and this untold history can give us a better understanding of recent developments and help us, as applied scientists, in future works.

Keywords History of probability · Axiomatic probability · Subjective probability

1 Introduction

In the beginning of the twentieth century, several scholars, although some of them working separately, were gathering efforts in an endeavor which ended with the

[1] We are using the term "axiomatization" as the setting of unproved *"basic statements about the concept (such as the geometry of the plane) to be studied, using certain undefined technical terms as well as the terms of classical logic."* [52, p. 9].

M. A. Diniz (✉) · S. Gallo
Federal University of São Carlos, São Carlos, Brazil
e-mail: marciodiniz@ufscar.br

S. Gallo
e-mail: sandro.gallo@ufscar.br

A. Polpo et al. (eds.), *Bayesian Inference and Maximum Entropy Methods in Science and Engineering*, Springer Proceedings in Mathematics & Statistics 239, https://doi.org/10.1007/978-3-319-91143-4_14

axiomatization,[1] of the three main interpretations of the concept of probability. The frequentist interpretation won the day, and based its foundations on the works of Richard von Mises. The subjective or personalistic interpretation based its developments on the works of Frank Plumpton [43] and, especially, Bruno [20]; and the logic or logicist school, advocated initially by [27, 31], was fully axiomatized by [15], in the philosophical field, and [17], for the applied sciences.

We decided to turn our attention to the origins of axioms in the subjective side, as the origins of axiomatic probability in the frequentist school is well documented.[2] Therefore, starting with the problem of how to measure subjective probabilities, the main goal of this note was to search for the first *explicit* uses of the definition of subjective probability using betting odds or ratios, i.e., using the Dutch book argument, as it is presently known. These efforts can be, in the future, part of a larger project, in which the primary goal will be, in the spirit of [25], to identify and understand the specific events that lead the subjective approach toward axiomatization. In particular, we would like to understand why the definition of probability using the Dutch book argument was used explicitly only in the same period when the other schools were also looking for axioms.

In our search, we have found two authors prior to Ramsey and de Finetti that used the mentioned definition: Émile Borel, in an article of 1924, and Jean-Baptiste Estienne, a French army officer, in a series of four articles published in 1903 and 1904. As far as we know, the ideas proposed by Borel date from the early 1920s and, regarding Estienne, probably from the late 1890s. We tried to identify, in the references given by Borel and Estienne, inspirations and quotations common to Ramsey and de Finetti in order to determine, in the literature on probability of the beginning of the last century, at least some elements that point to specific events that lead to the referred axiomatization.

To the best of our knowledge, the genesis of the axiomatic approach in the subjective school was not traced yet, and our contribution is a start that will, eventually, fill this gap when joined with future work. The untold history can give us a better understanding of recent developments and help us, as applied scientists, to *"find some connection between this abstract entity which satisfies certain mathematical stipulations and the pragmatic content, the real meaning, of the important statements of scientific and social intercourse which contain the word "probability" or one of its synonyms."*[3]

2 Historical Context

In the late nineteenth century, pure mathematicians of different fields were searching for suitable axioms for their areas, and probability was no exception. [4, p. 98] wrote that probability theory would be ranked among the pure sciences only if *"the*

[2] See [1, 46] and references therein.

[3] [34, p. 4].

principles upon which its methods are founded should be of an axiomatic nature." This was written decades before Hilbert presented his 23 open problems at the International Congress of Mathematicians in Paris in 1900. The sixth problem was to treat axiomatically, based on the model of geometry, those parts of physics in which mathematics already played an important role, especially mechanics and, as Hilbert classifies, probability.[4]

Even though applied probability was broadly disseminated, mathematicians were not satisfied with the foundations of probability calculus,[5] once its whole nature seemed to be concerned with concepts that lie outside mathematics: events, trials, randomness, probability. As [41] wrote, *"one can hardly give a satisfactory definition of probability."*

After some tentative replies to Hilbert's challenge, these concerns were completely satisfied by Kolmogorov's book. Probability was regarded, mainly, as a physical or statistical concept, in the sense that represented physical properties of objects or tendencies of aleatory events to present stable relative frequencies in the long run. These were the ideas proposed by [47, 48][6] and [44], and which explicitly inspired Kolmogorov's system.[7]

This great advance, however, reinforced, among pure mathematicians, the idea that *"the subjectivistic theory of probability remained pretty much of a philosophical curiosity. None of those for whom probability theory was a means of livelihood or knowledge paid much attention to it."*[8]

Nevertheless, we now know that some scholars were seriously engaged in this curiosity. The articles of [20, 43] are recognized as the most important ones, followed by the synthesis of [45]. Working independently, Ramsey and de Finetti started by tackling the fundamental issue of how to measure subjective probabilities, and the idea was to use betting odds or ratios.[9]

---

[4] See [26, p. 454]. To explain what he meant by axioms for probability, Hilbert cited Georg Bohlmann, who named the rules of total and compound probability as axioms rather than theorems, [3].

[5] Probability was not regarded as an interesting research topic by pure mathematicians. As an example of this feeling, see below the remarks made by Camille Jordan about the probability lessons he had to teach at the *Polytechnique*. The only exception was, perhaps, the Russian school in St. Petersburg led by Markov and Tchebychev.

[6] See [50] for an English translation.

[7] In a footnote to §2, "The Relation to Experimental Data," of his book, [33, p. 3], mentioned that "*In establishing the premises necessary for the applicability of the theory of probability to the world of actual events, the author has used, in large measure, the work of R. v. Mises, pp. 21–27.*" See [49]. Therefore, although Kolmogorov's approach is strictly mathematical, i.e., can be adopted regardless of the interpretation given to the axioms, the frequentist school rapidly embraced it.

[8] [34, p. 15].

[9] [43, p. 31] says: *"The subject of our inquiry is the logic of partial belief, and I do not think we can carry it far unless we have at least an approximate notion of what partial belief is, and how, if at all, it can be measured."* On his side, [20, pp. 302–303] says *"Now it is a question of* **measuring** *subjective probability, that is, to translate in the determination of a number, our degree of uncertainty about a given sentences; this is the first problem that presents when one wants to establish the calculation of probabilities according to the subjectivistic conception."*

The genesis of this idea, however, was not new, and may be traced back to the origins of probability calculus. Cardano (1501–1576) does not mention probability explicitly, but implicitly says that stakes should be placed according to probabilities.[10] Thus, at least since Cardano, gamblers, and scholars knew how to compute stakes from probabilities when they wanted to engage in a fair bet. The other direction (how to compute probabilities from placed stakes) did not show up until the early twentieth century, the period when our search for explicit uses of the Dutch book argument finds the earlier instances. We now proceed to examine more closely the found examples, providing also some brief biographical information about their authors.

3 Jean-Baptiste Estienne

Jean-Baptiste Eugène Estienne was born in 1860 at Condé en Barrois (now Les Hauts-de-Chée), a small village between Reims and Nancy. He was admitted to the *École Polytechnique* in 1880, graduating in 1882 as 131st of his year. Also, in 1882, he won the first prize for mathematics in the *concours général*.[11]

He joined the French army as a second lieutenant in 1883, serving with the artillery from 1884. Having studied ballistics, he presented his first statistical application to the subject in a work entitled *"Étude sur les erreurs d'observation,"* presented to the *Académie des Sciences* in 1890.[12] This work stimulated the introduction of modern indirect firing methods.

Promoted to captain of the 1st Artillery Regiment in 1891, he began to develop telemetric instruments and, in 1902 he was made squadron commander of the 19th Artillery Regiment. He promoted the development of precision instruments for the technical artillery section in Paris, and the use of telephonic connections to enable the artillery to switch targets quickly. During this period, he presented other contributions at the *Académie* and articles that were published in journals like the *Comptes Rendus de l'Académie de Sciences* and the *Revue d'artillerie*. The *"Essai sur l'art de conjecturer"* is a series of four articles published by the *Revue d'Artillerie* in 1903 and 1904.

At that time,Estienne was already reputed to be one of the most competent and progressive officers in France, and one of the founders of modern artillery. He was involved in the creation of the French air force and other artillery methods, most notably tanks, during World War I. For this reason, he is known as *père des chars*, or father of tanks, to this day.

---

[10]See [40, p. 202]: *"So there is one general rule, namely, that we should consider the whole circuit [the sample space], and the number of those casts which represents in how many ways the favorable result can occur, and compare that number to the remainder of the circuit, and according to that proportion should the mutual wagers be laid so that one may contend on equal terms."*

[11]Academic competition held every year between senior high school students.

[12]See [22]. Bertrand and Jordan were the chairs of the session when he presented the work.

3.1 Essai sur l'art de Conjecturer

Estienne starts with the classical definition of probability, that assumes all the possible cases — of some random experiment, we would say in modern days — are equally probable. However, he adds, this equality of probabilities is not a mathematical fact provable through reason or experience, but it depends on the personal appreciation of each individual, and says that *"I may have the right not to recognize the equality of two chances admitted by others"*. This definition of probability is called by Estienne **mathematical probability**. He also warns that the narrow limits imposed by the definition of mathematical probability lead to inconsistencies when such a theory was applied *"to facts absolutely foreign to its object"*.

A more applicable concept would be that of what he calls **vulgar probability**, which we can safely identify with a degree of belief or subjective probability. Therefore, *"the vulgar probability of an uncertain fact always exists in the mind of man,"* i.e.,it could be applied to every uncertain fact we face, and not only those where the possible cases can be enumerated and judged as equally probable. In the beginning of the second section, he presents the definition of the betting price or quotation, *la cote*, given below.

> *"Definition of quotation –An individual expresses the degree of probability that he attributes, rightly or wrongly, to the happening of an uncertain event, by the fraction $\frac{a}{a+b}$, a being the quantity he is willing to bet against b that the event will happen. The number $\frac{a}{a+b}$ is the quotation of the event, for the considered individual."*

Estienne mentions that the quote does not depend on the units in which a and b are measured, suggesting that a small amount should be used in order to avoid problems relating quotes and material welfare.

Then,he proceeds to establish the general rules the calculi with quotations should obey. As primary result, he points out that the individual should base his betting prices on the common sense — *le bon sens* — in order to avoid contradictions when assessing such prices. This implies, for instance, that a quote should not be greater than one and that the quotes of certain, or believed as certain, events should be one.

The first main principle proved from the definition — the principle of total quotes —, is merely the law of addition for the union of disjoint events. From this principle, he remarks that the quotation on some event, when added to the quotation on the negation of the same event, must be one. As a second remark, he mentions that the *"mathematical probability, when it exists, is a special case of the quotes"*, deriving the classical definition as a particular case of the quotation calculus when the possible cases are judged equally probable.

The second principle — the principle of composed quotes —, is the law of multiplication for the intersection of events. As a corollary of this principle, he derives Bayes' rule. The second section is closed remarking that the principles that rule the quotation calculus are the same that rule probability calculus and that the difference between the *vulgar* and the *mathematical* definitions of probability are just superficial. Therefore, Estienne adds, since probability calculus is also based on the same principles demonstrated valid for the quotation calculus, one can concentrate on the interpretation of the obtained results, since analytical results are plentifully provided. In the sequel of the essay, Estienne illustrates the developed concepts with applications in games of chance and ballistics.

4 Émile Borel

Félix Édouard Justin Émile Borel was born in 1871 in Saint-Affrique, department of Aveyron in Southern France.[13] In 1889[14] he applied for the *École Polytechnique* and the *École Normale Supérieure*.[15] Being qualified as the top candidate for both, he chose to attend the later, beginning his studies the same year. In 1893, even before finishing his doctorate, Borel was appointed lecturer at the University of Lille.

In 1897 he returned to the *École Normale* also as lecturer, and in 1904 was appointed to the University of Paris at Sorbonne where, in 1909, he became full professor of the chair of theory of functions, specially created for him. In 1920,he also became full professor of the chair of probability calculus and mathematical physics, which he held until his retirement in 1941. Between 1910 and 1920 he was also Assistant Director of the *École Normale*.

His first paper on probability was published in 1905. In 1906, he started *La Revue du mois*, a monthly magazine in which he defined in a specific manner the new scientific program he intended to follow thereafter, the *"practical worth of the calculus of probabilities."* The *Revue* counted with contributions of several scholars of diverse fields.[16] In 1909 Borel published the paper that became a classical work in the mathematical theory of probability, on "denumerable probabilities", from which the Borel-Cantelli Lemma originated.

During World War I, Borel was called by Paul Painlevé, a mathematician that started a political career after the Dreyfus case, to become head of the Directorate for inventions in the service of national defense. When Painlevé was minister of war from March to September 1917, Borel became director of the ministry's technical services, and when Painlevé rose to prime minister, at the most tragic moment of World War I, in the autumn of 1917, Borel became general secretary of the government, entrusted with all missions, including those connected with the scientific aspects of the war.[17]

[13]Distant 120 km from Montpellier.

[14]The same year he won the first prize for mathematics in the *concours général*, the same prize won by Estienne in 1882.

[15]Preparatory school of teachers for the high school level.

[16]Poincaré and Karl Pearson, for instance, wrote contributions about probability and statistics.

[17]This allows one to conjecture that Borel and Estienne eventually met.

Borel's political activities did not prevent him from dedicating time to scientific projects. In the early 1920s, he started a new publication of which he was the only editor: the *Traité du Calcul des Probabilités et de ses Applications*, published in four volumes and eighteen fascicles[18] between 1925 and 1939. This was an important work, which fell rapidly into relative obscurity after World War II, partly due to Kolmogorov's book, but also due to the publication of other books with a style closer to pure mathematics.

In his later life, Borel became deputy for the department of Aveyron from 1924 to 1936, Minister for the Navy in 1925, mayor of Saint-Affrique from 1929 to 1941 and from 1945 to 1947.

4.1 Borel's Review

Borel [10], "*A propos d'un traité de probabilités*," is a review of [31]. After discussing some aspects of Keynes' book that are not relevant to our study, Borel turns to the question of how to measure probability, a topic covered by Keynes on Chap. III, "The measurement of probabilities" of his book. Keynes' view was that some probabilities could be compared, but not explicitly measured or attached to it a specific value.

With opposing view, Borel suggests a procedure to measure probabilities that is analogous to the setting of prices of goods bought and sold in a given market. He says, on page 332: *"it seems that the method of betting allows, in majority of the cases, a numerical evaluation of probabilities which has exactly the same character as the evaluation of prices by exchange transactions."*

Being aware of the technical problems related to bets and money prizes, he suggests that, if we do not want *"to take into account the attraction or the reluctance caused by bets, I would be able to offer the choice between two bets giving the same advantages in case of gain. Paul says it will rain tomorrow; I admit that we agree on the precise meaning of this statement and offer him, at his choice, to receive 100 francs if his statement is true, or to receive 100 francs, if he obtains 5 or 6 throwing a die. In this second case, the probability of receiving 100 francs is equal to one third; if he prefers to receive 100 francs in case his prediction is accurate, it is because he attributes to this prediction a probability greater than one third. [...] The same method applies to all verifiable sentences and allows the numerical evaluation of probabilities with a precision quite comparable to the precision with which the prices are evaluated."*

He reminds that *"for probabilities as for prices, [...] sentimental reasons will intervene to distort the numerical evaluation"*, but adds that *"Even if the exceptional cases were, which I do not think, more numerous for probabilities than for market prices, it would still be possible to establish a mathematical theory applicable to all probabilities numerically evaluated and this theory would have a very large scope of application, as well as economic theories have an importance that is not diminished*

[18]Five of them written by Borel himself.

by the fact that there are values, such as the conscience of a judge, that are not for sale."

5 Common Influences

In this section, we search for ideas or authors that may have influenced Estienne or Borel. We start looking for references given in [20, 43].

Ramsey [43] does not provide any previous influence or reference for his approach, saying only that the *"old-established way of measuring a person's belief is to propose a bet, and see what are the lowest odds which he will accept."*, [43, p. 172].

On his side, [20, p. 303] explicitly mentions that the idea of using betting odds to measure subjective probabilities was based on an observation due to [2, p. 27]. In the referred page, Bertrand says that the probability of some event for a person is p if that person was prepared to exchange the consequences attached to the happening of the event to identical consequences attached to a drawing from an urn with a probability equal to p. To illustrate the concept, Bertrand remarks that, if I accept that the probability for the doctor coming upon being called is 9/10 and is 1/3 for a successful treatment, the probability of cure through treatment is, "for me", as Bertrand highlights, $9/10 \times 1/3 = 3/10$. From this, he concludes that one's probability evaluations must follow the rules of total and composite probability — the actual subject of the section where we find these lines — but nothing more is elaborated from these ideas.

Finetti [21, p. 6, footnote], also wrote later that Bertrand introduced subjective probabilities *"for the sole purpose of opposing them with other 'objective probabilities'."*[19]

5.1 Estienne's Background

Estienne [23] also refers to [2], quoting several examples of this book, but also implicitly. Since Bertrand taught at the *École Polytechnique* from 1854 to 1894,[20] it is interesting to review some facts about the courses Estienne probably took during his years there, i.e., the early 1880s.

In those years, Bertrand and Camille Jordan took turns teaching the Analysis course, an annual course which should cover the main topics on probability theory in three lessons, and it was given to second-year students.[21] Bertrand taught the classes of students that started on even years and Jordan taught the classes that started on

[19] Actually de Finetti mentioned page 24 in [21], probably a typo since in [20] he mentions page 27. The error was kept in the translation published in [34].

[20] [16, p. 61].

[21] [16, p. 63].

odd years.[22] Therefore, since Estienne started his studies in 1880, he was probably taught by Bertrand.[23]

After 1885, the *Polytechnique* professors receive strict instructions and probability was explicitly inserted in the syllabus[24] but before that there was a lot of freedom regarding the syllabus.[25] This fact and some remarks made by Jordan about the same period,[26] make us wonder that his students, before 1885, were not introduced to probability.

However, even Jordan's students apparently had contact with basic probability and some of its applications. These were provided by the mandatory ballistics course, taught at the *École d'application de l'artillerie et du génie*, in Fontainebleau, by captain Esprit Pascal Jouffret.[27]

This course had as part of the syllabus the use of probabilistic methods to ballistics. From [12, p. 40], we know that Jouffret's course was based on four articles published on the *Revue Maritime et Coloniale* in 1873 and 1874.[28] In [29], he labels the classical definition of probability as *mathematical probability*, to distinguish it from the *philosophical probability*, as he calls it, in a way similar to Estienne, that distinguished mathematical probability from *vulgar* probability.

Jouffret says that philosophical probability is based on the induction and analogy and that these probabilities do not seem expressible by numbers, even though "*they are imposed on us with more or less force, according to the degree of naturalness and simplicity, their concordance with notions we already have, etc.*"[29] Just after this remark, he returns to the classical definition and, to illustrate the concept, considers some event A, which has f favorable and equiprobable "cases", being N the total number of also equiprobable cases. Then the probability of A, p, is f/N, and of its complement, q, is $1 - p$, leading him to conclude that it is a **bet** of p against $1 - p$, or of $p/(1 - p)$ against one, that event A will happen.

The use of betting odds to illustrate probabilities computed under the classical assumption was not new, being extensively used by [2, 36, 37], but with early examples given in [14, 35], just to mention French authors. In his subjective view, [23, 61, p. 407] partly agrees with such an illustration, but adds: *"Laplace was certainly entitled to express his degree of personal conviction by the offering of such a bet, if his calculations authorized him, but he abused the authority of his geniality by*

[22][39, pp. 10–11].

[23]In fact, Jordan taught in 1888 — see "Camille JORDAN: Leçons de Probabilités à l'Ecole Polytechnique (1888),"available at http://www.jehps.net/decembre2009.html — which gives us evidence that Estienne was, indeed, taught by Bertrand.

[24]That is why we have class notes of Jordan's 1888 lecture notes.

[25]See [16, p. 63].

[26]According to [19, p. 205], Jordan wrote in 1894: *"I would like to see [...] disappear without regret [from the course of analysis] the three lessons that I devote to the calculus of probabilities"*.

[27]See [19] and [39, p. 11].

[28]See [28–30].

[29][29, p. 10].

claiming to impose his figure on the universality of men."[30] However, none of the above authors explicitly used betting odds to *define* probability in a way at least similar to Estienne.[31]

5.2 Borel's Background and Former Works

As mentioned above, Borel was approved at both the *École Polytechnique* and the *École Normal Supérieure*. He chose the second, and this was the most appropriate path to enter the closed French academic world at the end of the nineteenth century. According to [18], probability was not taught at the *École Normal*, but from [13, p. 287], we know that Borel attended Poincaré's course at the Sorbonne.[32] Therefore, we may say that Borel was fully acquainted with [41][33] which, alongside [2], were the most important textbooks on probability at the beginning of the twentieth century.

In his doctoral thesis, [5] studied series that were known to diverge on a dense set of points on a closed curve and hence, it was thought, could not be continued analytically into the region bounded by the curve. Borel then discovered that the set of points where divergence occurred, although dense, can be covered by a countable number of intervals with arbitrarily small total length, i.e., the first version of the presently called Heine–Borel theorem.

As remarked by [46], this discovery led Borel to a new theory of measurability for subsets of the [0, 1] interval. The same approach was used by Borel's former student at the *École Normale*, Henri Lebesgue, as the basis of a new theory of integration in [38]. The first article Borel wrote using measure theory on probability was [6]. This article explains how the presently called Borel measure allows one to extend and make more precise the calculation of geometric probabilities. The results obtained allowed him to state that *"the new theory justified Poincaré's intuition that a point chosen at random from a line segment would be incommensurable with probability 1,"* Shafer and Vovk [2006, p. 16].

---

[30]He goes on bringing an example: *"When the dimensions of a rectangular field are given the values 200m and 300m, we are forced to admit that the field has [an area of] 6 hectares; it would be a mistake to claim that the numeral 6 is imposed by geometry on those who would not have first agreed on the accuracy of the dimensions."* This echoes [2, p. 28]: *"If it is alleged that it is impossible to measure in figures the probabilities of which we are speaking, the objection would be as unfounded as if, evaluating the length of a field of rectangular appearance at 300m and the width at 100m, to add, irrespective of any verification, that such measures, however doubtful they may be, and these assessments assign to the field an area of three hectares."*

[31]When Estienne took Bertrand's course, in 1881–1882, the book of his professor has not yet been published. In the preface, [2, p. v] mentions that the book was based on the course he taught at the *Collège de France*. Bertrand also mentions Jouffret, as "Jauffret" [2, p. xxxvi], citing his colleague's illustration of the law of large numbers using an example from ballistics.

[32]Poincaré was full professor of the chair of probability theory and mathematical physics, being eventually succeeded by Borel [51, p. 36].

[33]Like [2], entitled *Calcul des Probabilités*. A second edition was published in 1912.

According to [13], Borel gave his first course on probability calculus at the Sorbonne in 1908-1909. His textbook, *Éléments de la theorie des probabilités*, was published in 1909, the same year he published another paper on the subject, [8]. This article was important because it strengthened the connection between measure theory and probability. In it, Borel proves a version of the law of large numbers — the Borel-Cantelli lemma — for a denumerable sequence of independent trials. He poses the question as a problem of geometrical probability: the fraction of ones in the binary expansion of a real number chosen at random from [0, 1] converges to 1/2 with probability 1. Being related to number theory, the paper called the attention of pure mathematicians [51, p. 57], [13, p. 289] and was crucial to the ongoing research that resulted in Kolmogorov's book.

Mentioning the importance of this 1909 article, [24, p. 54], recalls also the introduction of countable additivity in probability theory: *"It was at the moment when Mr. Borel introduced this new kind of additivity into the calculus of probability — in 1909, that is to say — that all the elements needed to formulate explicitly the whole body of axioms of probability theory came together. [...] This is what Mr. Kolmogorov did. This is his achievement."*

Although involved with the axiomatization of the objective interpretation of probability, Borel inherited from Poincaré a double position on the meaning of probability. [42], a popular exposition on science and philosophy, admits the subjective character of probability, but also states the objectivity of statistically stable phenomena observed in nature.

In [7], a note written for the *Revue du Mois*, he differentiates objective and subjective probabilities.[34] The difference, for him, is one of degree: when there is a situation where the probabilities are far away from 1 and 0, probability has a subjective value in the sense that some action has to be taken, and it is up to the individual to decide [51, p. 44].

Borel's early concept seems odd when compared with the present meaning given to subjective probability, but in his 1924 review of Keynes' book, we find a developed notion. There he accepts Keynes' idea that a probability is relative to a body of knowledge, and accepts the modern subjective character of probabilities proposing the method of betting discussed above. This view was revisited in [11], the last fascicle of the series *Traité du calcul des probabilités et ses applications*.[35]

---

[34]In [9, pp. 226–227], he wrote that *"It is not a difference of nature that separates the objective probability from the subjective probability, but only a difference of degree. A result from probability calculus deserves to be called objective when the probability becomes large enough to be confounded with practical certitude."*

[35]In [11], in pages 84 through 86 he explains the method of betting and in the last section (Conclusion and probability of a single trial), he reviews the argument and mentions [21]. In 1928 Borel help was important in the establishment of the *Institut Henri Poincaré*, a research institution devoted to probability theory and mathematical physics, where several lectures were held in the 1930s, including de Finetti's, presented in 1935 and published in 1937.

6 Final Remarks

Our search for early explicit uses of betting odds to define subjective probabilities led us to two French authors of the beginning of the twentieth century, both influenced by important scholars of the period — Joseph Bertrand and Henri Poincaré — when the French school still admitted a dual interpretation of the meaning of probability: subjective and objective.

Regarding our main characters, we believe our report allows one to see Estienne as the progressive and pragmatical man from the military, advocating the subjective definition of probability, and Borel as the traditional scholar, a true heir of Poincaré, accepting both the objective and subjective views. While Borel only suggested the argument as a way to elicit probabilities, Estienne presented a formulation that today one clearly recognizes as precursory to ideas completely formalized by Ramsey and de Finetti. He explicitly used an individual's assessed betting price as the definition of subjective probability; recognized *le bon sens* — which de Finetti called "coherence" — to avoid contradictions in the offered prices as providing the mathematical basis for the rules of probability, and derived the sum and product rules of probability calculus. From this perspective, we can say that Estienne's efforts were much closer to a rigorous axiomatization than the suggestions made by Borel.

These remarks made us speculate a little more about Estienne's views. After reading several scholars that wrote about probability during the nineteenth and beginning of the twentieth century, we distinguished two (almost) opposing groups: one, more purist, advocating mathematical rigor above everything; the other, more pragmatic, looking for applications in any scientific field. Some scholars are easily qualified in one of the two groups, but some were, perhaps, at the intersection.

For us, it is not wrong to say that Estienne and Borel were at this intersection. Although it is not easy to see Borel at both groups, we can also qualify Estienne as a dualist. He wanted to define the concepts clear and precisely, but once this was done, one would be entitled to use probability in whatever applications were needed.

As possible developments of this research, we propose, first, a detailed review of the literature of probability theory of the late nineteenth century and of the historiography of gambling and betting on sports. A more detailed study of the academic environment of mathematics and probability in Europe and U.S. at the end of the nineteenth century is also recommended. A second possible extension is to search and understand the reasons of the described axiomatization, as well as for the time period of which happened. In this direction, it would be relevant to study the parallel concept of utility and how it helped to promote the full axiomatization of subjective probability in [43, 45], and the advent of game and decision theory.

References

1. Barone, J., Novikoff, A.: A history of the axiomatic formulation of probability from Borel to Kolmogorov: Part I. Arch. Hist. Exact Sci. **18**, 123–190 (1978)
2. Bertrand, J.: Calcul des Probabilités. Gauthier-Villars (1889)
3. Bohlmann, G.: Lebensversicherungs-mathematik. In: Encyklopädie der Mathematischen Wissenschaften, vol. 1(2), pp. 852–917. Teubner (1901)
4. Boole, G.: XIII. On the conditions by which the solutions of questions in the theory of probabilities are limited. Philos. Mag. **8**, 91–98 (1854)
5. Borel, É.: Sur quelques points de la théorie des fonctions. Annales Scientifiques de l'École Normale Supérieure **12**, 9–5 (1895)
6. Borel, É.: Remarques sur certaines questions de probabilité. Bulletin de la Société Mathématique de France **33**, 123–128 (1905)
7. Borel, É.: La valeur pratique du calcul des probabilités. Revue du mois **1**, 424–437 (1906)
8. Borel, É.: Les probabilités denombrables et leurs applications arithmétiques. Rendiconti del Circolo Matematico di Palermo **27**, 247–270 (1909)
9. Borel, É.: Le Hasard. Libraire Félix Alcan (1914)
10. Borel, É.: A propos d'un traité de probabilités. Revue Philosophique de la France et de l'Étranger **98**, 321–336 (1924)
11. Borel, É.: Valeur pratique et philosophie des probabilités. In: Traité du Calcul des Probabilités et de ses applications, vol. IV, fasc. 3, pp. 1–182. Gauthier-Villars (1939)
12. Bru, B.: Problème de l'efficacité du tir à l'école d'artillerie de Metz. Mathématiques et Sciences Humaines **136**, 29–42 (1996)
13. Bru, B.: Statisticians of the Centuries, Chapter 'Émile Borel', pp. 287–291. Springer (2001)
14. Buffon, G.-L.L.: Essai d'artithmétique morale (1777). In: Ouevres complètes de Buffon, vol. 9, pp. 373–444. Pourrat Frères (1835)
15. Carnap, R.: Logical Foundations of Probability. University of Chicago Press (1950)
16. Courtebras, B.: À l'école des probabilités : une histoire de l'enseignement français du calcul des probabilités. Presses universitaires de Franche-Comté (2006)
17. Cox, R.: Probability, frequency and reasonable expectation. Am. J. Phys. **14**, 1–13 (1946)
18. Crépel, P.: De Condorcet à Arago:l'enseignement des probabilités en France de 1786 à 1830. Bulletin de la Société des Amis de la Bibliothèque de l'École Polytechnique **4**, 29–81 (1989)
19. Crépel, P.: La formation polytechnicienne 1794–1994, chapter 'Le calcul des probabilités: de l'arithmétique sociale a l'art militaire', pp. 197–215. Dunod (1994)
20. de Finetti, B.: Sul significato soggettivo della probabilità. Fund. Math. **17**, 298–329 (1931)
21. de Finetti, B.: La prévision: ses lois logiques, ses sources subjectives, translated in [34]. Annales de l'Institut Henri Poincaré **7**, 1–68 (1937)
22. Estienne, J.B.: Étude sur les erreurs d'observation. Comptes Rendus Hebdomadaires des Séances de l'Académie des Sciences **110**, 512 (1890)
23. Estienne, J.B.: Éssai sur l'art de conjecturer. Revue d'artillerie **61**, 405–449, 62:73–117, 64:5–39, 65–97 (1903–1904)
24. Fréchet, M.: Exposé et discussion de quelques recherches récentes sur les fondements du calcul des probabilités. In: Wavre, R. (ed.) Colloque Consacré a la Théorie des Probabilités: Les fondements du calcul des probabilités, number 735 in Actualités Scientifiques et Industrielles, pp. 23–55. Hermann (1938)
25. Hacking, I.: The Emergence of Probability: A Philosophical Study of Early Ideas about Probability. Cambridge University Press, Induction and Statistical Inference (1975)
26. Hilbert, D.: Mathematical problems. Bull. Am. Math. Soc. **8**, 437–479 (1902)
27. Jeffreys, H.: Theory of Probability. Clarendon Press (1939)
28. Jouffret, E.P.: Étude sur l'établissement et l'usage des tables du tir. Revue d'artillerie **2**, 370–389, 3:52–72, 208–227, 324–340, 386–401 (1873a)
29. Jouffret, E.P.: Sur la probabilité du tir et la méthode des moindres carrés. Revue maritime et coloniale **38**, 5–30 (1873b)

30. Jouffret, E.P.: Sur la probabilité du tir et la méthode des moindres carrés. Revue maritime et coloniale **42**, 665–680, 986–1021, 43:131–152 (1874)
31. Keynes, J.M.: A Treatise on Probability. Macmillan (1921)
32. Kolmogorov, A.N.: Grundbegriffe der Wahrscheinlichkeitsrechnung. Springer (1933)
33. Kolmogorov, A.N.: Foundations of the Theory of Probability, Chelsea, 2nd edn. (transl. of [32]) (1956)
34. Kyburg, H.E., Smokler, H.E.: Studies in Subjective Probability. Wiley (1980)
35. Lacroix, S.F.: Traité élémentaire du calcul des probabilités. Courcier (1816)
36. Laplace, P.S.: Théorie Analytique des Probabilités. Courcier (1812)
37. Laplace, P.S.: Essai Philosophique sur les Probabilités. Bachelier (1814)
38. Lebesgue, H.: Sur une généralisation de l'intégrale définie. Comptes Rendus Hebdomadaires de Séances de l'Académie des Sciences **132**, 1025–1028 (1901)
39. Meusnier, N.: Sur l'histoire de l'enseignement des probabilités et des statistiques. Eletron. J. Hist. Probab. Stat. **2**, 1–20 (2006)
40. Ore, O.: Cardano. Princeton University Press, The Gambling Scholar (1953)
41. Poincaré, H.: Calcul des Probabilités. Gauthier-Villars (1896)
42. Poincaré, H.: La Science et l'hypothèse. Ernest Flammarion (1902)
43. Ramsey, F.P.: The Foundations of Mathematics and other Logical Essays, Chapter 'Truth and Probability (1926)', pp. 156–198. Routledge & Kegan Paul (1931)
44. Reichenbach, H.: The Theory of Probability. University of California Press (1949)
45. Savage, L.J.: The Foundations of Statistics. Wiley (1954)
46. Shafer, G., Vovk, V.: The sources of Kolmogorov's Grundbegriffe. Stat. Sci. **21**, 70–98 (2006)
47. Venn, J.: The Logic of Chance. McMillan (1866)
48. von Mises, R.: Wahrscheinlichkeit. Springer, Statistik und Wahrheit (1928)
49. von Mises, R.: Wahrscheinlichkeitsrechnung und ihre anwendung in der statistik und theoretischen physik. Leipzig (1931)
50. von Mises, R.: Probability, Statistics and Truth. William Hodge and Co., translation of [48] (1939)
51. von Plato, J.: Creating Modern Probability. Cambridge University Press (1994)
52. Wilder, R.L.: Introduction to the Foundations of Mathematics. Wiley (1965)

Model Selection in the Sparsity Context for Inverse Problems in Bayesian Framework

Mircea Dumitru, Li Wang, Ali Mohammad-Djafari and Nicolas Gac

Abstract The Bayesian approach is considered for inverse problems with a typical forward model accounting for errors and *a priori* sparse solutions. Solutions with sparse structure are enforced using heavy-tailed prior distributions. The particular case of such prior expressed via normal variance mixtures with conjugate laws for the mixing distribution is the main interest of this paper. Such a prior is considered in this paper, namely, the Student-t distribution. Iterative algorithms are derived via posterior mean estimation. The mixing distribution parameters appear in updating equations and are also used for the initialization. For the choice of mixing distribution parameters, three model selection strategies are considered: (i) parameters approximating the mixing distribution with Jeffrey law, i.e., keeping the mixing distribution well defined but as close as possible to the Jeffreys priors, (ii) based on the prior distribution form, fixing the parameters corresponding to the form inducing the most sparse solution and (iii) based on the sparsity mechanism, fixing the hyperparameters using the statistical measures of the mixing and prior distribution. For each strategy of model selection, the theoretical advantages and drawbacks are discussed and the corresponding simulations are reported for a 1D direct sparsity application in a biomedical context. We show that the third strategy seems to provide the best parameter selection strategy for this context.

Keywords Inverse problems · Gaussian scale mixtures · Sparsity enforcing Parameter selection

---

M. Dumitru (✉) · L. Wang · A. Mohammad-Djafari · N. Gac
Laboratoire des signaux et systèmes, CNRS – CentraleSupélec – Université Paris-Saclay, 3, 91192 Gif sur Yvette, Rue Joliot-Curie, France
e-mail: Mircea.Dumitru@lss.supelec.fr

L. Wang
e-mail: Li.Wang@lss.supelec.fr

A. Mohammad-Djafari
e-mail: Mohammad-Djafari@lss.supelec.fr

N. Gac
e-mail: Nicolas.Gac@lss.supelec.fr

A. Polpo et al. (eds.), *Bayesian Inference and Maximum Entropy Methods in Science and Engineering*, Springer Proceedings in Mathematics & Statistics 239, https://doi.org/10.1007/978-3-319-91143-4_15

1 Introduction

In this paper, we compare three model selection strategies for a particular context of the Bayesian approach for inverse problems. More precisely, we consider a linear model describing the forward problem and the available prior information about the sparse structure of the unknown. The sparse structure is modeled via heavy-tailed priors ($\mathscr{P}$), well known in the literature for enforcing sparsity [1–3]. The particular class of priors considered in this article is the zero-mean normal variance mixtures. The unknowns are estimated using the Posterior Mean (PM) estimation via Variational Bayesian Approximation (VBA), [4, 5]. Typically, the initialization of the derived iterative algorithm is done using the hyperparameters [6], i.e., the mixing distributions ($\mathscr{M}$) parameters. Therefore, the model selection is a crucial step in such algorithms. In this specific context, three different strategies for model selection are considered and compared for the particular Student-t prior case. We consider the linear forward model,

$$\boldsymbol{g} = \boldsymbol{H}\boldsymbol{f} + \boldsymbol{\varepsilon}, \tag{1}$$

where $\boldsymbol{g}$ represents the $N \times 1$ observed data, $\boldsymbol{H}$ represents a $N \times M$ measurement matrix, $\boldsymbol{f}$ represents the unknown *sparse* signal and $\boldsymbol{\varepsilon}$ accounts for measurement *and* modeling errors. In this paper, the sparsity is accounted in the Bayesian hierarchical prior models framework, using sparsity enforcing prior distributions to model f_j, $j \in \{1, \ldots, M\}$, [1, 7]. For computational reasons, we consider heavy-tailed distributions expressed via zero-mean normal variance mixtures with conjugate priors as mixing distributions,

$$\begin{cases} p(f_j \mid v_{f_j}) = \mathscr{N}(f_j \mid 0, v_{f_j}) \\ p(v_{f_j} \mid \boldsymbol{\xi}_f) = \mathscr{M}(v_{f_j} \mid \boldsymbol{\xi}_f) \end{cases}, \tag{2}$$

where $\boldsymbol{\xi}_f$ represents the parameters of the mixing distribution. In this article, the Inverse Gamma distribution ($\mathscr{I}\mathscr{G}$), corresponding respectively to the two parameters Student–t ($\mathscr{S}t$) distribution will be considered for simulations results and comparisons between the different model selection strategies considered. However, the framework is general and can be used for other sparsity enforcing priors expressed as normal variance mixture, e.g., the Normal-Inverse Gaussian ($\mathscr{N}\mathscr{I}\mathscr{G}$) distribution and the Variance–Gamma ($\mathscr{V}\mathscr{G}$) distribution. The *nonstationary independent Gaussian uncertainties model* is assumed with conjugate priors modeling the variances,

$$\begin{cases} p(\varepsilon_i \mid v_{\varepsilon_i}) = \mathscr{N}(\varepsilon_i \mid 0, v_{\varepsilon_i}) \\ p(v_{\varepsilon_i} \mid \boldsymbol{\xi}_\varepsilon) = \mathscr{M}(v_{\varepsilon_i} \mid \boldsymbol{\xi}_\varepsilon) \end{cases}, \tag{3}$$

where $\boldsymbol{\xi}_\varepsilon$ represents the parameters of the mixing distribution. For the derived iterative algorithms, the parameters of the *posterior* mixing distributions $\widehat{\boldsymbol{\xi}}_f$ and $\widehat{\boldsymbol{\xi}}_\varepsilon$ modeling variances v_{f_j} and v_{ε_i}, have updating expressions depending on $\boldsymbol{\xi}_f$ and $\boldsymbol{\xi}_\varepsilon$.

The model selection, i.e., the choice of *prior* mixing parameters $\boldsymbol{\xi}_f$ and $\boldsymbol{\xi}_\varepsilon$ is therefore crucial in the context of non-supervised algorithms. In practice, such algorithms can be obtained by considering noninformative prior mixing distributions, i.e., considering the Jeffreys prior as the mixing distribution (more exactly, conserving the *conjugate prior* setting, using the conjugate prior with parameters values $\boldsymbol{\xi}_f$ and $\boldsymbol{\xi}_\varepsilon$ such that the corresponding mixing prior is close to Jeffreys prior). This approach was successfully used in [8].

Two other model selection approaches accounting for the sparsity particular context and the specific sparsity enforcing priors used are considered. The first one is based on the form of the prior distribution, i.e., a model selection strategy considering the parameters for which the prior distribution is as concentrated as possible around the mean. For the second one, we first show that the variance of the posterior distribution $\mathrm{Var}_{[\mathscr{P}]}$, modeling $\boldsymbol{f}$ is linked with the expectation of the $\mathscr{M}$ distribution $\mathrm{E}_{[\mathscr{M}]}$, modeling the corresponding variance $\boldsymbol{v}_f$. More precisely

$$\mathrm{Var}_{[\mathscr{P}]}\left(f_j\right) = \mathrm{E}_{[\mathscr{M}]}\left(v_{f_j}\right). \tag{4}$$

Second, we consider a small variance for the prior distribution, i.e., $\mathrm{Var}_{[\mathscr{P}]}\left(f_j\right) = \varepsilon \searrow 0$ in order to impose a model that is concentrating the points f_j around the zero-mean. Clearly, doing this, via Eq. (4), the expectation of the mixture distribution has the same value and doing the same for the mixture distribution variance, $\mathrm{Var}_{[\mathscr{M}]}\left(v_{f_j}\right) = \omega \searrow 0$ will impose a sparse structure for $\boldsymbol{v}_f$, with small values v_{f_j} corresponding to the small values f_j and *significant* values v_{f_j} corresponding to the *significant* values f_j. A sparse structure is therefore enforced by:

1. considering a heavy-tailed prior distribution.
2. setting a small variance for prior distribution, $\mathrm{Var}_{[\mathscr{P}]}\left(f_j\right) = \varepsilon \searrow 0$.
3. enforcing a sparse structure for $\boldsymbol{v}_f$ by setting a small variance also for the mixing distribution $\mathrm{Var}_{[\mathscr{M}]}\left(v_{f_j}\right) = \omega \searrow 0$.

The rest of the paper is organized as follows. Section 2 is introducing the general hierarchical prior model, setting the context of the particular class of sparsity enforcing prior used, and presenting the normal variance mixtures considered during paper and their behavior depending on the parameters. The corresponding PM algorithms are developed in Sect. 3. Empirical evaluations of performances and comparisons between the results corresponding to the two approaches for modeling the hyperparameters are presented in Sect. 4. Conclusions are drawn in Sect. 5.

2 Hierarchical Prior Models Based on the Normal Variance Mixtures

The framework of the hierarchical prior model discussed in this paper, Fig. 1, is based on the sparsity enforcing prior distributions expressed as marginals of normal variance mixtures and nonstationary independent Gaussian uncertainties (noise) model with conjugate priors modeling the variances. The posterior distribution writes

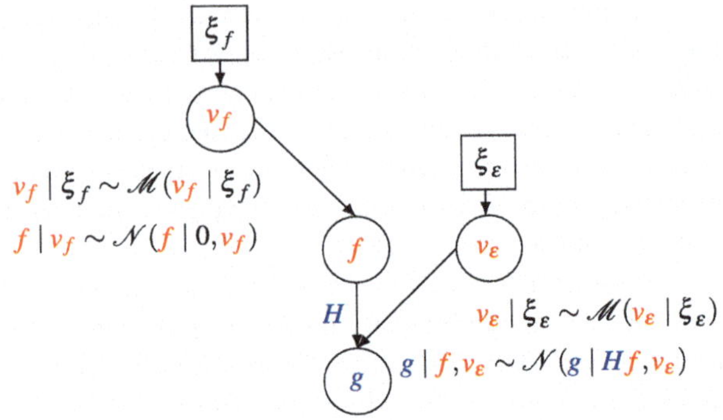

Fig. 1 Hierarchical Prior Model for Forward Model, Eq. (1)

$$p(\boldsymbol{f}, \boldsymbol{v}_\varepsilon, \boldsymbol{v}_f \mid \boldsymbol{g}) \propto \mathcal{N}(\boldsymbol{g} \mid \boldsymbol{H}\boldsymbol{f}, \boldsymbol{v}_\varepsilon)\, \mathcal{N}\left(\boldsymbol{f} \mid \boldsymbol{0}, \boldsymbol{v}_f\right) \mathscr{M}(v_{f_j} \mid \boldsymbol{\xi}_f)\, \mathscr{M}(v_{\varepsilon_i} \mid \boldsymbol{\xi}_\varepsilon). \tag{5}$$

In this specific framework, the product of the two conditional distributions $p(\boldsymbol{g} \mid \boldsymbol{f}, \boldsymbol{v}_\varepsilon) = \mathcal{N}(\boldsymbol{g} \mid \boldsymbol{H}\boldsymbol{f}, \boldsymbol{v}_\varepsilon)$ and $p\left(\boldsymbol{f} \mid \boldsymbol{v}_f\right) = \mathcal{N}\left(\boldsymbol{f} \mid \boldsymbol{0}, \boldsymbol{v}_f\right)$ is common to the posterior distribution, while the differences are induced by the choice of the mixing distributions $p(v_{f_j} \mid \boldsymbol{\xi}_f) = \mathscr{M}(v_{f_j} \mid \boldsymbol{\xi}_f)$ and $p(v_{\varepsilon_i} \mid \boldsymbol{\xi}_\varepsilon) = \mathscr{M}(v_{\varepsilon_i} \mid \boldsymbol{\xi}_\varepsilon)$. We consider in the following the particular case of the Student–t prior expressed as a normal variance mixture.

2.1 Inverse Gamma Mixing Distribution

In Eq. (2), the Inverse Gamma is considered as the mixing distribution, $\mathscr{M}(v_{f_j} \mid \boldsymbol{\xi}_f) = \mathscr{IG}\left(v_{f_j} \mid \alpha_f, \beta_f\right)$, with the probability density function given by:

$$\mathscr{IG}\left(v_{f_j} \mid \alpha_f, \beta_f\right) = \frac{\beta_f^{\alpha_f}}{\Gamma\left(\alpha_f\right)} v_{f_j}^{-\alpha_f - 1} \exp\left(-\frac{\beta_f}{v_{f_j}}\right), \quad \alpha_f > 0,\ \beta_f > 0, \tag{6}$$

where $\Gamma(\cdot)$ denotes the Gamma function. The corresponding hyperparameters are $\boldsymbol{\xi}_f = \left(\alpha_f, \beta_f\right)$, and the corresponding prior $p\left(f_j \mid \alpha_f, \beta_f\right)$ is a two-parameter $\mathscr{S}t$ distribution:

$$p(f_j \mid \alpha_f, \beta_f) = \frac{\Gamma(\alpha_f + \frac{1}{2})}{\sqrt{2\pi\beta_f}\,\Gamma(\alpha_f)} \left(1 + \frac{f_j^2}{2\beta_f}\right)^{-(\alpha_f + \frac{1}{2})} = \mathscr{S}t(f_j \mid \alpha_f, \beta_f). \tag{7}$$

$\alpha_f = \beta_f = \nu_f/2$ corresponds to the standard $\mathcal{S}t$ form, [6]. The expectation of mixing distribution $\mathcal{I}\mathcal{G}$ (equal to the variance of the $\mathcal{S}t$ distribution) and the variance of the mixing distribution $\mathcal{I}\mathcal{G}$ are given by

$$\mathrm{E}_{[\mathcal{I}\mathcal{G}]}\left(v_{f_j}\right) = \mathrm{Var}_{[\mathcal{S}t]}\left(f_j\right) = \frac{\beta_f}{\alpha_f - 1}; \ \mathrm{Var}_{[\mathcal{I}\mathcal{G}]}\left(v_{f_j}\right) = \frac{\beta_f^2}{\left(\alpha_f - 1\right)^2\left(\alpha_f - 2\right)}, \tag{8}$$

with $\alpha_f > 1$ for the first equality and $\alpha_f > 2$ for the second one. This model gives the possibility to consider a heavy-tailed distribution to model the sparse structure of $\boldsymbol{f}$. It is expressed via the Normal distribution and a conjugate prior, which has great computational advantages, guaranteeing the same family distributions for the posterior distributions. The choice of the (prior) parameters α_f and β_f plays a crucial role. Different approaches can be considered to choose the parameters.

1. An approach based on the prior distribution form, imposing a small value for β_f and a large value for α_f. Establishing how small or how large the parameters should be set is difficult.
2. *Close to* Jeffreys prior, setting both parameters close to zero. As before, the same difficulty of establishing *how close to zero* the parameters should be fixed is encountered.
3. Using Eq. (8) to fix the parameters depending on the mixing and prior distribution moments. The relation between α_f and β_f parameters and the moments ε and ω is given by

$$\alpha_f = 2 + \frac{\varepsilon^2}{\omega} \ ; \ \beta_f = \varepsilon\left(1 + \frac{\varepsilon^2}{\omega}\right). \tag{9}$$

   This model selection is based on the data characteristics, i.e., the sparse structure. However, the same difficulty appears for establishing *how small* should ε and ω be.

Table 1 resumes the three strategies considered for model selection.

3 PM Estimation via VBA

The PM estimation is considered via Variational Bayesian Approximation (VBA). The posterior distribution is first approximated with a separable one,

$$p(\boldsymbol{f}, \boldsymbol{v}_f, \boldsymbol{v}_\varepsilon \mid \boldsymbol{g}) \approx q(\boldsymbol{f}, \boldsymbol{v}_f, \boldsymbol{v}_\varepsilon \mid \boldsymbol{g}) = q_1(\boldsymbol{f}) \prod_{j=1}^{M} q_{2j}(v_{f_j}) \prod_{i=1}^{N} q_{3i}(v_{\varepsilon_i}), \tag{10}$$

by minimizing the Kullback–Leibler divergence. Proportionality relations for each separable distribution are obtained. It can be shown that $q_1(\boldsymbol{f})$ is a multivariate Normal distribution,

Table 1 Student-t sparsity enforcing prior: model selection strategies

Parameters	Mixing distribution	Prior distribution	Moments
$\alpha_f \searrow 0;\ \beta_f \searrow 0$	$\mathscr{IG}\left(v_{f_j} \mid \alpha_f, \beta_f\right)$	$\mathscr{S}t\left(f_j \mid \alpha_f, \beta_f\right)$	Not defined
Simulates the Jeffreys prior and has the advantages of a conjugate prior distribution but is difficult to measure *how close* to 0 should the two parameters should be chosen.			
$\alpha_f \nearrow \infty;\ \beta_f \searrow 0$	$\mathscr{IG}\left(v_{f_j} \mid \alpha_f, \beta_f\right)$	$\mathscr{S}t\left(f_j \mid \alpha_f, \beta_f\right)$	Not defined
Advantages of a conjugate prior law, parameters chosen in accordance with their influence on the distribution form but difficult to measure *how close* to ∞ (0) should α_f (β_f) be fixed			
$\alpha_f = 2 + \frac{\varepsilon^2}{\omega};\ \beta_f = \varepsilon\left(1 + \frac{\varepsilon^2}{\omega}\right)$	$\mathscr{IG}\left(v_{f_j} \mid \alpha_f, \beta_f\right)$	$\mathscr{S}t\left(f_j \mid \alpha_f, \beta_f\right)$	Defined
Advantages of a conjugate prior law, parameters chosen in accordance with data structure, considers the moments of the $\mathscr{IG}$ and $\mathscr{S}t$ laws; same difficulties as above for ε and ω			

$$q_1(\boldsymbol{f}) = \mathcal{N}\left(\boldsymbol{f} \mid \widehat{\boldsymbol{f}}, \widehat{\boldsymbol{\Sigma}}\right); \quad \widehat{\boldsymbol{f}} = \left(\boldsymbol{H}^T \widetilde{\boldsymbol{V}}_\varepsilon \boldsymbol{H} + \widetilde{\boldsymbol{V}}_f\right)^{-1} \boldsymbol{H}^T \widetilde{\boldsymbol{V}}_\varepsilon \boldsymbol{g}, \quad \widehat{\boldsymbol{\Sigma}} = \left(\boldsymbol{H}^T \widetilde{\boldsymbol{V}}_\varepsilon \boldsymbol{H} + \widetilde{\boldsymbol{V}}_f\right)^{-1}, \tag{11}$$

using the notations

$$\widetilde{v}_{\varepsilon_i} = \left\langle v_{\varepsilon_i}^{-1} \right\rangle_{q_{3i}(v_{\varepsilon_i})}, i \in \{1, \ldots, N\}; \widetilde{\boldsymbol{v}}_\varepsilon = \left[\ldots \widetilde{v}_{\varepsilon_i} \ldots\right]^T; \widetilde{\boldsymbol{V}}_\varepsilon = \operatorname{diag}\left[\widetilde{\boldsymbol{v}}_\varepsilon\right],$$

$$\widetilde{v}_{f_j} = \left\langle v_{f_j}^{-1} \right\rangle_{q_{2j}(v_{f_j})}, j \in \{1, \ldots, M\}; \widetilde{\boldsymbol{v}}_f = \left[\ldots \widetilde{v}_{f_j} \ldots\right]^T; \widetilde{\boldsymbol{V}}_f = \operatorname{diag}\left[\widetilde{\boldsymbol{v}}_f\right]. \tag{12}$$

In general, $q_{2j}(v_{f_j})$ and $q_{3i}(v_{\varepsilon_i})$ belong to the same family as the $\mathcal{M}$ distribution. In particular, for the $\mathcal{S}t$ prior, $q_{2j}(v_{f_j})$ are $\mathcal{IG}(v_{f_j} \mid \alpha_f + \frac{1}{2}, \widehat{\beta}_{f_j})$ and $q_{3i}(v_{\varepsilon_i})$ are $\mathcal{IG}(v_{\varepsilon_i} \mid \alpha_\varepsilon + \frac{1}{2}, \widehat{\beta}_{\varepsilon_i})$ distributions, with the analytical expressions of β parameters given by:

$$\widehat{\beta}_{f_j} = \beta_f + \frac{1}{2}\left(\widehat{\boldsymbol{\Sigma}}_{jj} + f_j^2\right); \widehat{\beta}_{\varepsilon_i} = \beta_\varepsilon + \frac{1}{2}\left[\boldsymbol{H}_i \widehat{\boldsymbol{\Sigma}} \boldsymbol{H}_i^T + \left(g_i - \boldsymbol{H}_i \widehat{\boldsymbol{f}}\right)^2\right] \tag{13}$$

The parameters corresponding to the multivariate Normal distribution are expressed via $\widetilde{\boldsymbol{V}}_f$ and $\widetilde{\boldsymbol{V}}_\varepsilon$ (and by extension all elements forming the two matrices $\widetilde{v}_{f_j}$, $j \in \{1, 2, \ldots, M\}$ and $\widetilde{v}_{\varepsilon_i}$, $i \in \{1, 2, \ldots, N\}$, Eq. (12)). The following relation holds:

$$\left\langle x^{-1} \right\rangle_{\mathcal{IG}(x \mid \alpha, \beta)} = \frac{\alpha}{\beta} \tag{14}$$

4 Simulation Results

The forward model Eq. (1) is considered for a 1-D application in biology, where a short time series of gene expressions is modeled as

$$g(t) = \sum_{j=1}^{M} f_{1j} \cos\left(\frac{2\pi}{p_j} t\right) + f_{2j} \sin\left(\frac{2\pi}{p_j} t\right) + \varepsilon(t), \tag{15}$$

where $p_j \in [8, \ldots, 32]$, $t = 1, \ldots, N$, and the objective is to find $\boldsymbol{f}_1 = \left[f_{1j}, \ldots, f_{1M}\right]$ and $\boldsymbol{f}_2 = \left[f_{2j}, \ldots, f_{2M}\right]$. This relation can be written as:

$$\boldsymbol{g} = \boldsymbol{H}\boldsymbol{f} + \boldsymbol{\varepsilon} = \left[\boldsymbol{H}_1 | \boldsymbol{H}_2\right] \begin{bmatrix} \boldsymbol{f}_1 \\ \boldsymbol{f}_2 \end{bmatrix} + \boldsymbol{\varepsilon} = \boldsymbol{H}_1 \boldsymbol{f}_1 + \boldsymbol{H}_2 \boldsymbol{f}_2 + \boldsymbol{\varepsilon}. \tag{16}$$

The objective is a precise estimation of the periodic component (PC) vectors ($\boldsymbol{f}_1$ and $\boldsymbol{f}_2$, considered between 8 and 32 hours) corresponding to a short (relative to the a priori dominant period) signal $\boldsymbol{g}$ (considered for four days, sampled every hour).

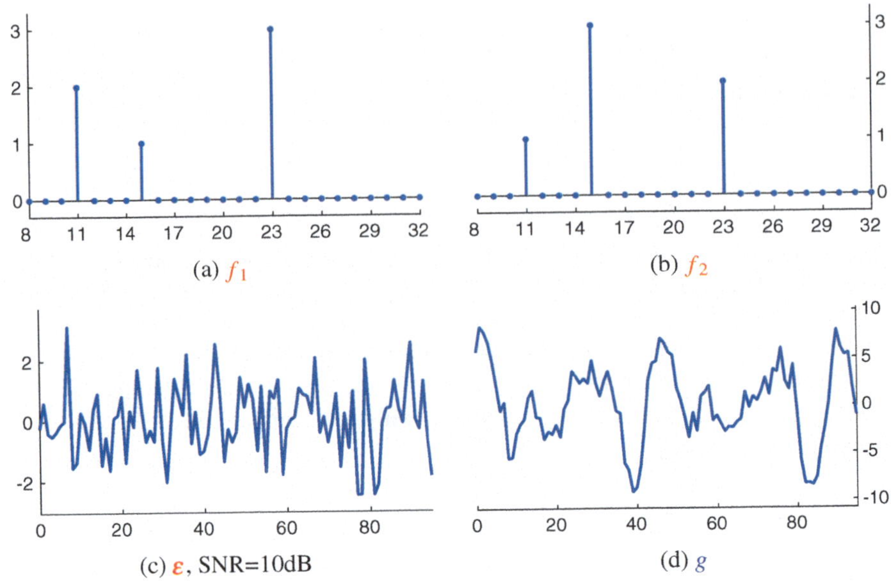

Fig. 2 Synthetic data: 4-days noisy signal $\boldsymbol{g}$ sampled every hour and corresponding PC vectors $\boldsymbol{f}_1$ and $\boldsymbol{f}_1$

The biological prior information is the reduced number of nonzero periods, i.e., the sparse structure of the PC vector. More details about the application and limitations of the classical methods can be found in [6, 9].

The synthetic PC vectors $\boldsymbol{f}_1$ and $\boldsymbol{f}_2$ are presented in Fig. 2a, b, the corresponding biological signal in Fig. 2c and the added noise (corresponding to 10 dB) in Fig. 2c. We consider the PM iterative algorithm, corresponding to the Student-t prior model. The three different model selection strategies are considered. The algorithm initialization is done via the hyperparameters, i.e., the $\mathscr{M}$ distribution parameters, modeling, respectively v_f and v_ε. In this paper, we are interested in testing which model selection strategy is most adequate in enforcing the sparsity, for this specific prior model and this specific application. The results discussed in this section are limited to this aspect. For this purpose, the simulations are done assuming $E_{[\mathscr{M}]}(v_\varepsilon)$ and $Var_{[\mathscr{M}]}(v_\varepsilon)$ to be known, and then the noise hyperparameters, namely the $\mathscr{M}$ parameters modeling v_ε, are derived, depending on the noise model, via Eq. (9).

The estimated PC vectors $\widehat{\boldsymbol{f}}_1$ and $\widehat{\boldsymbol{f}}_2$, corresponding to the $\mathscr{S}t$ prior, are compared with synthetic data, in Fig. (3).

1. Results corresponding to the Jeffreys prior model selection are reported in Fig. 3a, b. The results correspond to $\alpha_f = \beta_f = 10^{-3}$. Clearly, the estimation results are far from the synthetic input. More important, for this model selection, the prior distribution does not enforce sparsity very well. Indeed, for the first periods of the PC vector the estimation is rather good with the zero values well estimated, but for the second part of the PC vector the sparsity is not enforced.

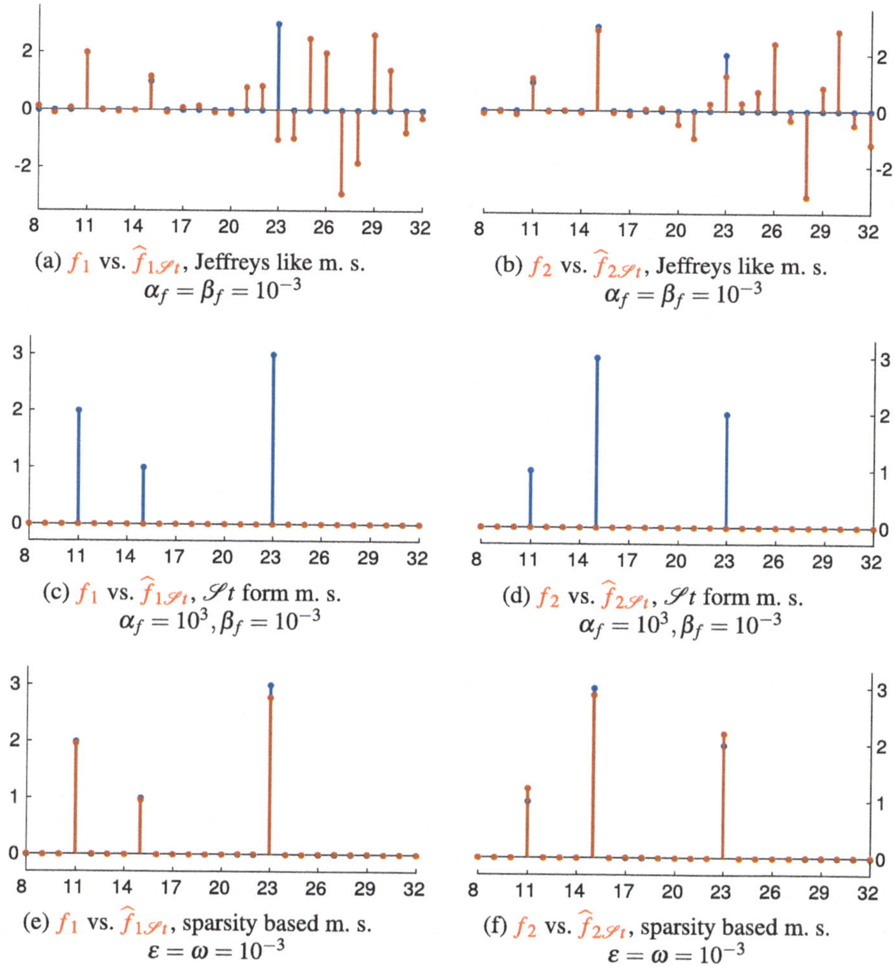

(a) f_1 vs. $\widehat{f}_{1\mathscr{S}t}$, Jeffreys like m. s. $\alpha_f = \beta_f = 10^{-3}$

(b) f_2 vs. $\widehat{f}_{2\mathscr{S}t}$, Jeffreys like m. s. $\alpha_f = \beta_f = 10^{-3}$

(c) f_1 vs. $\widehat{f}_{1\mathscr{S}t}$, $\mathscr{S}t$ form m. s. $\alpha_f = 10^3, \beta_f = 10^{-3}$

(d) f_2 vs. $\widehat{f}_{2\mathscr{S}t}$, $\mathscr{S}t$ form m. s. $\alpha_f = 10^3, \beta_f = 10^{-3}$

(e) f_1 vs. $\widehat{f}_{1\mathscr{S}t}$, sparsity based m. s. $\varepsilon = \omega = 10^{-3}$

(f) f_2 vs. $\widehat{f}_{2\mathscr{S}t}$, sparsity based m. s. $\varepsilon = \omega = 10^{-3}$

Fig. 3 Comparison between the data f_1, f_2 (blue) and $\widehat{f}_{1\mathscr{S}t}$, $\widehat{f}_{2\mathscr{S}t}$ (red). Three model selection strategies, $\mathscr{S}t$ prior model: Jeffreys like, ($\alpha_f = \beta_f = 10^{-3}$, Fig. 3a, b), based on $\mathscr{S}t$ form, ($\alpha_f = 10^{-3}$, $\beta_f = 10^{-3}$, Fig. 3c, d) and sparsity based, ($\alpha_f = 2 + \frac{\varepsilon^2}{\omega}$, $\beta_f = \varepsilon\left(1 + \frac{\varepsilon^2}{\omega}\right)$, $\varepsilon = \omega = 10^{-3}$, Fig. 3e, f)

2. Results corresponding to the model selection based on the $\mathscr{S}t$ prior distribution is reported in Fig. 3c, d. The results correspond to $\alpha_f = 10^3$, $\beta_f = 10^{-3}$. In this case, the result is too sparse: all PC vector values are estimated as zero values. Those particular values correspond for a *strong* prior, which does not account for data. We will see that this model selection strategy can lead to very good estimation results. In particular, for $\alpha_f = 10^1$, $\beta_f = 10^{-1}$ both estimations $\widehat{f_1}$ and $\widehat{f_2}$ are precise.

Table 2 m. s. for $\mathcal{S}t$ prior model: qualitative estimation depending on the numerical values for α_f and β_f parameters.

	Jeffreys like $\alpha_f = \beta_f = 10^{-k}$	$\mathcal{S}t$ form $\alpha_f = 10^k$; $\beta_f = 10^{-k}$	Sparsity mechanism $\varepsilon = \omega = 10^{-k}$, Eq. (9)
$k = 1$	✗	✓	✓
$k = 2$	✗	✗	✓
$k = 3$	✗	✗	✓
$k = 4$	✗	✗	✗
$k = 5$	✗	✗	✗

3. Finally, the results corresponding to the model selection based on the sparsity mechanism are reported in Fig. 3e, f. The results correspond to $\varepsilon = \omega = 10^{-3}$ (see Eq. (9)).

The reconstruction results corresponding to different hyperparmeters values, for each of the three model selection strategies, are reported in Table 2. We consider $k \in \{1, 2, 3, 4, 5\}$ and the following values corresponding to each model selection strategy:

1. for Jeffreys like model selection, we consider $\alpha_f = \beta_f = 10^{-k}$.
2. for model selection based on the $\mathcal{S}t$ form, we consider $\alpha_f = 10^k$; $\beta_f = 10^{-k}$.
3. for model selection based on sparsity mechanism, α_f and β_f are defined via Eq. (9), using $\varepsilon = \text{Var}_{[\mathcal{S}t]} = \text{E}_{[\mathcal{IG}]}$ and $\omega = \text{Var}_{[\mathcal{IG}]}$. We consider $\varepsilon = \omega = 10^{-k}$.

Table 2 reports the quality reconstruction not in the sense of some numerical measure, like L_1 or L_1 reconstruction errors but rather if the results are as sparse as the synthetic inputs $\boldsymbol{f}_1$ and $\boldsymbol{f}_2$. We notice that the model selection strategy based on the sparsity mechanism, for the $\mathcal{S}t$ prior model is more flexible. Good results can be achieved using the $\mathcal{P}$ distribution form. In this case, having to set a small value for one hyperparameter (β_f) and a significant value for the other (α_f) is a difficult task, since each of both hyperparameters are influencing the $\mathcal{P}$ distribution form, enforcing sparsity to much. Model selection strategy based on the sparsity mechanism is the statistical measures of the unknown of the model and its corresponding variance, which generally can be approximately inferred in each application. Moreover, some preliminary results are indicating a strong influence of ε and a week influence of ω in the model selection, so reducing it to only one parameter.

5 Conclusion

In the Bayesian framework, using heavy-tailed distributions in order to enforce sparsity, we have compared the PM iterative algorithms reconstruction results corresponding to a specific sparsity enforcing law (Student-t) corresponding to three different model selection strategies in terms of sparsity enforcing.

The model selection strategy based on the sparsity mechanism can lead to good results in terms of sparsity enforcing and seems to be more flexible than the other two strategies considered. This model selection strategy is based on the assignment of the hyperparameters using the statistical measures of the prior and mixing distributions, ε and ω. The interval for selecting ε and ω is rather large in all three cases. Some preliminary results indicate a week dependency between ω the reconstruction results in terms of sparsity enforcing.

The sparsity is not enforced when the model selection strategy based on Jeffreys priors is used. The model selection strategy based on the prior distribution form can give good reconstruction results in term of sparsity enforcing but in this case, the interval seems to be rather small.

Clearly, the results strongly depend on the application and on the specific prior law (induced by the choice of the mixing distribution). For future work, those strategies will be compared for other sparsity enforcing distributions and other applications. Also, a key concept is the *sparsity rate* (SR). Another perspective of this work is to study a possible relation between the SR and the model selection, more precisely a link between the model selection and SR.

We mention that in this paper, we have measured the reconstruction results in terms of sparsity enforcing. Evidently, this is just the first the step in a much more detailed analysis, accounting also for different reconstruction measures, like L_1, L_2 reconstruction errors, false positives, etc. This paper reports the preliminary results corresponding to the best model selection strategies for the Student-t prior model in terms of sparsity enforcing.

References

1. Mohammad-Djafari, A.: EURASIP J. Adv. Signal Process. **52** (2012). https://doi.org/10.1186/1687-6180-2012-52
2. Babacan, S.D., Nakajima, S., Do, M.N.: IEEE Trans. Signal Process. **14**, 2906 (2014). https://doi.org/10.1109/TSP.2014.2319775
3. Dobigeon, N., Hero, A.O., Tourneret, J.Y.: IEEE Trans. Image Process. **18**, 2059 (2009). https://doi.org/10.1109/TIP.2009.2024067
4. Smídl, V., Quinn, A.: The Variational Bayes Method in Signal Processing. Signals and communication technology, 1st edn. Springer, Heidelberg (2006). https://doi.org/10.1007/3-540-28820-1
5. Beal, M.J.: Variational Algorithms for Approximate Bayesian Inference. Ph.D. thesis, University College London (2003)
6. Dumitru, M.: A Bayesian Approach for Periodic Components Estimation for Chronobiological Signals. Ph.D. thesis, Université Paris-Saclay (2016)
7. Mohammad-Djafari, A., Dumitru, M.: Digit. Signal Process. **47**, 128 (2015). https://doi.org/10.1016/j.dsp.2015.08.0
8. Orieux, F., Giovannelli, J.F., Rodet, T.: J. Opt. Soc. Am. A **27**(7), 1593 (2010). https://doi.org/10.1364/JOSAA.27.001593
9. Dumitru, M., Mohammad-Djafari, A., B.S, S.: EURASIP J. Bioinf. Syst. Biol. **3** (2016). https://doi.org/10.1186/s13637-015-0033-6

Sample Size Calculation Using Decision Theory

Milene Vaiano Farhat, Nicholas Wagner Eugenio and Victor Fossaluza

Abstract Decision Theory and Bayesian Inference have an important role to solve some common problems in research and practice in the medical field. These decisions may be from different natures and can consider several factors, such as the cost to carry out the study and each sample unit and, especially, the risks for the patients involved. Here, the estimation of sample size calculation considers the cost of sampling units and clinically relevant size of the credible interval for difference between groups. By fixing a probability to the HPD region, the Bayes' Risk is calculated for each sample size possible and it is chosen the optimal sample size, that minimizes the risk. In addition, a second solution is presented by setting the amplitude of the credible interval, leaving its probability free. It is considered a Normal distribution for data with unknown mean and fixed variance (Normal prior) and the case where both mean and variance are unknown (Normal-Inverse Gamma prior). It is presented as a solution considering the statistical distribution of sufficient statistics. In scenarios with no analytical solutions, the optimal sample sizes are presented using Monte Carlo methods.

Keywords Bayesian statistics · Decision Theory · Sample size

1 Introduction

Clinical research is a branch of health science that determines the safety and efficacy of drugs, devices, diagnostic products and treatment regimens intended for human use. These can be used for prevention, treatment, diagnosis, or relief of the symptoms

M. Vaiano Farhat (✉) · N. Wagner Eugenio (✉) · V. Fossaluza (✉)
IME-USP, São Paulo, Brazil
e-mail: milene.farhat@usp.br

N. Wagner Eugenio
e-mail: neugenio@usp.br

V. Fossaluza
e-mail: victor.ime@gmail.com

A. Polpo et al. (eds.), *Bayesian Inference and Maximum Entropy Methods in Science and Engineering*, Springer Proceedings in Mathematics & Statistics 239, https://doi.org/10.1007/978-3-319-91143-4_16

of an illness. The positive and well-established results of these surveys are used by health professionals in clinical practice.

Both in research and in clinical practice, it is common for the medical professionals to come across a situation where they need to make decisions. These decisions may be of various natures, such as which is the best clinical trial design and how the sample should be composed or what type of treatment is best for each type of patient. In addition, these decisions must take into account several factors: the cost to carry out the study and each sample unit and, especially, the risks to the patients involved.

The Decision Theory provides a mathematical basis for rational decision-making, taking into account the "usefulness" (or "losses") of each possible action [1–4]. Although [5–7] and other authors have already suggested the use of Decision Theory in the clinical area, the lack of mathematical or statistical knowledge on the part of many medical researchers make its use limited. A very consistent way to apply Decision Theory is through Bayesian Inference.

The purpose of this paper is to use Decision Theory to solve some common problems in medical research and practice. Particularly, the problem of sample size calculation for the usual issues of the area was treated considering the cost of the study, the sample units and the amplitude of the difference between the groups, so that the detected difference is clinically relevant. The solution to a similar problem was presented in [8]. In addition, it was developed as a function in [9] for performing the calculations.

2 Bayesian Region Estimation

Let Θ be the parametric space (or *state space*) such that each $\theta \in \Theta$ represents a possible realization of the "state of nature", the unknown element of interest, and consider the *actions space* $\mathscr{D}$ where each element $d \in \mathscr{D}$ represents a possible action (or statement about θ) which the decision maker can choose. Also, suppose it is possible to perform an experiment where a realization $x \in \mathscr{X}$ from a random variable X, which carries information about θ, is observed. A prior and an experimental information are complemented by a new type of information that refers to the consequences of decisions in their interaction with the surrounding state of nature: the loss function (or, alternatively the utility function) $l : \Theta \times \mathscr{D} \to \mathrm{R}_+$ that, for each $\theta \in \Theta$ and $d \in \mathscr{D}$, associates the loss $l(\theta, d)$ caused by choosing d when θ is the realization of the state of nature.

Suppose that the experiment is performed and a particular result x is observed, obtaining the posterior distribution for $\theta, f(\theta|x)$. In this case, for each $x \in \mathscr{X}$, the optimal decision $\delta^*(x)$, also called *Bayes Decision*, is the $d_x \in \mathscr{D}$ action that minimizes the *posterior risk* function, defined by

$$r_x(d) = E[l(\theta, d)|x] = \int_\Theta l(\theta, d) f(\theta|x)\, d\theta.$$

It means, $r_x(d_x) = \inf_{d \in \mathscr{D}} r_x(d)$. The posterior risk $r_x(d_x)$ of the decision d_x is called by *Posterior Bayes' Risk*. One advantage of this approach is that it is possible to include in the loss function not only the risk of the decision but also the costs involved in the research. In particular, the problem of sample size calculation will be discussed under this approach. In this case, we will conduct a *pre posterior analysis*, that is obtain a optimal sample size n^* before observing x, minimizing the *risk function*

$$\rho(\delta) = E\left[l(\theta, \delta(X))\right] = E\left[E[l(\theta, \delta(X))|X]\right] = \int_{\mathscr{X}} \int_{\Theta} l(\theta, d) f(\theta|x)\, d\theta\, f(x)\, dx.$$

Note that the function $\delta : \mathscr{X} \to \mathscr{D}$ which minimizes $\rho(\delta)$ is equivalent almost surely to decision $\delta^*(x) = d_x$ for each x in $\mathscr{X}$.

The **interval estimation** (or **estimation by regions** when the parametric space dimension is greater than one) is an decision problem where the aim is to estimate a subset of the parametric space $d \in \mathscr{D}$ that contains the parameter with high probability, where the decision space $\mathscr{D}$ is a σ-algebra of Θ sets. The loss function, in this case, should penalize more the sets that do not contain "the real" $\boldsymbol{\theta}$ as well as very large sets, because they bring a few information about $\boldsymbol{\theta}$. In this way, consider $\lambda(d)$ the Lebesgue measure of the decision d, $I(\boldsymbol{\theta} \in d)$ the function which indicates whether Θ belongs to the set d and $c(n)$ a function that represents the costs involved in the survey for a sample size n. For now, let's consider $c(n) = cn$, where c is a previously fixed value. Thus, it will be considered a loss function that takes into account the cost the size of the region and the pertinence of $\boldsymbol{\theta}$

$$l(\boldsymbol{\theta}, d) = \lambda(d) - k\mathrm{I}(\boldsymbol{\theta} \in d) + c(n), \quad k \in \mathrm{R}_+.$$

The smallest region of the parametric space δ^* that contains the parameter $\boldsymbol{\theta}$ with probability γ is called **HPD region** *(High Posterior Density)*.

Definition: We say that $d_x \in \mathscr{D}$ is an HPD region of probability γ if:
(i) $f(\boldsymbol{\theta}^*|\boldsymbol{x}) \geq f(\boldsymbol{\theta}|\boldsymbol{x}), \forall \boldsymbol{\theta}^* \in d_x, \forall \boldsymbol{\theta} \notin d_x$.
(ii) $P(\boldsymbol{\theta} \in d_x|\boldsymbol{x}) = \gamma$.

It follows that the *optimal decision* d_x with respect to the loss $l(\boldsymbol{\theta}, d)$ against a prior $f(\boldsymbol{\theta})$ is the action $d_x \in \mathscr{D}$ which minimizes the *posterior risk* function. Thus,

$$r_x(d_x) = \inf_{d \in \mathscr{D}} E_{\boldsymbol{\theta}|\boldsymbol{x}}[l(\boldsymbol{\theta}, d)|\boldsymbol{x}] = E_{\boldsymbol{\theta}|\boldsymbol{x}}[\lambda(d_x) - k\mathrm{I}(\boldsymbol{\theta} \in d_x) + cn|\boldsymbol{x}].$$

For each sample size n, it is possible to calculate the Bayes' Risk $\rho_n(\delta^*)$. In this paper, we want to find the optimal sample size, that is, the value n^* which presents the lowest Bayes' risk.
Result: For each $\boldsymbol{x} \in \mathscr{X}$, the optimal decision d_x in relation to the loss $l(\boldsymbol{\theta}, d)$ versus a prior $f(\boldsymbol{\theta})$ is a HPD region.

Demonstration (continuous case):

$$r_x(d) = E[l(\boldsymbol{\theta}, d)|\boldsymbol{x}] = \int_{\Theta} [\lambda(d) - k\mathrm{I}(\boldsymbol{\theta} \in d)d\mathrm{P}(\boldsymbol{\theta}|\boldsymbol{x}) = \int_{\Theta} \mathrm{I}(\boldsymbol{\theta} \in d)d\boldsymbol{\theta} - \int_{\Theta} k\mathrm{I}(\boldsymbol{\theta} \in d)d\mathrm{P}(\boldsymbol{\theta}|\boldsymbol{x})$$

$$= \int_d 1 d\boldsymbol{\theta} - \int_d kf(\boldsymbol{\theta}|\boldsymbol{x})d\boldsymbol{\theta} = \int_d (1 - kf(\boldsymbol{\theta}|\boldsymbol{x}))d\boldsymbol{\theta} = \int_{\Theta} (1 - kf(\boldsymbol{\theta}|\boldsymbol{x}))\mathrm{I}(\boldsymbol{\theta} \in d)d\boldsymbol{\theta}.$$

So that the integral is minimized in the set d_x such that

$$1 - kf(\boldsymbol{\theta}|\boldsymbol{x}) \leq 0 \;\Leftrightarrow\; f(\boldsymbol{\theta}|\boldsymbol{x}) \geq \frac{1}{k} \;\Rightarrow\; d_x = \left\{\boldsymbol{\theta} \in \Theta : f(\boldsymbol{\theta}|\boldsymbol{x}) \geq \frac{1}{k}\right\}. \qquad \blacksquare$$

In the Bayesian literature, the set d_x is also called *Credible Region* γ.

3 Optimal Sample Size for Normal Distribution

It was considered the case in which the population mean is unknown and its variance is known and the bivariate case, where both are unknown. In order to calculate the sample size we can think of two possible cases: to establish a probability of the HPD region, with the purpose of obtaining the optimal decision with a predefined probability, as well as the case where it is desired to fix the length of the credible interval, a common case in clinical research.

3.1 Unknown Mean μ

Lets $\boldsymbol{X}_n = X_1, \ldots, X_n$ be a random sample, that is, $X_1, \ldots, X_n$ are random variables conditionally independent and identically distributed belonging to the same space $\mathscr{X} = \mathrm{R}$, such that each $X_i|\mu \sim \mathrm{N}(\mu, \sigma^2)$, with variance σ^2 fixed. Consider $\mu \in \Theta$ the parameter such that each μ represents a possible "state of nature", and let us consider that μ have a prior distribution $\mu \sim \mathrm{N}(m, v^2)$. In this way,

$$f(\boldsymbol{x}|\mu) = \frac{1}{\sqrt{2\pi\sigma^2}} e^{-\frac{(x-\mu)^2}{2\sigma^2}} \quad \text{and} \quad f(\mu) = \frac{1}{\sqrt{2\pi v^2}} e^{-\frac{(\mu-m)^2}{2v^2}}.$$

The Theorem of [1] states that after observing $\boldsymbol{X} = \boldsymbol{x}$, the posterior distribution for μ is normal with mean m_x and variance v_x^2, with

$$m_x = \frac{\sigma^2 m + nv^2\bar{x}}{\sigma^2 + nv^2} \quad \text{and} \quad v_x^2 = \frac{\sigma^2 v^2}{\sigma^2 + nv^2}.$$

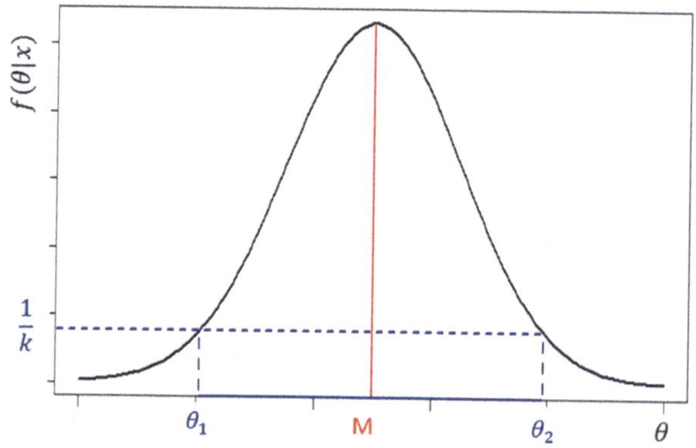

Fig. 1 Example of HPD region for normal distribution

It is known that $\bar{X}_n = \frac{1}{n}\sum_{i=1}^{n} X_i$ is a *sufficient statistic* for μ, so $f(\mu|\boldsymbol{x}) = f(\mu|\bar{x})$. In addition, $\bar{X}_n \sim N\left(m, \frac{\sigma^2}{n}\right)$, and henceforth, for all calculations will be used the statistic $\bar{X}_n$ in place of the entire sample $\mathbf{X}_n = X_1, ..., X_n$ (Fig. 1).

3.1.1 Credible Interval with Fixed Probability γ

Given the preestablished condition of $\mathrm{P}(\mu \in d|\bar{x}) = \gamma$, $\gamma \in (0, 1)$, it is reasonable to consider a loss function $l : \mathscr{D} \times \Theta \to \mathrm{R}$ which takes into account only the size of the region d and the cost of observations. Thus, the loss function will be given by

$$l(\mu, d) = \lambda(d) + cn.$$

As is usual in statistical practice, let us consider $\gamma = 0.95$. In our example, we have that the posterior distribution $\mu|\bar{x} \sim N(m_x, v_x^2)$ is symmetrical, we can have μ_1 and μ_2 reminding that $Z = \frac{\mu - m_x}{v_x} \mid \bar{x} \sim N(0, 1)$ and let

$$\mathrm{P}(\mu_1 \le \mu \le \mu_2 \mid \bar{x}) \;\Rightarrow\; \mathrm{P}\left(\frac{\mu_1 - m_x}{v_x} \le Z \le \frac{\mu_2 + m_x}{v_x}\right)$$

$$\Rightarrow\; \frac{\mu_i \pm m_x}{v_x} = \Phi^{-1}\left(\frac{1-\gamma}{2}\right) \quad\Rightarrow\quad \mu_i = m_x \pm \Phi^{-1}\left(\frac{1-\gamma}{2}\right) v_x.$$

It follows that the size of the interval d_x is given by

$$\lambda(d_x) = \mu_2 - \mu_1 = 2v_x\,\Phi^{-1}\left(\frac{1+\gamma}{2}\right).$$

Hence, the action $d_x \in D$ which minimizes the *posterior risk* is an HPD interval, that is, the smallest region with probability 0.95. In this way, the Bayes' risk of $\delta^*(X) = [\mu_1(X), \mu_2(X)]$ is given by

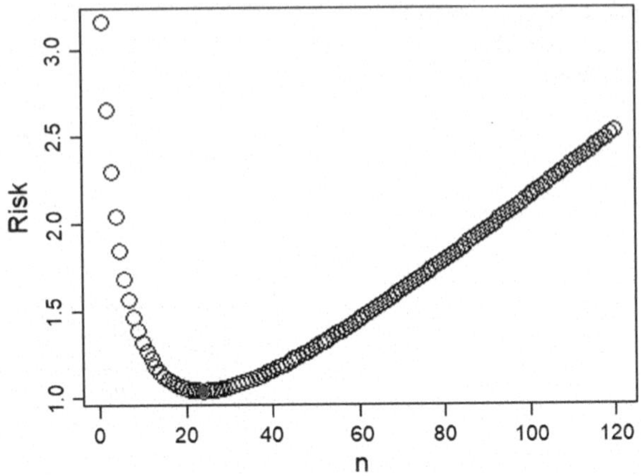

Fig. 2 Graphic of Bayes' risk by sample size with fixed probability γ

$$\rho_n(\delta^*) = E_{\mu,\bar{X}}\left[l(\mu,\delta^*(X))\right] = \int_{\mathscr{X}}\int_{\Theta} \lambda(d_x) f(\mu|\bar{x}) f(\bar{x})\, d\mu d\bar{x} + cn$$

$$= \int_{\mathscr{X}}\int_{\Theta} 2v_x\, \Phi^{-1}\left(\frac{1+\gamma}{2}\right) f(x,\mu)\, d\mu dx + cn \quad = 2v_x\, \Phi^{-1}\left(\frac{1+\gamma}{2}\right) + cn,$$

because $v_x^2 = \frac{v^2\sigma^2}{nv^2+\sigma^2}$. Analytically, the larger is the sample size n, the smaller is the first term of the Bayes' Risk. On the other hand, the cost increases with n.

To illustrate, the *Bayes' Risk* for the optimal decision δ^* for each sample size $n = 1, 2, \ldots, N$ after setting the probability of the HPD region at $\gamma = 0.95$, we use the software *R*. From the values obtained, we can find the optimal sample size n^* which presents the lowest Bayes' Risk.

Assuming that the population variance $\sigma^2 = 4$, the prior parameters are $m = 0$, $v = 1$, and the cost per observation is $c = 0.02$, the optimal sample size presenting the lowest Bayes' Risk was $n^* = 24$ (Fig. 2).

3.1.2 Credible Interval with Fixed Length $\lambda(d)$

Given the preestablished condition of $\text{P}(\mu \in d|\bar{x}) = \gamma$, $\gamma \in (0, 1)$, it is reasonable to consider a loss function $l : \mathscr{D} \times \Theta \rightarrow R$ which takes into account only the size of the region d and the cost of observations.

Under the condition that the interval's size is fixed $\lambda(d) = \varepsilon$, the loss only depends on the parameter μ belonging or not to the set d and the cost per n observations, cn. Thus, for $k \geq 0$,

$$l(\mu, d) = kI(\mu \notin d) + cn.$$

Then, it follows that the posterior risk is

$$\rho_n(\delta^*) = \int_{\mathscr{X}} \int_{\Theta} l(\delta(X), \mu) f(\mu|\bar{x}) f(\bar{x})\, d\mu d\bar{x} + cn = k \int_{\mathscr{X}} \mathrm{P}(\mu \notin d_x|\bar{x}) f(\bar{x})\, d\bar{x} + cn.$$

For $d = [\mu_1, \mu_2]$ with $\mu_2 - \mu_1 = \varepsilon$, recalling that

$$E_{\mu,\bar{X}}\left[k\mathrm{I}(\mu \notin d)\right] = kE_{\mu,\bar{X}}\left[\mathrm{I}(\mu \notin d)\right] = kE_{\bar{X}}\left[1 - \mathrm{P}(\mu_1 \leq \mu \leq \mu_2|\bar{x})\right] = kE_{\bar{X}}\left[r_x(d)\right].$$

For each $x \in \mathscr{X}$, the posterior risk $r_x(d)$ is minimized when $\mathrm{P}(\mu_1 \leq \mu \leq \mu_2|x)$ is maximum and this occurs in the HPD region. Thus, we can write

$$d_x = [\mu_1, \mu_2] = \left[m_x - \frac{\varepsilon}{2}\,;\; m_x + \frac{\varepsilon}{2}\right].$$

Therefore, the *Bayes' Risk* of the decision function δ^* is

$$\begin{aligned}
\rho(\delta^*) &= \int_{\mathscr{X}} \int_{\Theta} k\left[1 - \mathrm{I}\left(m_x - \frac{\varepsilon}{2} \leq \mu \leq m_x + \frac{\varepsilon}{2}\right)\right] f(\mu|\bar{x}) f(\bar{x})\, d\mu d\bar{x} + cn \\
&= \int_{\mathscr{X}} k\left[1 - \mathrm{P}\left(\frac{\mu_1 - m_x}{v_x} \leq z \leq \frac{\mu_2 + m_x}{v_x} \mid \bar{x}\right)\right] f(\bar{x})\, d\bar{x} + cn \\
&= \int_{\mathscr{X}} k\, 2.\Phi\left(-\frac{\delta}{2v_x}\right) f(\bar{x})\, d\bar{x} + cn = 2k\, \Phi\left(-\frac{\delta}{2v_x}\right) + cn.
\end{aligned}$$

Analytically, then, the larger is the sample size n, the smaller the variance of the posterior $v_x^2 = \frac{v^2\sigma^2}{nv^2+\sigma^2}$ and consequently, the smaller the probability of μ not being in the HPD region. On the other hand, the cost increases with n.

To illustrate the *Bayes' Risk* to the optimal decision δ^* for each sample size $n = 1, 2, \ldots, N$ with a fixed length $\lambda(d) = \varepsilon$, we use software *R*. We can find the optimal sample size n^* which presents the lowest Bayes' Risk.

Using as an example, the same values as in the previous case ($\sigma^2 = 4$, $m = 0$, $v^2 = 1$ e $c = 0.02$) and $k = 4$, the optimal sample size that presents the lowest Bayes' risk was $n^* = 16$ (Fig. 3).

3.2 Unknown Mean and Variance

Let $\boldsymbol{X}_n = X_1, \ldots, X_n$ be a random sample, that is, $X_i|\mu, \sigma^2 \sim \mathrm{N}(\mu, \sigma^2)$ are conditionally independent and identically distributed (c.i.i.d) variables. Consider $\boldsymbol{\theta} = (\mu, \sigma^2)$ the parameters in $\Theta = \mathrm{R} \times \mathrm{R}_+$. Let us consider, a prior, $\sigma^2 \sim \mathrm{GI}(a, b)$, $a, b > 0$ and $\mu|\sigma^2 \sim \mathrm{N}(m, \sigma^2/v)$, $m \in \mathrm{R}$ and $v > 0$. In this case, the joint probabil-

Fig. 3 Graphic of Bayes' risk by sample size with fixed interval length ε

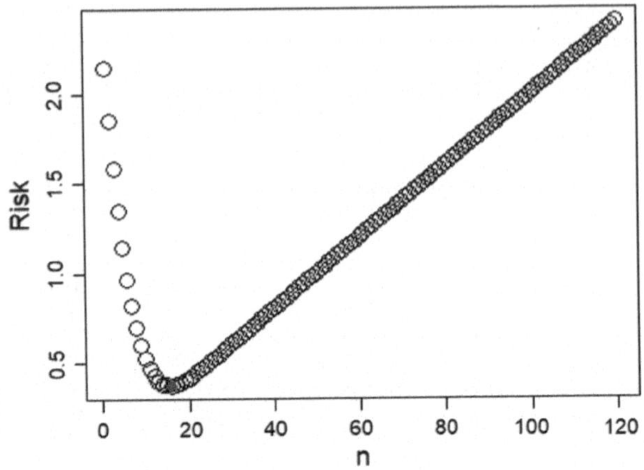

ity distribution is called **Normal-Inverse Gamma**, $NIG(m, v, a, b)$. The *sufficient statistics* here are $\bar{x} = \frac{1}{n}\sum x_i$ and $s^2 = \frac{1}{n-1}\sum(x_i - \bar{x})^2$ and is known that the posterior is also a NIG distribution:

$$\mu, \sigma^2|\bar{x}, s^2 \sim \text{NIG}(m_x, v_x, a_x, b_x), \quad m_x \in \text{R}, v_x, a_x, b_x > 0,$$

with

$$f(\mu, \sigma^2|\bar{x}, s^2) = \frac{\sqrt{v_x}}{\sqrt{2\pi\sigma^2}} \frac{b_x^{a_x}}{\Gamma(a_x)} \left(\frac{1}{\sigma^2}\right)^{a_x+1} e^{-\frac{2b_x+v_x(\mu-m_x)^2}{2\sigma^2}},$$

where $m_x = \frac{(vm+n\bar{x})}{(v+n)} + \frac{nv(\bar{x}-m)^2}{2(v+n)}$; $\quad v_x = v + n$; $\quad a_x = a + \frac{n}{2}$; $\quad b_x = b + \frac{n-1}{2}s^2$.

In this case, the loss function are similar to the univariate case:

$$l(\boldsymbol{\theta}, d) = \lambda(d) - k\text{I}(\boldsymbol{\theta} \in d) + c(n), \quad k \in \text{R}_+,$$

It follows that, for each $x \in \mathscr{X}$ the optimal decision $\delta^*(x) = d_x$ in relation to the loss $l(\boldsymbol{\theta}, d)$ against the prior $f(\mu, \sigma^2)$, is an HPD region again. Figure 5 presents an example of HPD regions with different probabilities.

3.2.1 Credible Region with Fixed Probability γ

Given the preestablished probability $\text{P}(\boldsymbol{\theta} \in d|\bar{x}) = \gamma$, it is reasonable to consider the loss function $l : \mathscr{D} \times \Theta \to \text{R}$ just with the terms related to region's size and cost of observations.

$$l(\boldsymbol{\theta}, d) = \lambda(d) + c(n).$$

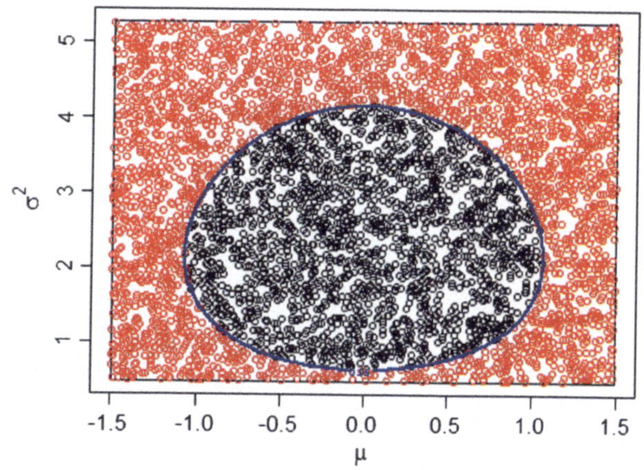

Fig. 4 Hit or Miss process to find the region measure

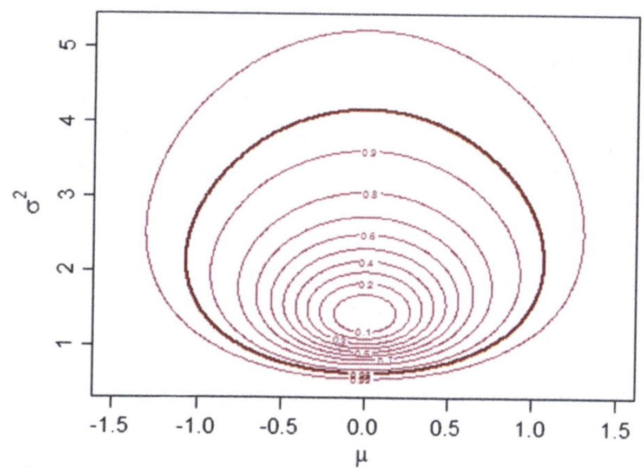

Fig. 5 HPD regions from a Normal-Inverse gamma posterior distribution

To illustrate the *Bayes' risk* for the optimal decision δ^* for each sample size $n = 1, 2, \ldots, N$, we fixed the probability of the HPD region in 0.95. To obtain $\lambda(\delta^*)$, we apply the Monte Carlo process of Hit or Miss by establishing a rectangle that encompasses the HPD region of interest (Figure 4). From the values obtained, we can find the optimal sample size n^* that presents the lowest Bayes' risk.

Assuming that the prior parameters are $m = 0$, $v = 1$, $a = 2$, $b = 3$ and the cost function $c(n) = 0.3n$, we use the Monte Carlo method to find the optimal sample size $n^* = 18$ (Fig. 6). We also tested using the cost function $c(n) = log(n)$ and the result was approximately $n^* = 86$.

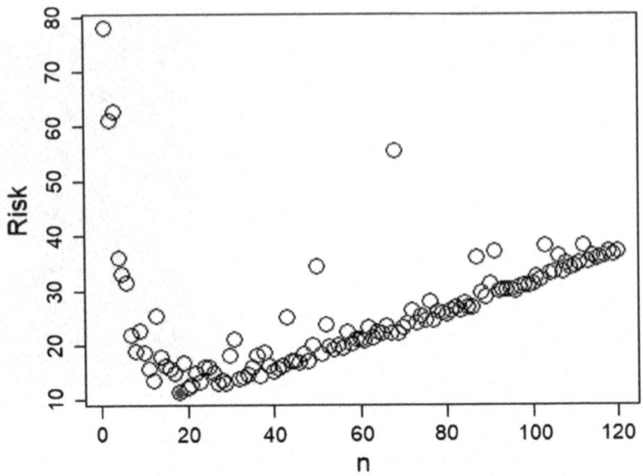

Fig. 6 Graphic of sample size by the Bayes' risk

References

1. DeGroot, M.H.: Optimal Statistical Decisions. MacGraw-Hill, New York (1970)
2. Lindley, D.V.: Making Decisions. Wiley, New York (1985)
3. Parmigiani, G., Inoue, L.: Decision Theory: Principles and Approaches. Wiley, New York (2009). ISBN 978-0-471-49657-1
4. Raiffa, H., Schlaifer, R.: Applied Statistical Decision Theory. Harvard University, Boston (1961)
5. Pauker, S.G., Kassirer, J.P.: The threshold approach to clinical decision making. New Engl. J. Med. **302**(20), 1109–1117 (1980)
6. Schwartz, J.K.W.B., Gorry, G.A., Essig, A.: Decision analysis and clinical judgment. Am. J. Med. **55**(4), 459–472 (1973)
7. Watts, N.T.: Clinical decision analysis. Phys. Ther. **69**(7), 569–576 (1989)
8. Fossaluza, V., Silva, P.V.: Optimal sample sizes for comparison of proportions using fbst. AIP Conference Proceedings, vol. 1490, pp. 153–159 (2012). http://scitation.aip.org/content/aip/proceeding/aipcp/10.1063/1.4759599
9. R Core Team.: R: A Language and Environment for Statistical Computing. R Foundation for Statistical Computing, Vienna, Austria (2013). http://www.R-project.org/

Utility for Significance Tests

Nathália Demetrio Vasconcelos Moura and Sergio Wechsler

Abstract The range of possible readings among and within the statistical inference, in addition to the relevance of these in the applied context, justify the extensive literature analyzing and comparing the main methodologies. However, the fact that each approach is built upon their own structures, varying even the spaces in which they are evaluated, limit the conclusions to the specified scenarios. As a solution for that, in the context of hypotheses tests, we work with the decision theory, which provides a unique language to incorporate the logic of each existent philosophy. For such, after discussing the main points of the frequentist and Bayesian inference, the main approaches are presented, particularly regarding to precise hypotheses, and then unify by the decision-theoretic viewpoint. Additionally, by through this perspective we analyze, interpret and compare the loss functions of some precise approaches, in the context of significance tests.

Keywords Significance tests · Decision theory · FBST · Loss function · Bayes Fisher

1 Introduction

The main goal of Statistical Inference is to answer about random phenomena based on the available information. For such, it is possible to work with different paradigms, including likelihood-based, fuzzy, among others, with the Frequentist approach being by far the most used. For this school, the probability of an event is given by the limit of the relative frequencies, being such frequencies represented by an entity called parameter, defined according to infinite and hypothetical repetitions of the associated experiment. Particularly relevant, the parameter is responsible for specifying the

N. D. V. Moura (✉) · S. Wechsler
University of São Paulo, São Paulo, Brazil
e-mail: nathdvs@ime.usp.br

S. Wechsler
e-mail: sw@ime.usp.br

A. Polpo et al. (eds.), *Bayesian Inference and Maximum Entropy Methods in Science and Engineering*, Springer Proceedings in Mathematics & Statistics 239, https://doi.org/10.1007/978-3-319-91143-4_17

behavior of the referenced random variable. Nevertheless, dealing with such limit as a fixed quantity, despite unknown, imposes some difficult to analysis. For instance, the need for an infinite sequence of repetitions of the experiment, carried out under the same conditions, or the violation of the Likelihood Principle.

To circumvent such limitations, we have the option of extending the analysis to the Bayesian understanding. In this, by looking at the parameter, the entity of interest, as a latent random entity, we obtain a harmonious reading with the way that uncertainty is commonly used. And the laws of probability being the structure according to which a coherent individual must express his uncertainty. Besides, the axioms of coherence [1], presupposed for such approach, are: simple, interpretable, and intuitive.

However, in practice, there are applications working with different readings, particularly with regard to the Hypothesis Tests, and even more to the Precise Hypothesis case. As a solution, we will address the Hypothesis Tests in a single language: decision theory, representing the main logics and objectives through the respective loss functions. Additionally, by through this perspective, we analyze, interpret and compare the loss functions of some precise approaches, in the context of significance tests.

2 Decision Theory

Aiming to structure a methodology that helps us choose the best action taking into account our objectives, circumstances and knowledge, we have the **decision theory**. In this, the action to be taken admits values in the decision space $\mathscr{D}$, and is influenced by the results of an entity involving uncertainty, called Ω. So, given the preferences of the decision agent, given by the loss function $L(\cdot)$ in relation to the possible consequences ($\mathscr{D} \times \Omega$), we get the optimal choice.

For this, we look for the decision so that the associated loss is minimal. However, since the choice must be made without knowledge of the state of nature, we assign probability to the set Ω, which is therefore seen as a random variable. Thus, we estimate its behavior through the understanding that the decision agent has on the parametric space, represented by the probability distribution $\pi(\theta)$, called priori.

Considering also the case where the decision agent has access to a sample x, where the respective random variable X has its sigma-algebra of subsets of the sample space (χ) indexed by Ω, the decision agent starts to contemplate the knowledge of such evidences. And the action is specified according to a **decision rule** δ,

$$\begin{aligned} \delta_\pi : \chi &\rightarrow \mathscr{D} \\ x &\mapsto \delta_\pi(x). \end{aligned} \tag{1}$$

Thus, by means of the **priori expected loss**, or **risk against the priori**,

$$R_\pi(\delta) = E_\pi[L(\delta(X), \Theta)] = \int_\Omega \int_\chi L(\delta(x), \theta) f(x|\theta)\, \pi(\theta)\, dx\, d\theta. \tag{2}$$

Therefore, we seek for the strategy that minimizes the risk in relation to $\pi(\theta)$, so that $\delta_\pi^* = arg\ \min_{\delta \in \mathscr{D}} R_\pi(\delta)$. And, if the order of integration in (2) is alterable, the δ_π^* is equivalent to finding the rule that minimizes the expected loss a posteriori, or **risk against the posteriori** π, that is,

$$r_{\pi(\cdot|x)}(\delta) = \mathbb{E}_{\pi(\cdot|x)}[L(\delta(X), \Theta)] = \int_\Omega L(\delta(x), \theta)\pi(\theta|x)\, d\theta. \quad (3)$$

3 Hypothesis Testing

Hypothesis tests have the purpose of indicating the most plausible scenario among a collection of conjectures. However, it is usual to work with only two premises, so that they configure a partition of the parametric space Ω. Typically named as null and alternative, we have, respectively: $H_0 : \Theta \in \Omega_0$ and $H_1 : \Theta \in \Omega_1$.

In theoretical terms, the procedures are specified by a function φ, defined in class $\{\varphi : \chi \to \{0, 1\}\}$, so we decide by H_0 if $\varphi = 0$, e H_1 otherwise. Having further that the value of φ is determined by means of a Rejection Region, such a subset of the sample space is mathematically given by: $\varphi^{-1}(\{1\}) = \{x \in \chi : \varphi(x) = 1\}$. Regarding the specification of the hypotheses, there are two types of errors that can occur. The error of type I is given by $\alpha(\varphi) = \mathrm{P}[\varphi(X) = 1|\Theta \in \Omega_0]$, which occurs when we incorrectly label the alternative hypothesis as true. On the other hand, the type II error is defined by $\beta(\varphi) = \mathbb{P}[\varphi(X) = 0|\Theta \in \Omega_1]$, in relation to the null hypothesis.

Additionally, in the frequentist context, it is usual to still work with the Power Function, or Power of Test. Such quantity associates the probability of rejecting the null hypothesis at each value of Θ. Then, we define the size of the test, given by the supreme power function, considering only the values of $\Theta \in \Omega_0$, i.e., $\alpha = \sup_{\Theta \in \Omega_0} \mathbb{P}_\varphi[\varphi = 1|\theta]$. Finally, we call the value α_0 the significance level, if this is the upper limitation for the other test sizes. Whereas, in the Bayesian context, we work directly with the posteriori probabilities of the hypotheses. Now we describe the main approaches.

3.1 *Fisher and p-Value*

The most widespread reading in relation to hypothesis testing, refers to the philosophy of Sir Karl Popper, disseminated in the statistics area by Sir Ronald Fisher. According to this, a hypothesis can never be proven by an empirical study. However, a counterexample is sufficient for its negation. In hypothesis testing, such a premise implies that we consider inductive reasoning, so regardless of the amount of evidence in favor of the premise in question, it should not be accepted [2]. Although it is not necessary to indicate whether we are dealing with the null or alternative hypothesis, it is usual to specify H_0, so this is the one that we attach the greatest importance.

For the context discussed, the descriptive level of observed significance (or p-value), introduced by Pearson, is presented as an appropriate tool. This is because the p-value searches from the unobserved samples for evidence at odds with the null hypothesis, considering for such, that the related experiment is fixed. Thus, by ordering the sample space given by H_0, we examine the probability of obtaining samples as extreme as that observed. However, this metric has a number of undesirable characteristics, such as its magnitude being dependent of sample size, or the difficulty of interpretation, since the conditional definition $\mathbb{P}(x|\Omega_0)$ is summarily intuited as a conditional probability $\mathbb{P}(\Omega_0|x)$. In any case, the principle of seeking evidence against H_0, instead of evaluating both hypotheses, is diffused to the point of having a specific class of tests, called **significance tests**.

3.2 Neyman–Pearson and Likelihood Ratio

The perspective advocated by Jerzy Neyman and Egon Pearson (N–P) complements the frequentist scenario regarding hypothesis testing. For this reading, we shall initially consider that the test consists of simple hypotheses, that is, $H_0 : \Theta = \theta_0$ and $H_1 : \Theta = \theta_1$. Thus, there is a critical region given in function of the ratio of probabilities evaluated in the respective subspaces Ω, that is, $\lambda(X) = f(X|\theta_0)/f(X|\theta_1)$. However, given the impossibility of simultaneously controlling the two errors involved, the analysis is limited to the family consisting of the significance level tests α_0. Formally, for $k \geq 0$,

$$\varphi^*(x) = \begin{cases} 1 \text{ se } \lambda(x) < k \\ 0 \text{ se } \lambda(x) > k. \end{cases} \tag{4}$$

In case one of the assumptions is compound, say H_1, we restrict the domain to the Uniformly Most Powerful (UMP) tests. In general, terms, to extend the analysis with some guaranteed properties, it will always be necessary to continue applying restrictions in the domain of tests. Additionally, there are some undesirable characteristics, like the imbalance between errors I and II when the sample size increases, sometimes reaching the inversion of the initially specified match. DeGroot reread the question from a broader perspective, working with the minimization of the linear combination of errors. And later, Pericchi and Pereira [3] generalized the idea, by weighing the likelihoods, obtaining a globally optimal test, plus a balance between the specified errors and the sample size.

3.3 Bayes and Conditional Measures

In contrast to the frequentist theory, which bases its conclusions on samples and unobserved events, Bayesian Inference presents conclusions derived directly from the parametric space. Thus, we can indicate a premise with greater chance of occurrence through the posteriors ratio (denominated Bayes Factor) and the loss function used, that is,

$$\frac{\mathbb{P}(\Theta \in \Omega_0|x)}{\mathbb{P}(\Theta \in \Omega_1|x)} = \frac{\mathbb{P}(X|\Theta \in \Omega_0)\,\mathbb{P}(\Theta \in \Omega_0)}{\mathbb{P}(X|\Theta \in \Omega_1)\,\mathbb{P}(\Theta \in \Omega_1)} \geq k(L(d, \Theta)). \tag{5}$$

This reasoning is interesting, since it contemplates not only the acceptability of an isolated hypothesis, but also the circumstances of the said complement, without priorities.

4 Precise Hypotheses

Hypothesis tests have an important special case: when the conjecture of interest has Lebesgue measure zero, also known as precise hypothesis. The best-known example is the case where the parametric space is defined in the real line: $H_0 : \Theta = \theta_0$ versus $H_1 : \Theta \neq \theta_0$, circumstance which we will give emphasis.

The absence of probability in Ω, does not result in mathematical restrictions in the frequentist approach. However, in the context of significance tests, the constraint of the subspace Ω_0 assigns particular importance to the structure of Popper, given the limitation of the hypothesis in relation to the parametric space as a whole. Whereas in the Bayesian context, if the priori distribution on Ω is continuous, the posterior probability of the subset Ω_0 will be zero, invalidating the usual approaches, justifying, therefore, the development of other criteria. Following we introduce the main criteria.

4.1 Jeffreys

The Jeffreys Test, the most widespread approach, circumvented the problem of the posterior probability of Ω_0 by imposing the specification of a priori with positive probability for H_0. Thus, we started to have $\pi_{\theta_0}(\theta, \zeta)$ defined according to a combination of probabilities: $(1 - \zeta)\pi(\theta)$ to Ω_1 and ζ to Ω_0. Such rebalancing is not a problem if it is, in fact, the analyst's opinion. However, as usually it is only a practical palliative for a mathematical limitation, we violate the principle of coherence, in addition to requiring a greater amount of evidence against H_0 to enables its rejection.

4.2 FBST

In order to develop a Bayesian significance test that holds the coherence assumptions, Pereira and Stern [4] introduced the Full Bayesian Significance Test (FBST). Although this test is feasible for applications in different spaces, its contribution is more expressive in the context of the precise hypotheses, as it is developed based on the principle of least surprise, aiming for evidence in favor of the null hypothesis.

For such, it sorts the parametric space according to the posteriori probability, and seeks the θ^* that belongs to the region of H_0 and its density is maximum. Then, we form the tangent set to the null hypothesis, configured by all points with density lower than the obtained θ^*. Formally,

Definition 1 For the tangent set $T(x) = \{\theta : \pi(\theta|x) > \sup_{\Omega_0} \pi(\theta|x)\}$ the FBST evidence measure in favor of H_0 is: $EV(\Omega_0, x) = 1 - \int_{T(x)} \pi(\theta|x) d\theta$.

For high values of $EV(\Omega_0, x)$, or *e-value* as it is also known, θ_0 will be among the most likely points a posteriori, and will favor the null hypothesis. Additionally, this approach presents advantages as: intuitive logic, geometric interpretation, consistency, and invariance under one-to-one parameter transformations.

5 Loss Function

In order to approach the tests of significance according to a single language, we work with decision theory. For this, we consider the space of decisions $\mathscr{D}$, given by $\{d_0, d_1\}$, where d_i denotes the action of accepting the hypothesis $H_i : \Theta \in \Omega_i$, with $i \in \{0, 1\}$, and losses L_0 and L_1, respectively. In addition, assuming that there is a differentiated posture in relation to the null hypothesis, the decision is presented in relation to the H_0, this is, d_1 is read as rejection of H_0. Besides, that conservative behavior is incorporated into the analysis through the loss function. Thus, for a sample x, we will have

$$\varphi_\pi(x) = \begin{cases} d_0 \text{ if } \dfrac{\pi(\Theta \in \Omega_0|x)}{1 - \pi(\Theta \in \Omega_0|x)} > \dfrac{L_0}{L_1} \\ d_1 \text{ if } \dfrac{\pi(\Theta \in \Omega_0|x)}{1 - \pi(\Theta \in \Omega_0|x)} < \dfrac{L_0}{L_1}, \end{cases} \tag{6}$$

and the conclusion will be given according to the already known Factor of Bayes. Note that assigning randomness to the set Ω does not invalidate the generalization of the analyzes, since the interest is in replicating the philosophy of each approach and not the system itself. Considering this perspective, follows the description of the FBST, and the Popper's perspective (essence of the p-value).

Table 1 Loss function of the FBST test

	Accept H_0	Reject H_0
$\theta \notin T(x)$	b	a
$\theta \in T(x)$	$b + c\,[\mathbb{I}\{\theta \in T(x)\}]$	0

Table 2 Risk for some cases of evidence

$EV(\Omega_0, x)$	$r_{\pi(\cdot\|x)}(d_0)$	$r_{\pi(\cdot\|x)}(d_1)$
0	$b + c$	0
0.5	$b + 0.5\,c$	$0.5\,a$
1	b	a

5.1 FBST and Madruga et al.

By making use only of the information contained in the posterior density, the FBST has been classified as full Bayesian since its genesis. However, only in the work of Madruga et al. [5] this measure was analyzed according to the decision theory, being obtained by minimizing the loss function given from positive a, b and c (Table 1).

Note that, in this case, unlike classical theory, we consider a broader class, wherein the observed sample is also incorporated into the loss function. Thus, assuming that the tangent space of the FBST is defined from the sample, we have, from the minimization of $L(d, \theta, x)$, that the acceptance of the hypothesis H_0 will occur if, and only if, $EV(\Omega_0, x) > \frac{b+c}{a+c}$.

In practical terms, the loss function is structured in order to evaluate the tangent set, that is, we describe our expectations regarding the information brought by the sample. However, once it is a measure of significance, the penalty associated with rejection of the null hypothesis when $\theta \in T(x)$ is smaller. For a better understanding, follows some examples (Table 2).

Usually, by specifying the loss function according to the real values of the parametric space, without worrying about our knowledge or the results of the sample, we are working with the penalty related to the phenomenon, rather than the study itself. Because such scenario is a simplification, we fail to incorporate our preferences regarding the reflexes of the study, such as psychological, financial, and even social issues. Whereas, when dealing with losses according to what we are actually going to obtain, and not with a utopian scenario of absolute knowledge, we have a more realistic reading, as well as a full analysis of the situation.

5.2 Popper and Rice

Aiming at a loss function that reflected Popperian philosophy, Rice [6] reread the decision space so that d_0 represents the decision "Nothing to declare", while for d_1

Table 3 Effects for different values of γ

γ	Penalty for reporting	Penalty for not reporting	Inverse of variability
0.01	10	0.10	99
0.1	3.16	0.32	9.00
0.3	1.83	0.55	2.33
0.5	1.41	0.71	1.00
0.7	1.20	0.84	0.43
0.99	1.01	0.994	0.01

we provide results indicating the rejection of H_0. Thus, by choosing to make such a statement, the conclusions are presented by means of an estimate $\hat{\theta}$, and the said loss is evaluated by the usual quadratic loss. On the other hand, by omitting the results, the loss occurs in terms of no longer known unfoldments and how informative they might be. This loss is represented proportionally to the distance between θ and θ_0. Thus, both losses are maintained on the same scale, allowing the decision agent to specify his/her opinion to the relation between them, by means of a factor γ, hence nothing to declare implies that $\gamma^{1/2}\,(\theta_0 - \theta)^2$, and the rejection of H_0 implies $\gamma^{-1/2}\,(d - \theta)^2$. Therefore, according to Bayes rule, we will report results ($\varphi_\pi^R(x) = d_1$) if,

$$R(\Omega_0, x) = \frac{\mathbb{E}^2_{\pi(\cdot|x)}[\Theta - \theta_0]}{Var_{\pi(\cdot|x)}[\Theta]} \geq \frac{1-\gamma}{\gamma}. \tag{7}$$

In practical terms, we reject H_0 when the estimate is "far" from the tested value. Otherwise, we do not have conclusions. The hesitation related to the statements is represented by the value γ, so that the smaller this quantity, the more skeptical the analysis, and the penalty for reporting results will always be higher than the alternative. Noting also that the product of the weights is fixed, they follow the effects for different values γ (Table 3).

Thus, considering the correspondence between the units of imprecision of the estimate d and the loss inherent in the lack of results, the agent should look for the point of balance between such entities. It should also be noted that the conclusion is taken on the basis of the inverse of a measure of variability, similar to the square of the coefficient of variation, but with a focus on the a posteriori parametric space. Thus, when such a measure of variability is significant, we have indicatives about the lack of accuracy and, consequently, the results are not reported.

In the reading made by Rice, we identify several important gains in relation to Fisher's test: lack of assumptions about repeatability and stopping rules of the experiment, mandatory incorporation of the alternative hypothesis, coherence between rejection of the hypothesis and its subsets and even informational results for large samples.

Table 4 Example 1: $R(\Omega_0, x) \times EV(\Omega_0, x)$, $n = 10$

# Successes	Rice: $\varphi_\pi^R(x) = d_1$	FBST: $\varphi_\pi^E(x) = d_0$
0	7.989	0.020
1	1.664	0.123
2	0.173	0.450
3	0.065	1.000
4	0.728	0.471
5	2.080	0.156
6	4.294	0.036
7	7.865	0.006
8	14.040	0.001
9	26.624	0.000
10	64.716	0.000

6 FBST x Rice

Although both FBST and Rice methodologies share the same principle as *Onus Probandi*, where the defendant is to be presumed innocent, the Rice test has as its central concern whether or not to reject the hypothesis, while the Madruga et al. measures its consistency.

Despite the differences between the two approaches, when $\Omega \subset \mathbb{R}$ the tests are essentially equivalent in the frequentist sense of having a one-to-one relationship between test statistics, for such, follows an illustration.

Example 1 Consider a random sample i.i.d. of Bernoulli's, conditioned in the parameter θ, and the interest in testing the hypotheses $H_0 : \theta = 0.3 \times H_0 : \theta \neq 0.3$, assuming a priori Beta(1, 1). Thus, knowing that all the information of the 2^n possible results can be examined by means of the number of successes obtained, we have the following values for the statistics of Rice and FBST for a sample size 10 (Table 4).

Considering the same analysis for a sample size 1000, we can see in the Fig. 1 the coherence between both results, by means of a negative association, as expected, since we are comparing opposite decisions. Note that we choose to report the results, regardless of the sample size, in the case of γ less than 0.015.

In the inferential context, where the goal is to learn about the parameter, not reporting results seems inappropriate. However, cautious behavior is nothing more than a characteristic of an philosophy, that is, a perspective that is perfectly valid for certain scenarios, such as when the contradiction of the tested hypothesis is not absolute, the sample presents itself as a too limited tool, or the analyst can simply prefer a more cautious stance.

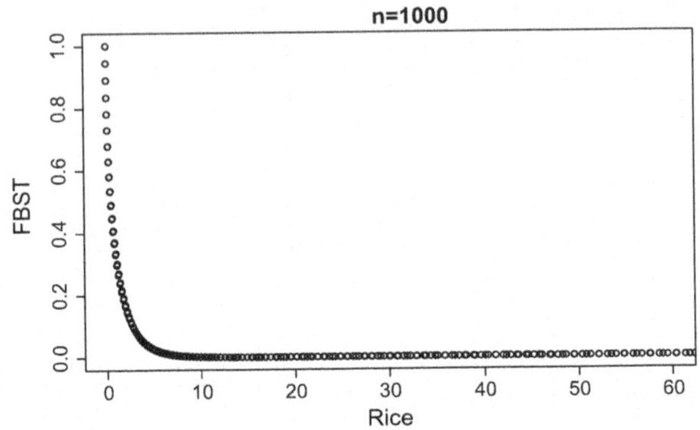

Fig. 1 Example 1: $R(\Omega_0, x) \times EV(\Omega_0, x)$, $n = 1000$

7 Conclusion

In this paper, we discuss the main approaches to hypothesis testing, particularly with regard to significance tests, and reading according to the statistical decision theory. From the point of view of coherence, we concluded that, between the approaches evaluated, we have two satisfactory options from the point of view of coherence: FBST and Rice. Finally, by comparing both readings, we obtain harmonious results with the respective proposals. Besides, they are consistent with each other. For future works, the proposal is to extend the Rice test to parametric spaces larger than one and analyze practical applications of the discussed approaches.

References

1. Kadane, J.B.: Principles of Uncertainty. Chapman & Hall, Boca Raton (2011)
2. Fisher, R.A.: The Design of Experiments. Oliver and Boyd, Edinburgh (1935)
3. Pericchi, L.R., Pereira, C.A.B.: Changing the paradigm of fixed significance levels: testing hypothesis by minimizing sum of errors type I and type II. Braz. J. Probab. Stat. **1310.0039** (2013)
4. Pereira, C.A.B., Stern, J.M.: Evidence and credibility: full bayesian significance test for precise hypotheses. Entropy **1**(4), 99–110 (1999). https://doi.org/10.3390/e1040099
5. Madruga, M.R., Esteves, L.G., Wechsler, S.: On the Bayesianity of Pereira-Stern tests. Sociedad de Estadistica e Investigacion Operativa **10**, 291–299 (2001)
6. Rice, K.: A decision-theoretic formulation of Fisher's approach to testing. Am. Statis. **64**(4), 345–349 (2010)

Probabilistic Equilibrium: A Review on the Application of MAXENT to Macroeconomic Models

Paulo Hubert and Julio M. Stern

Abstract The concept of equilibrium is central to many macroeconomic models. However, after the 2008 crisis, many of the most used macroeconomic models have been subject to criticism, after their failure in predicting and explaining the crisis. Over the last years, a response to this situation has been the proposal of new approaches to the study of macroeconomical systems, in particular, with the introduction of thermodynamics and statistical physics methods. In this paper, we offer a brief review of the application of the maximum entropy framework in macroeconomics, centered around the different interpretations of the equlibrium concept.

Keywords Maximum entropy · Macroeconomy · Equilibrium

1 Introduction

The classical example of equilibrium in dynamical systems comes from mechanics: the pendular system, composed of a point mass suspended by a chord, attached to a fixed platform. Put on movement by an initial impulse, the system, in the absence of attrition, enters an equilibrium state of perpetual and periodical motion. When attrition enters the picture, the system dissipates energy to the surroundings, and the new equilibrium is one of rest: the point mass evolves to the least-energy state, and remains there unless some new action is imposed upon it.

Many states of equilibrium in mechanics are states in which the system repeats itself, or stays at rest. These are situations in which our description of the system dynamics can dispense of an infinite time axis: all possible configurations present

P. Hubert (✉) · J. M. Stern
IME-USP, R. do Matao, 1010 - Vila Universitaria, São Paulo 05508-090, Brazil
e-mail: phubert@ime.usp.br

J. M. Stern
e-mail: jstern@ime.usp.br

A. Polpo et al. (eds.), *Bayesian Inference and Maximum Entropy Methods in Science and Engineering*, Springer Proceedings in Mathematics & Statistics 239, https://doi.org/10.1007/978-3-319-91143-4_18

themselves during a limited time interval. Therefore, it is much easier to make predictions about the future (or draw conclusions about the past) of a system in a state of equilibrium. Even further: once the system's singularities are known (for mechanical systems, these are the solutions to the equation $\dot{x} = 0$), the theory of differential equations allows one to consider the limiting behavior of a system (either forwards or backwards in time) in terms of its relation with these stationary points.

Macroeconomic theory has for a long-time drawn inspiration from mechanics and its methods [1]. Therefore, the concepts of equilibrium and limiting behavior of a system are central in many macroeconomical models. However, as we intend to argue, this is not a mere consequence of the use of rational mechanics' methods: equilibrium states are objects with an intrinsic epistemological interest. One important issue that arises, then, is the ontological interpretation one gives to the equilibrium and its attainment as a limiting behavior of the system's trajectories.

In this paper, we offer a brief review of the use of equilibrium concepts both in classical and contemporary macroeconomics, and rely on the relationship between rational mechanics and economy to discuss the recent application of statistical mechanical methods to macroeconomics. We argue, in the spirit of Kuhn, that macroeconomical science is in a state of exploration after a paradigm crisis, and analyze briefly some of the conceptual distinctions between classical and statistical equilibrium.

2 Equilibria in Classical Economics

Perhaps the most concrete example of the analogy between economical and mechanical systems is the Phillips' machine, or *MONIAC (Monetary National Income Analogue Computer)* (Fig. 1). It is an analogical computer that uses water flows to model the dynamic of income in an economy. The MONIAC, officially presented at the London School of Economics in 1949, was able to solve systems of up to nine simultaneous equations, with parametrizable coefficients [2, 3]. Once a dynamical equilibrium was reached the solution could be read in scales attached to the water tanks.

Regardless of the method of computation, however, economists are in general very fond of equilibria. We find mention of the term already in the seminal work of Debreu [4], in which he proposes a formalization of economical analysis based on the axiomatic method. There, he defines equilibrium in the following terms:

> *If the actions x_i, y_i satisfy the market equilibrium equality $x - y = w$, the economy is in equilibrium, i.e., every agent, given the price system and the actions of other agents, has no incentive to choose a different action, and the state of the economy is a market equilibrium. (Debreu* [4], *p. 79)*

Equilibrium is thus a state of affairs, a situation (described by a set of values for economic variables, plus a price system relating them) in which no agent has

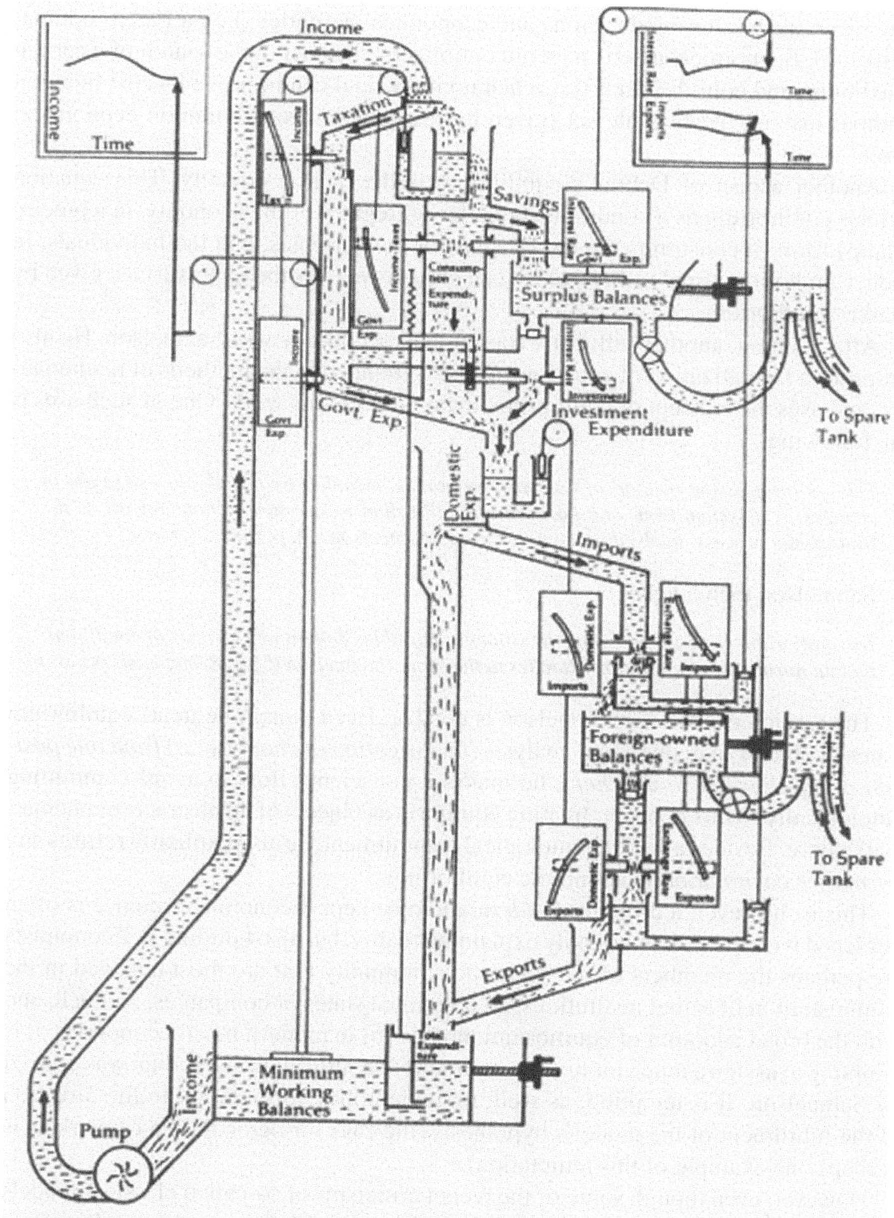

Diagram of the Phillips machine. Source: LSE Quarterly, Winter 1988, Nick Barr.

Fig. 1 Schematic drawing of the MONIAC [2]

any incentive to alter its decisions and economical activities (i.e., a Pareto optimal situation). Economical agents here are optimizers: they act in the search of a certain maximum, and equilibrium arises when no individual can improve its cost function without leaving the feasible set (given by the constraints of common economical life).

Another aspect of Debreu's equilibrium is the market equality. This equation defines equilibrium as a condition about the aggregates of the economy, in a precise (sharp) form: it constrains the total values for the variables, and the individuals, in their search for optimal utility points, can only move over the hypersurface given by market equilibrium.

After Debreu, another influent writer in macroeconomy was Samuelson. He also proposes a formalization of economic analysis founded in the methods of mathematics, and uses the concept of equilibrium many times in his work. One of such uses is the following:

> *This, in brief, is the method of comparative statics, meaning by this the investigation of changes in a system from one position of equilibrium to another without regard to the transitional process involved in the adjustment. (Samuelson* [5], *p. 8)*

Samuelson then adds:

> *By equilibrium is meant here only the values of variables determined by a set of conditions, and* ***no normative connotation attaches to the term****. (Samuelson* [5], *p. 8, emphasis is ours)*

The caution exerted by Samuelson is notable. Even though he treats equilibrium states as objects of economical analysis (*investigation of changes (...) from one position of equilibrium to another*), he makes a conscient effort to avoid committing ontologically to this concept, treating equilibria as objects of an abstract (mathematical) nature. Besides avoiding ontological commitment, he also explicitly refuses any *normative connotation* to economic equilibrium.

This is, however, a difficult *desideratum* to be kept. Economical science is often burdened with the task of not only explaining reality, but also building it. Economists are perhaps the members of the scientific community that are most engaged in the administration of actual institutions, be it national states or companies. As such, and with the broad adoption of equilibrium models [6] in modern macroeconomics, it is tempting to assign a much more concrete status to equilibrium states than was desired by Samuelson. It is tempting, as well, to guide policy formulation in the direction of the fulfillment of the model's hypothesis (the case for deregulation of markets is perhaps one example of this temptation).

However, even though some of the recent criticisms of so-called classical models direct their attacks precisely to the equilibrium concept [7, 8], some of the alternative frameworks proposed in the current literature still have it as a central epistemic notion. The reason for this attachment can perhaps be found in the epistemological analysis of Foerster and Luhmann.

3 Equilibria as Objects of Knowledge

In the work of Foerster [9], we find the following famous quotation about *eigenvalues*:

> *Eigenvalues have been found ontologically to be discrete, stable, separable, and composable, while ontogenetically to* ***arise as equilibria*** *that determine themselves through circular processes. (Foerster* [9], *p. 266, emphasis is ours)*

In his paper, Foerster analysis the "organization of sensorimotor interactions" of a cognoscent being with its environment. He proposes a model consisting of the alternate and recursive application of the operators *observation* and *coordination*: observation of new data triggers a coordination (behavior), which leads to a new observation and so forth. His eigenvalues (arising as equilibria) are the very objects of knowledge, coming to be in the interaction between being and environment.

A radical application and enlargement of Foerster's ideas can be found in the work of Luhmann [10]. He considers hierarchical models of recursive systems in a ladder of growing complexity, and uses this framework to describe not only the relation of one individual with its environment, but the very organization of human societies. In this same spirit, he analyzes the scientific endeavor by describing the collective work of scientists as a subsystem, horizontally differentiated from the broader system of human society. Successful scientific theories, then, would emerge as *eigenvalues* of a self-referent, recursive, dynamic system; theories, as collective objects of knowledge, share the same nature of Foerster's objects of understanding: they are also equilibria, and as such stable, discrete, limit states of a recursive process.

Under the light of Foerster and Luhmann's interpretation of the knowledge-building process, it becomes clearer why equilibrium states are useful objects inside theories. As eigenvalues of the studied system, these states are discrete, stable, separable; they are therefore much easier to name, classify, and study than the whole dynamics of the system. Equilibrium methods, in this sense, work like traps with which we take hold of a system of interest, in order to be able to describe it, make predictions about its future and inferences about its past.

This description of scientific theories introduces a strong recursive, self-referent aspect in epistemology. Equilibrium states and their attainment are objects of many sciences (physics and economics in particular), but in this framework, they also describe the process of scientific enquiry in itself. In this sense, when speaking about the nature of equilibria and limiting behavior, science is also talking about itself and its objects. In the very words of Luhmann:

> *The concept of self-referential systems can and must subsume science and one's own research. This requires taking leave of ontological metaphysics and apriority. Systems with built-in reflection are forced to forgo absolutes. And if science discovers this fact in the domain of its objects, the fact holds irrefutably for science, too (Luhmann* [10], *p. 485).*

This self-reference can also be found in the interpretation of statistical methods in physics, according to Jaynes [11]. For him, the maximum entropy solution represented the subjective (i.e., the scientist's) solution to an inference problem. However, it agrees with all measurements made in actual, thermodynamical systems in states of equilibrium (i.e., with the objective solution). Entropy, besides being interpreted as a measure of a system's quality, can also be interpreted as a measure of the scientist's knowledge about the system. What holds for the object, holds for the scientist as well.

4 The Change from One State of Equilibrium to Another

According to Kuhn, science advances in a dual fashion: continuously, in the periods of what he calls *normal science*, and discontinuously, when it is subject to a *paradigm shift* [12]. A paradigm, in this epistemology, is a scientific realization with two properties: if offers sufficiently unprecedented results, in order to attract an enduring group of participants and form a delimited (discrete) and stable research group. At the same time, it is sufficiently open to contain many unsolved problems in which this group can work (recursively), feeding itself in its own questions.

When normal science is taking place, Kuhn identifies two kinds of events: inventions and discoveries. The inventions are identified with *puzzle-solving* activities: new applications are developed, empirical evidence is accumulated, minor problems are solved (the ones that are not urgent, for their lack of solution is not enough to provoke a crisis).

Discovery, on the other hand, is associated to more extreme movements, caused by the presence of an *anomaly*:

> *Discovery begins with the awareness of an anomaly, that is, with the recognition that nature has in some way violated the paradigmatic expectations governing normal science. What follows is a more or less broad exploration of the area in which the anomaly has occurred. This work only stops after the paradigm's theory has been adjusted, in such a way that the anomalous has now turn into the expected. The assimilation of a new fact demands more than an additive adjustment of the theory. Until such an adjustment is completed - until the scientist has learnt to see nature in a different way, the new fact will not be considered completely scientific (Kuhn* [12], *p. 78, free translation from the Brazilian edition).*

In the realm of macroeconomy, the most recent and important anomaly was the financial crisis of 2007. The crisis, and subsequent depression, escaped entirely the models' predictions, and even their power of *post factum* explanation [13–16].

The economists reacted immediately: a thorough and deep theoretical exploration of monetary and financial systems (where it is consensual that the crisis originated) was launched. New approaches for economical modeling were proposed (Stock-flow consistent models, Agent-based models), ideas of less orthodox thinkers were revisited (maybe the most prominent example is the work of Minsky [17]), and methodological discussions were sharpened.

Olivier Blanchard, IMF's chief economist, makes a clear point about the necessity for theoretical exploration in postcrisis macroeconomics [18]:

> *Turning from policy to research, the message should be to let a hundred flowers bloom. Now that we are more aware of nonlinearities and the dangers they pose, we should explore them further theoretically and empirically—and in all sorts of models. This is happening*

already, and to judge from the flow of working papers since the beginning of the crisis, it is happening on a large scale. Finance and macroeconomics in particular are becoming much better integrated, which is very good news. (Blanchard 2014)

It seems reasonable, then, to describe the current moment of macroeconomical thinking in Kuhn's terms, and to say that macroeconomics is now going through a paradigm crisis: the change from one state of equilibrium to another. In these moments, as Kuhn points out (and Blanchard apparently agrees), it is fruitful to explore new methodological possibilities.

The application of maximum entropy methods is one of these possible explorations, one that is becoming frequent in the macroeconomics literature. In the next section, we briefly review a few papers on the subject.

5 The Statistical Equilibrium in Macroeconomics

We begin by analyzing the paper by Foley [19], which presents a direct application of the maximum entropy principle. Foley defines a space of transactions, in an economy with a certain (finite) number of goods. A transaction is a point in the space of goods, where each coordinate can represent demand (if it has a negative value) or supply (if it has a positive value) for that particular good. Agents are divided into categories, each category defined by a supply set, representing the totality of transactions which are at the same time feasible and desirable for agents belonging to that group.

He then defines an average excess demand measure, and by constraining this quantity to be 0 (i.e., applying the idea of market equilibrium) he obtains the maximum entropy distribution of agents inside each supply set. The statistical equilibrium he obtains (i.e., his maximum entropy distribution) is thus associated with the usual market equilibrium (zero excess demand), but instead of constraining the *total* excess demand to be 0, he constrains the *expected* excess demand to be 0, where this expectation is taken with respect to the MAXENT distribution.

Foley also points that the statistical equilibrium usually is not Pareto efficient. In other words, in an economy in equilibrium it is possible to find transactions between two or more agents that can improve both of their utility values. In a pure exchange model, the entropy associated to the Pareto equilibrium would be null (assuming convex utility functions), because all agents will be concentrated at the minimum cost point over the hypersurface of constant utility. In the statistical equilibrium model, however, agents can spread all along this hypersurface, even though with greater concentration around the minimum cost. In other words, the statistical equilibrium allows for horizontal inequality, even between individuals from the same class (i.e., agents with the same supply set), and, even further, an endogenous inequality that arises even between agents with identical initial resource allocation.

About this new possibility of horizontal inequality, he says:

The statistical equilibrium theory of markets is methodologically less ambitious than Walrasian competitive equilibrium theory. Walrasian theory seeks to predict the actual market outcome for every individual agent, while the statistical approach seeks only to characterize the equilibrium distributions of agents over outcomes, without predicting the fate of specific agents. (Foley [19] *pp.343–344).*

Another recent work applying thermodynamic methods to macroeconomics is the paper by Caticha and Golan [20]. In their model, there are a finite number of goods, and a finite number of agents. Goods can be seen as production and consumption goods simultaneously; an agent can be a producer and/or a consumer of any particular good. Each agent has an utility function that models its relative interest for different mixtures of goods. To each microstate of the economy (the specification of each agent's consumption and production functions, alongside with its utility function), there corresponds a macrostate: the total amount produced and consumed of each good, and the total utility of each agent. Again, as in Foley's work, the macrostates constraints are taken to be expected values, instead of exact, aggregate quantities.

By imposing an expected market closure constraint, a budget constraint on the agents, and fixating each agent's expected utility, they obtain the MAXENT distribution on the space of agents and goods, and use it to analyze the dynamics of an economy in state of (statistical) equilibrium.

Another line of application of the MAXENT framework has been the labor market. The paper published in Brazil by Soromenho [21], for one example, models the economy as composed of two kinds of agents: workers and firms. The goods of his economy are labor and currency, and the transactions considered are the exchange of a worker's labor for a firm's money. Workers can either be unemployed, or receive a wage for one unit of work. Firms have a wage budget that must be entirely spent. The wages are not fixed, meaning that, with a given fixed budget, each firm can hire more or less workers, paying a smaller or bigger wage respectively.

By imposing a viability condition (according to which the number of workers receiving wage i is the same number of workers employed by the firms with this same wage), the author obtains a MAXENT distribution. His method up to this point is quite similar to the methods we previously described. He differs, however, in the use he gives to the equilibrium distribution: he uses it to analyze a classical (nonstatistical) kaleckian model for a single good economy. He concludes that it is possible to replace the usual closure of the kaleckian model (exogenous markup for the wages, uniform distribution of employment between firms) by expected values for the same quantities (average wage, MAXENT distribution of employment) calculated with the MAXENT distribution, and still obtain the same results of the usual kaleckian model.

When applied to either the labor or the goods market, the maximum entropy framework is based on the idea of distributing a fixed quantity among individuals. The paper of Banerjee and Yakovenko [22] discuss the method precisely in this terms: in their paper, the economy is defined with three resources: money, income, and energy. The microstates of the economy are the sets of (either) goods in possession of each agent. Transactions occur between pairs of agents, and consist of the exchange of a constant amount of resources.

Agents are divided in classes, and the constraints are of a constant number of agents, and constant total amount of resources. With these constraints, the authors obtain the equilibrium distribution. Here, the statistical equilibrium receives a clearly ergodic interpretation:

> *After many transactions between different agents, we expect that a stationary probability distribution of money would develop in the system. (Banerjee and Yakovenko* [22], *p. 4)*

After obtaining the equilibrium distribution, the authors use it to analyze the situation in which two economies, with different equilibrium “temperatures” (which translates to different resource’s stocks *per capita*), start to interact. By postulating the validity of the second law of thermodynamics, they conclude that the flow of resources must be from the richer to the poorer countries.

5.1 Conclusion

The papers (very) briefly reviewed in the last section allow us to draw in broad lines what is the statistical equilibrium method, as applied to macroeconomics.

First of all, it begins with the modeling of individuals: a set of phase variables is chosen, and a model for the interactions between agents (exchanging of values for the phase variables) is proposed. From the start, it is recognized that *individual variability* might exist, even in equilibrium, and even between individuals which have exactly equal initial conditions. This recognition is the natural consequence of having a *probabilistic* solution for the model: a probability distribution over the possible microstates.

After the model for microstates is developed, in order to obtain a solution it is necessary to incorporate knowledge about macrostates as well. In physics, this is done by postulating conservation laws; in macroeconomy, the analogous idea is that of *market closure* (equality of supply and demand levels). This kind of constraint was used since the early works of Debreu and Samuelson; the main difference is that, in the statistical equilibrium framework, markets close only in *expected values*. This opens the possibility that real economies working outside market closure can still be analyzed by equilibrium methods. In the classical theory, at least in principle, the closure is a logical necessity in equilibrium, and any deviance from that breaks the model altogether (the recognition of this fact might have been one of the reasons that motivated Samuelson to insist upon the abstract nature of equilibrium solutions).

But besides these two points, we believe that the adoption of statistical equilibrium methods in macroeconomics can have another epistemological impact. This might be the case because, in the words of Blanchard [18]:

> *The techniques we use affect our thinking in deep and not always conscious ways. This was very much the case in macroeconomics in the decades preceding the crisis. The techniques were best suited to a worldview in which economic fluctuations occurred but were regular, and essentially self-correcting. The problem is that we came to believe that this was indeed the way the world worked. (Blanchard, 2014)*

In most papers we analyzed, statistical equilibrium receives an interpretation which can be linked to the ergodic school of thought in statistical mechanics [23]. Equilibrium is a concrete state of affairs, one that ought to be reached by the real economy, given that the interactions modeled are allowed to continue for a sufficient long period of time, and as long as the model assumptions are satisfied. The ergodic interpretation is then a very normative one, in the sense that it precognizes that, if the transactions and elements of the economy are such and such, the *free course of the economy will lead* to this and this situation. Uncertainty only enters the picture because individuals are unpredictable and do not exactly behave as the model says they do; if the economy is free to evolve for a sufficient long time, this unpredictability will "cancel out" and the predictions will be fulfilled.

In the subjective interpretation, on the other hand, statistical equilibrium solutions are understood as the most conservative (maximum entropy) probabilistic models for an economy, given the macro constraints we expect to be satisfied. It is explicitly a *tool* used by the scientist to describe what he believes would be the case if his description of the microstates is accurate and if his expectations about aggregates hold at least approximately. The uncertainty, in this case, is assumed *by the scientist*; he will be talking (probabilistically) about the current state and nature of a system, and not about its potential infinite evolution. Equilibrium solutions lose their normative status, as Samuelson wanted; or rather they cease to have a normative status over the real economies, to receive a normative status over the scientist's work.

In this sense, the subjective interpretation can work as a safeguard against the epistemological risk that Blanchard points out: the probabilistic model, derived as an equilibrium, will still be useful (and used) as a scientific and decision-making tool. However, the uncertainty of its predictions will be associated with the lack of complete knowledge about the system. Equilibrium will not be an ideal situation the economy might reach in the long run, but a practical state of affairs reached by macroeconomical science in the investigation of real economies. Knowing that, if the scientist again comes to believe that "this is indeed the way the world works," he might be not very much distant from the truth.

Acknowledgements The authors would like to acknowledge professor Gilberto Tadeu, at FEA-USP, for thoughtful comments on an earlier version of this work.

References

1. Mirowsky, P.: More Heat Than Light. Cambridge University Press, Cambridge (1995)
2. Swade, D.: Computer ressurrection: the bulletin of the Computer Conservation Society. The Phillips Economic Computer. http://www.cs.man.ac.uk/CCS/res/res12.htm. Accessed 13 Dec 2016
3. https://www.theguardian.com/business/2008/may/08/bankofenglandgovernor.economics. Accessed 13 Dec 2016
4. Debreu, G.: Theory of Value: An Axiomatic Analysis of Economic Equilibrium, 4th edn. Yale University Press, London (1971)

5. Samuelson, P.: Foundations of Economic Analysis, 9th edn. Cambridge Harvard University Press, Cambridge (1971)
6. Tovar, C.E.: BIS Working paper no. 258
7. Dixon, H., Creedy, J. (eds.): The foundations of economic thought. Equilibrium and Explanation. Blackwells, Oxford (1990)
8. Mirowsky, P.: Louvain Economic Review **55**(4), pp. 447–468
9. von Foerster, H.: Understanding Understanding: Essays on Cybernetics and Cognition. Springer, New York (2003)
10. Luhmann, N.: Social Systems. Stanford University Press, Stanford (1995)
11. Jaynes, E.T.: The Physical Review **106**(4), pp. 620–630
12. Kuhn, T.: A Estrutura das Revoluções Científicas, 9th edn. Perspectiva, São Paulo (2007)
13. https://www.gpo.gov/fdsys/pkg/CHRG-111hhrg57604/pdf/CHRG-111hhrg57604.pdf. Accessed 15 Dec 2016
14. Kocherlakota, N.: https://www.minneapolisfed.org/publications/the-region/modern-macroeconomic-models-as-tools-for-economic-policy. Accessed 15 Dec 2016
15. Blanchard, O.: https://piie.com/system/files/documents/pb16-11.pdf. Accessed 15 Dec 2016
16. Krugman, P.: http://krugman.blogs.nytimes.com/2016/08/12/the-state-of-macro-is-sad-wonkish/. Acessed 15 Dec 2016
17. Wray, L.R., Tymoigne, E.: The Levy Economics Institute, Working paper 543
18. Blanchard, O.: https://www.weforum.org/agenda/2014/10/olivier-blanchard-financial-crisis-macroeconomics/. Accessed 15 Dec 2016
19. Foley, D.K.: Journal of Economic Theory **62**, pp. 321–345
20. Caticha, A., Golan, A.: Physica A **408**, pp. 149–163
21. Soromenho, J.E.C.: Revista EconomiA **12**(3), pp. 407–425
22. Banerjee, A., Yakovenko, V.M.: New Journal of Physics **12**
23. Guttman, Y.M.: The Concept of Probability in Statistical Physics. Cambridge University Press, Cambrige (1999)

Full Bayesian Approach for Signal Detection with An Application to Boat Detection on Underwater Soundscape Data

Paulo Hubert, Julio M. Stern and Linilson Padovese

Abstract The problem of detecting a signal of known form in a noisy message is a long-studied problem. In this paper, we formulate it as the test of a sharp hypothesis, and propose the Full Bayesian significance test of Pereira and Stern as the tool for the job. We study the FBST in the signal detection problem using simulated data, and also using data from OceanPod, a hydrophone designed and operated by the Dynamics and Instrumentation Laboratory at EP-USP.

Keywords Underwater acoustics · Bayesian inference · Signal detection

1 Introduction

The problem of detecting the presence of a signal in a noisy sample can be stated as an inference problem where we compare two alternative hypothesis, H_0: *data is composed of noise only*, against H_1: *data is signal plus noise*. By *signal*, we understand a function of time, usually discretely sampled; data is, thus, a sequence of points indexed by a time variable.

One common application of signal detection is in telecommunications, where one intends to transmit a message through a noisy channel from a *transmitter* to a *receiver*; when the message is binary, the receiver must decide at each instant if a given (known) signal is present (in which case she assumes a 1 was transmitted) or

P. Hubert (✉) · J. M. Stern
IME-USP, R. do Matao, 1010 - Vila Universitaria, São Paulo 05508-090, Brazil
e-mail: phubert@ime.usp.br

J. M. Stern
e-mail: jstern@ime.usp.br

L. Padovese
EP-USP, Av. Prof. Luciano Gualberto, 380 - Butanta, São Paulo 05508-010, Brazil
e-mail: lrpadove@usp.br

A. Polpo et al. (eds.), *Bayesian Inference and Maximum Entropy Methods in Science and Engineering*, Springer Proceedings in Mathematics & Statistics 239, https://doi.org/10.1007/978-3-319-91143-4_19

absent (in which case she assumes a 0 was transmitted). In this kind of application, usually, the exact signal form is known both at the transmitter and the receiver, and the problem arises only because the channel is not ideal, i.e. it adds noise to the data that is collected at the receiver.

It is natural, in this situation, to postulate the model [1] $Y = \xi X + R$, where Y is the recorded noisy message, X is the particular signal form we are interested in, and R is noise, where by noise we understand whatever forms of random or non-random patterns besides the one that codifies the signal. The unknown parameter ξ is interpreted as a nonnegative gain factor. In this formulation, the problem of signal detection amounts to testing $H_0 : \xi = 0$ against $H_1 : \xi > 0$.

Testing hypothesis of equality, like the one defined above, is the main goal of the FBST (*Full Bayesian significance test*) [2] framework. In this work, we analyze the problem of signal detection as a sharp hypothesis test problem, and propose the FBST as the tool of choice for the job. We analyze both the simplest case, where the signal form is completely known at the receiver, and a more complicated version of the signal detection problem, namely the situation where the functional form of the signal is known, but not the values of the parameters that completely define it. After analyzing the performance of the FBST with simulated data, we apply it to the problem of detecting the presence of ships in soundscape data.

2 FBST for Signal Detection

2.1 Signal Known at the Receiver

We analyze first the problem of digital signal detection, which can be stated in the following terms: a transmitter sends a signal, modelled as a continuous function of time $x(t), t \in [0, T]$, through a noisy channel. The signal plus noise reaches a receiver, whose task is to analyze the message and decide whether the signal was or was not embedded in the message. The received message can be modelled as $y(t) = \xi x(t) + r(t), t \in [0, T]$, where $r(t)$ is noise, and ξ is a gain factor that represents the intensity of the signal (assumed constant for $t \in [0, T]$).

From a statistical point of view, the problem can be stated as the test of the sharp hypothesis $H_0 : \xi = 0$. Acceptance of the null hypothesis implies that no signal was present in the recording (i.e. a 0 was transmitted), whereas its rejection means that a signal was indeed present (a 1 was transmitted).

In many applications, specially in communications, the exact form of the signal is known both at the transmitter and the receiver. Given this information, the problem is greatly simplified, since our parametric space is only one-dimensional (ξ is the only unknown quantity, if one assumes known noise power).

In this section, we evaluate the performance of the *Full Bayesian significance test* (FBST) in this simpler version of the problem using simulated data. We consider, henceforth. a signal of the following form

$$x(t) = \sum_{h=1}^{m} A_h cos\,(2\pi h\omega t + \phi_h) \tag{1}$$

This form is the one of a sinusoidal wave with fundamental frequency ω and m harmonics. The A_h and ϕ_h represent each harmonic's amplitude and phase, respectively.

The choice for this particular form is motivated by our later application, namely the detection of ships in hydrophone recordings. The literature [3–6] of subaquatic acoustics suggests that the noise radiated by a moving ship is of the form in (1), plus broadband noise. We discuss this model in further detail in a later section.

For now, supposing that $\Theta = \{\omega, A_1, \ldots, A_m, \phi_1, \ldots, \phi_m\}$ is known, the problem of detecting this signal in a noisy recording can be modelled in the following way: the message at the receiver is given by $y(t) = \xi x(t) + r(t)$. We model the noise $r(t)$ as a Gaussian random variable with 0 db mean amplitude, and a variance of σ_r^2. The gain factor ξ is constrained to have values between 0 and 1.

We assume the message to be uniformly sampled at the receiver, at a sampling rate high enough to avoid aliasing problems. Thus our actual data is a set of N points $y[t_i], i = 1, \ldots, N$.

In this situation, and assuming a uniform prior in [0, 1] for ξ, and an improper prior for σ_r^2, the posterior distribution for ξ, given data $y[t_i]$, is

$$p(\xi|y, \Theta) = (2\pi\sigma_r^2)^{-N/2} exp\left[-\sum_{i=1}^{N} \frac{(y[t_i] - \xi x[t_i])^2}{2\sigma_r^2}\right] \tag{2}$$

Under $H_0 : \xi = 0$, the posterior is

$$p_{H_0}(\xi|y, \Theta) = (2\pi\sigma_r^2)^{-N/2} exp\left[-\sum_{i=1}^{N} \frac{y[t_i]^2}{2\sigma_r^2}\right] \tag{3}$$

In the FBST framework, the evidence against H_0 is defined as the integral of the posterior distribution over the surprise set, defined as the set of points, in the full parametric space, whose posterior values are higher than the maximum posterior under H_0. To calculate the evidence, then, we first need to obtain $\hat{p}_{H_0}$, the maximum posterior under H_0, which in this case is simply the value in (3) calculated at the maximum likelihood estimate for σ_r^2.

To calculate the integral of the full posterior, we use a traditional Metropolis-Hasting algorithm, with a uniform [0, 1] candidate distribution for ξ and an inverse gamma candidate for σ_r^2.

2.2 Simulated Data

To evaluate the FBST performance in the signal detection problem, we simulate a message with the following form: the signal has the functional form in (1), with

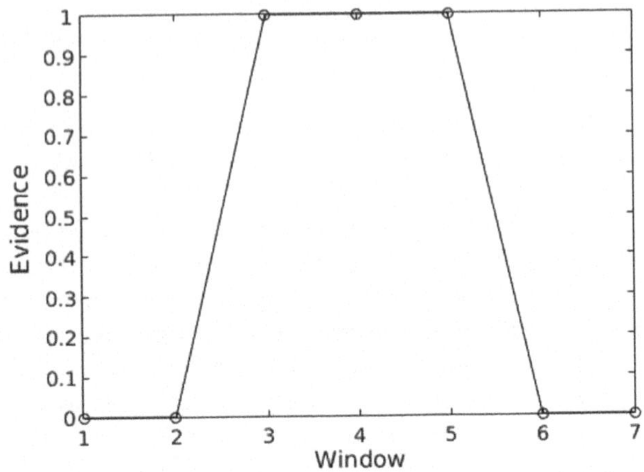

Fig. 1 Evidence values - known signal form

$m = 5, \omega = 60, A_i = \{0.005, 0.004, 0.003, 0.002, 0.001\}, \phi_i = \{-\pi, -\pi/2, 0, \pi/2, \pi\}$. We simulate a 7 s long signal, with $\xi = 0$ during the first and final 2 s, and $\xi = 1$ in the middle 3 s. We assume a sampling rate of 11025 Hz. We will use this same values for both cases (signal form completely known, and signal parameters unknown).

We simulate the message for four different SNR values: 0.9, 1.2, 1.5, 2 (SNR is here defined as the quotient between the deterministic signal's power, and noise power).

The results for the case where the signal form is completely known, and for the different SNR values, are shown in the Fig. 1. The results were the same, regardless of the SNR.

2.3 *Unknown Signal Parameters*

Now, we complicate matters a little bit further and assume that the signal form is known at the receiver, but not the parameters that fully define it. This situation might arise when, for instance, the receiver is not stationary with relation to the transmitter, or if the characteristics of the channel medium change over time.

Our model remains essentially the same as before, except that now the full posterior is 12-dimensional (the parameters are the gain factor, the fundamental frequency, the five amplitudes and five phases). In most real situations, however, there is strong prior information on the signal parameter's. We model this fact by imposing a Gaussian prior on ω, the fundamental frequency. The prior hyperparameters used were $\mu_\omega = 50, \sigma_\omega = 10$. Amplitudes and fundamental frequency are constrained to be positive, and phases lie in $[-\pi, \pi]$ by symmetry considerations. For these parameters and also for the signal noise variance, uninformative priors were adopted.

Given the prior distribution on ω, the new posterior has the form

$$p(\xi, \Theta|y) = (200\pi)^{-1} exp\left[-\frac{(\omega-50)^2}{200}\right] \times \tag{4}$$

$$(2\pi\sigma_r^2)^{-N/2} exp\left[-\sum_{i=1}^{N}\frac{(y[t_i]-\xi x[t_i])^2}{2\sigma_r^2}\right] \tag{5}$$

This time the evaluation of the posterior integral is not so straightforward; the parametric space is multidimensional, and there might me many local maxima in the posterior. Actually, the signal model we adopted is guaranteed to possess at least m local maxima, for $\omega = \omega_0, 2\omega_0, \ldots, m\omega_0$, where ω_0 is the true value of the fundamental frequency.

Given these characteristics of the problem, we choose to apply an evolutionary strategy to sample from the posterior, namely the DiffeRential Evolution Adaptive Markov Chain (DREAM) method of ter Braak and Vrugt [7]. This method consists in initializing a number of parallel Markov chains, which evolve dynamically by taking steps in a random direction given by the difference between one (or more) pair of chains. The method preserves ergodicity of the chain by application of the usual Metropolis acceptance ratio. This method is specially well suited for multimodal distributions; details can be found in [7, 8]. We use a version of the algorithm implemented in MATLAB by ourselves.

Again, the optimization step involved in the FBST calculation is immediate, since under H_0 the parametric space is one-dimensional and the maximum posterior is obtained by using the maximum likelihood estimate of σ_r^2. We apply the DREAM algorithm using 7 parallel chains (which as a side effect allows us to monitor the chain's convergence using, for instance, Gelman and Rubin's $\hat{R}$ statistic [9]), and sample 15.000 points after a burning period of 15.000. The results are shown in the Fig. 2.

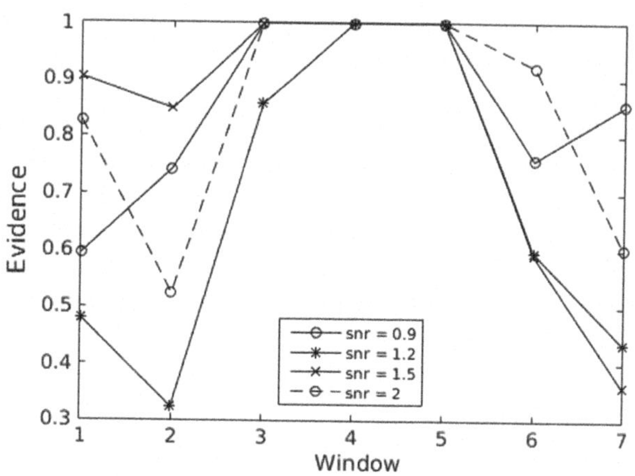

Fig. 2 Evidence values - unknown signal parameters

The method is thus very efficient in pointing out the signal's presence (rejecting a false H_0). However, it gives high values for the evidence against H_0 when it is true. This is caused by the generality of the model we adopted for our signal. We comment further on this fact in the next section.

3 Application to Soundscape Data

Soundscape data are audio recordings made by one or more hydrophones (acoustic recording devices that work underwater). This kind of data is used, among other things, to monitor the traffic of vessels (military or not) and to study the behaviour of marine species.

From the past 10 years, the Acoustics and Environment Laboratory (LACMAM) at EP-USP has been developing technology in the area of subaquatic acoustics. One of these technologies is the OceanPod [10], a hydrophone capable of 3-month continuous recordings, with a frequency band of 5–24 kHz.

One such hydrophone has been installed at a 20 m depth in the region of the *Laje de Santos* park, at the city of Santos in the Brazilian coast. This park is a marine preservation area, with the abundant presence of several marine species. The hydrophone recorded 3-months of sound before its retrieval by the LACMAM's team. The OceanPod mission has being repeated four times already, with a total of 1-year recording time.

In possession of these recordings, the laboratory has been using it with many different goals [11, 12]. One of these goals is to aid the development of a ship detection algorithm: since the park is a state preservation area, it is forbidden to fish in the park's area (actually, it is forbidden to even navigate through the park with fishing equipment inside the boat).

Nevertheless, given the abundance of fish in the park, many fishing boats disobey the park's regulation, specially at late hours of the night. Since the park's borders are at a 40 km distance from the coast, fiscalization is costly. Thus, the park administration, in a combined effort with the laboratory, intend to use the hydrophone's data to improve their fiscalization policies.

The problem of ship detection in sound signals is an old and much studied problem [4–6]. Recent work on the subject proposes the use of classification algorithms such as neural networks to identify the presence of ships. However, this kind of classification algorithm demands a large annotated sample, with which algorithms can be trained. This means that the researcher must either know beforehand the times of ships' passages, or else manually (auditively?) inspect the 3-month signal in order to separate and annotate samples. This is a demanding task, since the passage of boats is not very frequent. Also, listening to 3-month recordings of subaquatic sounds might be a rather dull job.

In order to help the preprocessing of these data, we propose in this paper a non-supervised classification algorithm, that can be run through the samples to select samples where it is at least highly likely that a ship was passing in the hydrophone's vicinity.

This task is similar to the detection problem we presented in the previous sections. However, this setting is a much more complicated one.

First of all, there is the functional form of the desired signal. As noted already, the literature of acoustic ship signature indicates that a ship's noise has mainly two components: a tonal component, given by a sinusoidal signal with a fundamental frequency and several harmonics, and a broadband noise component. The fundamental frequency of the tonal component, as received by the hydrophone, can vary depending on the ship's speed and direction of movement, and several other factors involving the ocean and wind conditions, the specific features of the ship's engine and propeller, etc. For the broadband noise, the situation is even worse, since no well-accepted functional form is known for this component, which is caused by many factors, including (but not limited to) cavitation effects from the propeller.

Even if we can find a suitable parametric model for the ship's noise signal, there's the problem of background noise in the recording. Noise, here, is taken to mean anything but the signal of interest; so it might include fish vocalizations, snapping shrimp and barnacle noise, the sound of waves, rain, etc. Worst of all, many biological sources of noise have precisely the same spectrum form as the tonal component of a ship's noise, namely a sequence of evenly spaced delta functions in the log-frequency domain.

As a first approach, we replicated the model used in the simulated data, where we test the presence of a tonal signal with m harmonics, against random, white noise. This approach failed miserably; the signal recorded by the hydrophone was far from being well described by a Gaussian noise component, and this first algorithm showed a profusion of false positives. It became then evident that a more precise model was needed.

To build this new model, we noticed an important difference in the spectrum taken from two different kinds of ships: Fig. 3 shows a spectrogram of the noise radiated by a large vessel, moving with close-to-constant speed, and at a large distance from the hydrophone. We see the equally spaced spectral lines almost parallel to the x-axis, mainly in the low frequency (20–500 Hz) band. It is known that low-frequency sounds are less attenuated by the ocean than high-frequency ones. Such sounds can then be detected at large distances, as is the case with the example below.

Figure 4, on the other hand, shows the spectrogram of a small vessel approaching the hydrophone with non-zero acceleration. The signal to noise ratio is much greater in this case, and also we see that the spectrum is distorted, showing negatively sloped lines in the spectrogram.

Since the actual goal of the analysis is to detect small, quicker vessels as the ones in Fig. 4, and since most biological acoustic signatures have the same functional form as (1), our next model thus incorporates the fact that *the fundamental frequency in*

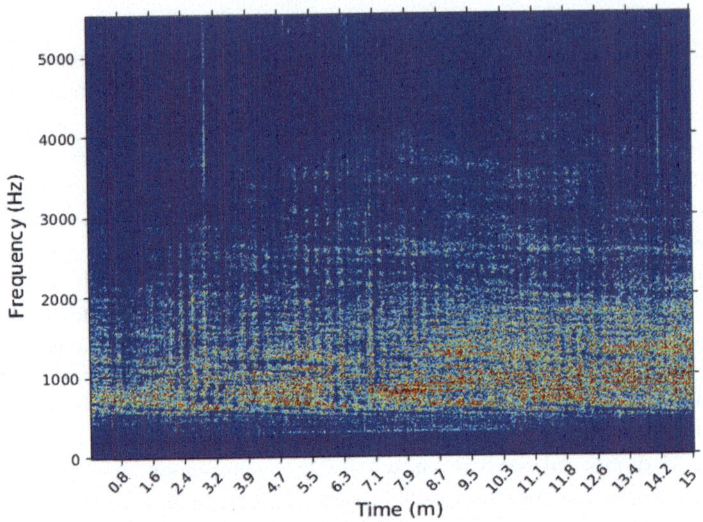

Fig. 3 Spectrogram for large boat

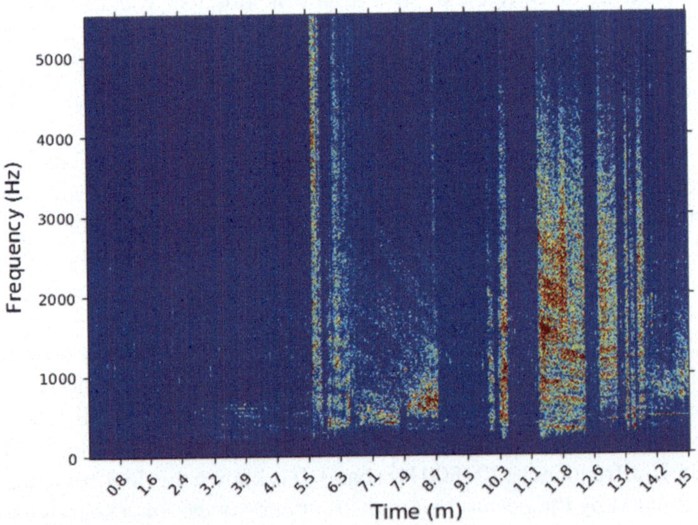

Fig. 4 Spectrogram for small boat

(1) *is time-dependent* for the kinds of events we want to detect, but constant to either large ships moving slowly and far away, or sounds of a biological nature. Our full model for the small vessel then becomes

$$x[t_i] = \sum_{h=1}^{m} A_i cos\ (2\pi h\omega(t_i)t_i + \phi_i) \tag{6}$$

where

$$\omega(t_i) = \omega_0 + \delta t_i \tag{7}$$

We use, thus, a linear function of time for the fundamental frequency of the ship's radiated noise. The full model then becomes

$$y[t_i] = x[t_i] + r[t_i] \tag{8}$$

and our new null hypothesis is $H_0 : \delta = 0$. This model, we expect, will differentiate between *statical sources* and *moving* ones. Incidentally, this might help us to detect more specific events, namely the *approximation* and *departure* of boats, rather than a stationary boat with engine turned on. In terms of aiding fiscalization in the park, this might be of greater interest than detecting any kind of ship-related events whatsoever.

Again, we model prior information available on the ship's fundamental frequency with a Gaussian prior, with $\mu_\omega = 40$, and $\sigma_\omega^2 = 25$. The reason for such a precise prior distribution is twofold: first of all, it helps to prevent the MCMC algorithm from wandering to much in the parametric space, helping it to avoid the inevitable local maxima at integer factors of the true fundamental frequency. Also, there is plenty of the literature in the subaquatic acoustic signature of small ships, and this literature points to fundamental frequencies usually in the range of 20–40 Hz.

Thus the posterior for this problem has the same form as in (4). To calculate the evidence against H_0, we first obtain the maximum posterior under H_0, applying a combination of the DREAM method and an interior-point optimization algorithm. We start 20 parallel chains, run it for a small number of iterations, and then apply the optimization algorithm to the maximum point of each chain. We then simulate again the 20 chains, using as starting values the maximum points, and repeat this procedure until convergence.

After obtaining the maximum posterior under H_0, we run the DREAM algorithm in the full parametric space to estimate the evidence value. We run the chains for 30.000 iterations, discarding the first 15.000.

There remains the choice of m, the number of fundamental harmonics. The number of harmonics in a ship's radiated noise can be as high as 20, if the signal has enough power. In our tests below, we apply the algorithm using $m = 7$ and $m = 10$.

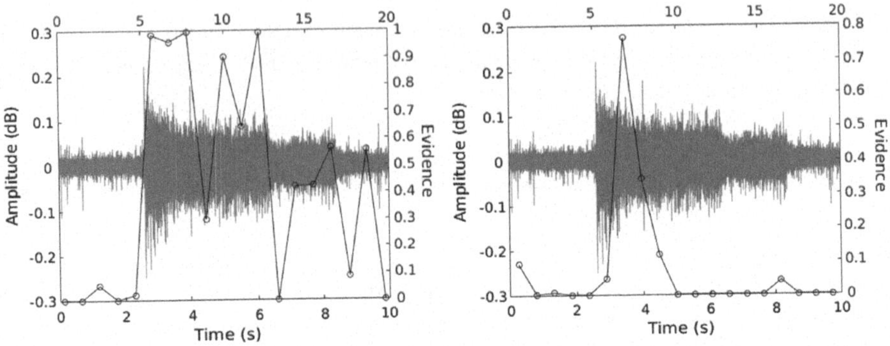

Fig. 5 Evidence values - $m = 7$ (LHS) and $m = 10$ (RHS)

3.1 Results

To test this model, we use a sample where it is known that small boats were passing.[1]

We calculate the evidence against H_0 (i.e. the evidence for the presence of a ship) using samples of 0.5 s. The hydrophone samples the signal at a 11.025 Hz, which gives us a total of 5.516 data points for each window. The results for a 10 s long signal, using non-overlapping 0.5 s samples, are shown in Fig. 5.

As we see in the left-hand side of Fig. 5, the problem of false positives was greatly reduced with the new model using $m = 7$. The first non-zero value for the evidence of a ship's passage is in the window starting at second 3, during which the signal first appears. After that, the evidence stays high for 3 s, falling to 0 again when the signal power falls considerably. After that we see evidence for the signal presence rise again. There is one possible false positive at the window between 8, 5 and 9 s, but auditive inspection of the signal shows a small occurrence of the engine sound at this point.

Using $m = 10$, the sensitivity of the test drops, and we see positive evidence for the presence of a ship only around second 3 in right-hand side of Fig. 5. This is due to the high dimensionality of the parametric space under H_0 (22 parameters) which allows for high posterior values under H_0.

Finally, we also applied our model in a sample of a large vessel, the one represented in the spectrogram Fig. 3. As expected, the evidence given by our model was less than 0.01 in all 0.5 s window extracted from that signal. This is another indication of the potentiality of our method in the detection of small vessels against other events, specifically against bigger boats in cruising speed.

[1] This was possible since touristic boats are allowed in the park for diving visits, and we happen to know that during weekends they are likely to be near the park.

4 Final Remarks and Future Work

The first goal of this paper was to evaluate the performance of the FBST framework in the task of signal detection. Being specially designed to calculate the evidence for sharp hypothesis, the FBST is a natural choice of tool for this job. Using simulated data, we confirmed that the FBST is a promising technique to be used in this kind of problems.

Our second goal was to apply the framework to a real data bank, namely audio recordings from a hydrophone. In this case, we wanted to design an algorithm that was able to preprocess subaquatic acoustic data, indicating sections of the audio that are highly likely to register the passage of small vessels. As we saw in the last section of the paper, by specifying a proper model for the ship's radiated noise, we obtained promising results with the FBST: the evidence values for our models are specially suited to detect the presence of rapidly moving vessels at a small distance from the hydrophone, which meets well the practical requirements for the problem at hand.

Also, the proposed framework is flexible: the models for the signal form can be modified to reflect different kinds of events, and the number of harmonics in the model can be used to adjust the evidence values for different values of signal power. Prior information can be incorporated easily in the model to adjust it for specific kinds of vessels, particularly by the previous estimation of the fundamental low-frequency of ships of interest using pre-annotated samples.

There are a few drawbacks, however; first of all, the algorithm relies on a MCMC technique, which in turn demands a large computing time until convergence. The computation of evidence for a 0.5 s window, with a model of 17 parameters, took roughly 15 min to complete (including both optimization and integration steps) in a Pentium Quadricore 1.6 GHz, 8 Mb RAM home computer, and with a serial algorithm running in a single core. However, since this method is aimed as a preprocessing tool, this is not a very serious drawback, and there are many ways to improve the performance of the algorithm, which we intend to investigate further on future works.

References

1. Daly, R.F., Rushforth, C.K.: IEEE Transaction on Information Theory **11**(1), pp. 70–76
2. Pereira, C.A.B., Stern, J.M.: Entropy **1**, pp. 99–110
3. Ogden, G.L., et al.: The Journal of the Acoustical Society of America, **129**, p. 3768
4. Chung, K.W., Sutin, A., et al.: Advances in Acoustic and Vibration **2011**
5. Kozaczka, E., Grelowska, G.: Acta Physica Polonica A **119** pp. 1009–1012
6. McKenna, M.F., Ross, D., et al.: J. Acoust. Soc. Am. **131**(1), pp. 92–103
7. ter Braak, C., Vrugt, J.: Statistics and Computing **18**(4), pp. 435–446
8. Vrught, J.: Environmental Modelling and Software **75**, pp. 273–316
9. Brooks, S.P., Gelman, A.: Journal of Computation and Graphical Statistics **7**(4), pp. 434–455
10. http://www.lacmam.poli.usp.br/Submarina.html. Accessed 13 June 2017
11. Padovese, L.: Marine Pollution Bulletin **105** pp. 65–72
12. Sánchez-Centriz, I., Padovese, L.: Ecological Information **38** pp. 31–38

Bayesian Support for Evolution: Detecting Phylogenetic Signal in a Subset of the Primate Family

Patricio Maturana Russel

Abstract The theory of evolution states that the diversity of species can be explained by descent with modification. Therefore, all living beings are related through a common ancestor. This evolutionary process must have left traces in our molecular composition. In this work, we present a randomization procedure in order to determine if a group of five species of the primate family, namely, macaque, guereza, orangutan, chimpanzee, and human, has retained these traces in its molecules. First, we present the randomization methodology through two toy examples, which allow to understand its logic. We then carry out a DNA data analysis to assess if the group of primates contains phylogenetic information which links them in a joint evolutionary history. This is carried out by monitoring a Bayesian measure, called marginal likelihood, which we estimate by using nested sampling. We found that it would be unusual to get the relationship observed in the data among these primate species if they had not shared a common ancestor. The results are in total agreement with the theory of evolution.

Keywords Phylogenetic signal · Randomization · Marginal likelihood · Nested sampling

1 Introduction

The theory of evolution states that the diversity of species can be explained by descendants with modification. Darwin [3] was able to provide evidence in favor of his theory, despite the limitations at that time. Nowadays, technology is a powerful tool which allows to generate a huge quantity of evidence in favor of this theory. The

---

P. Maturana Russel (✉)
Department of Statistics, University of Auckland, Private Bag 92019, 1142 Auckland, New Zealand
e-mail: p.russel@auckland.ac.nz

A. Polpo et al. (eds.), *Bayesian Inference and Maximum Entropy Methods in Science and Engineering*, Springer Proceedings in Mathematics & Statistics 239, https://doi.org/10.1007/978-3-319-91143-4_20

support comes from different areas, for instance, Molecular Biology, Paleontology, Biogeography, Biochemistry, and Phylogenetics. The present article is located in the latter which is the study of the evolutionary relationship among groups of organisms based typically on molecular sequencing data.

As in any other field, in phylogenetics data analysis is performed mainly under two statistical approaches: Frequentist and Bayesian. The latter has gained ground in phylogenetics due to its flexibility to deal with large dataset with the complex evolutionary models. Studying a particular Bayesian measure, the probability that the data have been generated from a tree-like evolutionary model, we asses whether the patterns of evolution in the molecular sequencing data (DNA) could reasonably arise due to chance. In other words, if the theory of evolution was right, the sequence alignments should contain information which connects the species from where the DNA was taken. If it is so, we asses if these patterns can be due to chance acting alone.

To evaluate if these patterns emerged from the molecular data is due to chance, we use a method known as *randomization*. This method allows to detect if the data contain nonrandom information that links the species in a common evolutionary history. It performs by comparing a statistic obtained from the data to the distribution of the same statistic obtained from a set of functional data, generated randomly from the original one, which consequently does not contain any phylogenetic signal. If the data support evolution, their information should be significant enough to be differentiated for that one obtained just by chance. This technique was already proposed by [1] in a nonparametric framework.

This article aims to show in a practical way how the evolution theory is supported for a logical method as it is randomization by studying a Bayesian quantity: the marginal likelihood. First, two toy examples are presented as means to understand the method and then an application on a real dataset which contains part of the primate family is given in order to detect phylogenetic signal. The description of the statistical methods and phylogenetic models are omitted but the respective references are given.

2 Randomization

Randomization is a method used to assess the effect of certain factor or treatment on a variable of interest. This is carried out by studying the properties of the distribution of a statistic calculated from randomized datasets. Each of these functional datasets is generated by randomly assigning the observations to the factor/treatment, i.e., the experimental units are relabeled. The new data will not show any effect of the factor on the variable. The factor is obviated and any difference between its levels is caused by chance. This is analogous to shuffling playing cards to eliminate any kind of intervention.

The method compares the statistic of the original data with the distribution of the same statistic of the randomized data. Such statistic, for example, can be mean, median, mode, or variance. This method does not need to make any assumption about the population, it just works with the data to make inferences. Assumptions such as normality or equal variances. The following example helps to understand the method.

2.1 *Toy Examples*

Consider that we have the marks of a test for 10 students differenced by the method of study (A or B) to which the students were randomly allocated. The marks are presented in percentage and are shown in Table 1. The objective is to determine which of the methods of study is more effective. Both examples are developed at the same context but they will differ in the dataset. They could have been treated analytically, but to illustrate randomization in a general way we have used simulations. They just have didactic purposes and clear patterns have been arbitrarily assigned.

2.1.1 Example 1

Clearly, method A presents higher marks than method B (see Table 1, Example 1). This can be also noticed comparing their means (see Table 2). Apparently, method A is better than B. But can this be due to chance acting alone? Randomization can give us an idea.

We generate a new dataset where each mark is assigned randomly to either method A or B. The number of marks per method is set to 5, as in the original dataset. Then,

Table 1 Data for the toy examples

Example 1		Example 2	
Method A	Method B	Method A	Method B
92.5	55.2	18.49	94.35
99.8	32.0	70.24	12.92
75.8	49.6	57.33	83.34
82.4	68.3	16.81	46.80
93.2	69.3	94.38	55.00

Table 2 Means for each method according to the example. "Difference" depicts the subtraction between the means of Method A and B

	Method A	Method B	Difference
Example 1	88.74	54.88	33.86
Example 2	51.45	58.48	−7.03

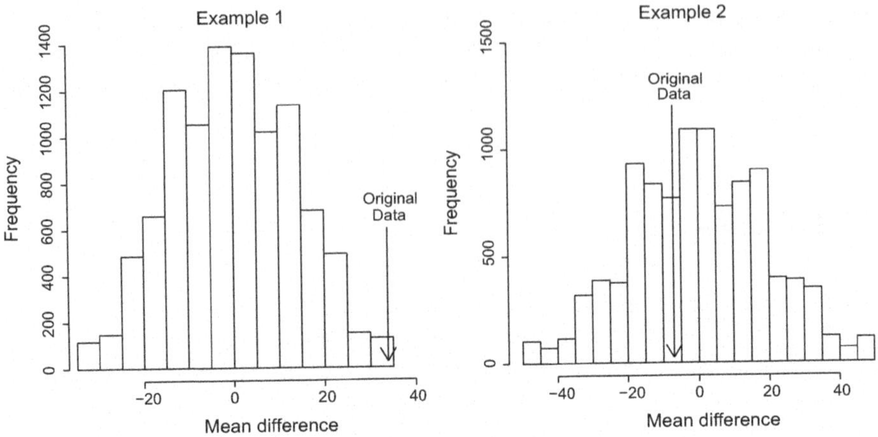

Fig. 1 Distribution of the mean of the randomized datasets for the toy examples. In Example 1, the observed difference is unlikely to have happened under chance acting alone. On the other hand, in Example 2, this difference could have been just due to chance and nothing to do with which method was used

the difference between the means is calculated and registered. This procedure is repeated 10,000 times. The mean differences are plotted in Fig. 1.

We can see that the mean difference is around zero. This is expected because the difference in means is just due to chance. The effect of the method has been obliterated. The observed difference, that was calculated from the original data, is 33.86 and located in the right extreme of the distribution. In case that chance is acting alone, it would be unusual to get an observed difference as big as that observed in the data. Assuming a well-designed experiment, we conclude that method A effectively yields better results than B on average.

2.1.2 Example 2

Now consider the data given in Table 1 for Example 2. In this case, both methods yield apparently similar results. The difference between their means is just 7.03 (see Table 2, Example 2). But again, can this be due to chance acting alone? To give an answer we repeat the procedure in Example 1. The results are shown in Fig. 1.

The distribution of the differences between the means of method A and B for the randomized datasets has its center around zero and is relatively symmetric. Similar characteristics were found in Example 1 because the potential effect of the method of study has been wiped out in both examples. The observed difference −7.03 is near its center. When the chance is acting alone, this difference is highly probable, unlike Example 1, where the difference in means was unusual under chance acting alone. Thus, assuming a well-designed experiment, we could claim that the methods of study yield similar results, on average, and the observed difference is just due to chance acting alone.

In these cases, we compared the effect of the method of study on the mark mean, but we could have studied any other characteristic, for instance, standard deviation, median, or a specific probability. In strict rigor, the comparison should be carried out by using an appropriate statistical test, for instance, a t-test. In the next case, we will study the probability of the data given the model in order to detect phylogenetic signal in a molecular dataset of five primates.

3 Phylogenetic Analysis

Now, we apply the same concept in order to analyze if a molecular dataset of a group of primates has information about their common evolutionary history. This is a subset of a dataset which has been previously analyzed in the literature [7]. This subset contains five kinds of primates: macaque, guereza, orangutan, chimpanzee, and human. The alignment corresponds to mitochondrial DNA which has length of 15,727 sites. To wit, the DNA is composed by four nucleobases: adenine (`A`), cytosine (`C`), guanine (`G`), and thymine (`T`). An extract of the data is shown in Fig. 2. The relationship among these species is uncontroversial and can be visualized as the tree shown in Fig. 3. Human and chimpanzee share a more recent common ancestor. This makes them more closely related. Orangutan is also part of this clade, but with a farer ancestor. Macaque and guereza form another clade. All the species are connected through their most recent common ancestor, which is located in the root of the tree (left vertex of the tree) (Fig. 3).

In order to eliminate any kind of correlation in the dataset, we permute each site generating a new dataset. In other words, each site is reordered randomly. For instance, site 2 = `(C,T,C,T,T)` displayed in Fig. 2, can be permuted as `(T,C,T,T,C)`. The theory of evolution [3] states that all organisms are related through common ancestors. So, if the data were generated by a tree, they should contain this information, unlike in case the data are randomized.

In the previous examples, the mean difference was studied, but now we will study the probability of the data given the model, which will be referred to from now as *marginal likelihood*. Phylogenetic deals with very small probability values, so it is

	1	2	3	4	5	6	7	8	9	10	11	12	13	14	15	16	17
Human	C	C	T	A	A	A	A	C	C	C	G	C	C	A	C	A	T
Chimpanzee	C	T	T	A	A	A	A	C	C	C	T	C	C	A	C	T	T
Orangutan	C	C	T	A	A	A	A	C	C	C	T	C	C	A	C	A	T
Guereza	C	T	C	A	A	A	A	C	C	C	G	C	A	A	C	C	T
Macaque	C	T	T	G	A	A	A	C	C	C	T	C	A	A	C	A	T

Fig. 2 Extract of the mitochondrial DNA for five species of primates. Each column represents a site

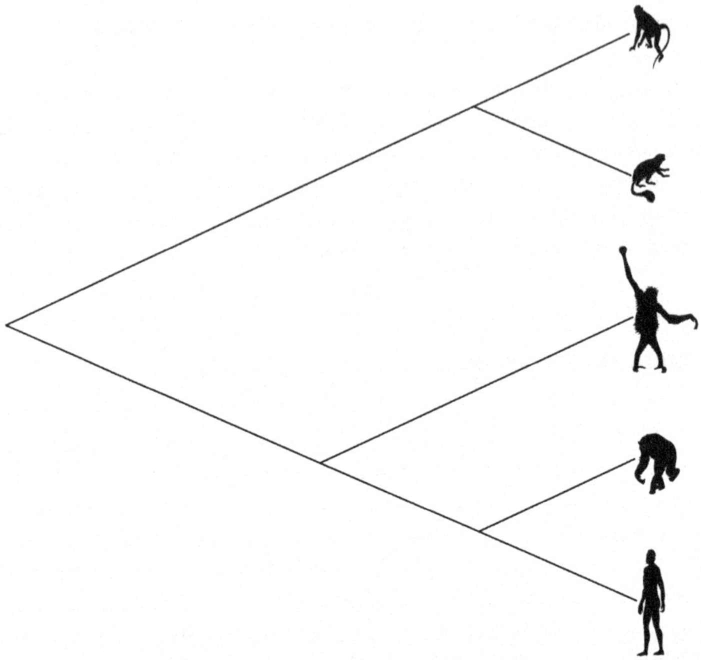

Fig. 3 Evolutionary relationship among five members of the primate family. From the top: macaque, guereza, orangutan, chimpanzee, and human. These species are related via common ancestry

convenient to work with log values. The evolutionary relationship among the species is modeled by the tree, which is displayed in Fig. 3. This tree represents the factor to be tested in this analysis, similar to the method of study that was tested in the previous example. We describe the evolutionary process along the tree assuming a GTR+Γ_4 model, which is the most general time reversible model. A good readable material about these models is given in [9]. The prior distributions on the parameters involved in the model are defined in Appendix 1.

The calculation of the marginal likelihood is a challenging problem in phylogenetics, even in simple models. Therefore, it requires a numerical approximation. Here, we estimate it via Nesting Sampling [8], algorithm introduced to phylogenetics by [4]. Details of the estimation process are given in Appendix 2.

We generate 1000 randomized datasets and calculate, for each one, their log-marginal likelihoods. Also, we estimate this quantity for the original dataset. The results are shown in Fig. 4 and the descriptive statistics in Table 3.

The estimates for the randomized data fluctuate between −53484 and −51675 with a mean of −51737. On the other hand, the observed log-marginal likelihood estimate is −49658 (with a standard deviation of 0.73). This is located at the right side of the distribution of the log-marginal likelihoods for the randomized datasets, approximately, 26 standard deviations away from the mean.

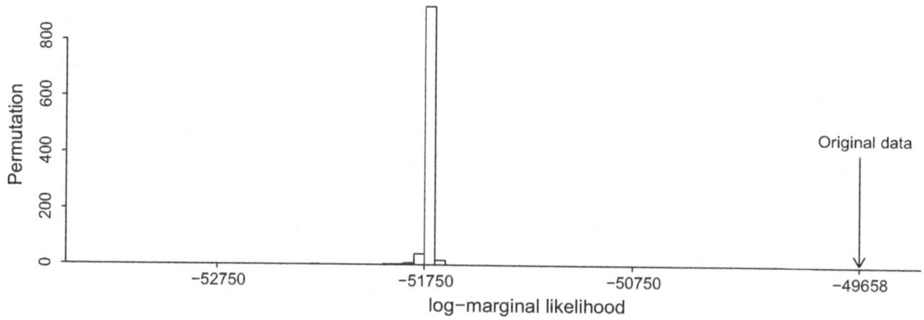

Fig. 4 Log-marginal likelihood of the observed data compared to the distribution of this quantity obtained from randomized datasets. The observed log-marginal likelihood is much higher than that one would expect under chance acting alone. The information contained in the molecular data of this primate family is more highly probable of being obtained due to common ancestry than just due to chance

Table 3 Descriptive statistics for the estimated log-marginal likelihoods from the randomized datasets

	Minimum	Mean	Std. Dev.	Maximum
Randomized data	−53484	−51737	80.03	−51675

Following the reasoning of Example 1, we conclude that it would be unusual that an observed log-marginal likelihood would be as large as the one observed in the data when chance is acting alone. The probability that the original data has been generated by the tree structure is much higher than the randomized datasets have. This means that the patterns in the DNA are more likely to be explained by the tree-like structure than just to occur due to chance. In other words, the data contain phylogenetic information that cannot be explained only by chance. The mitochondrial DNA has retained the common evolutionary history of these species, and our analysis has shown that it would have been highly unlikely to obtain this disposition of the bases in the data as a result of pure chance. This is evidence which supports the tree structure behind the evolutionary history of these 5 species of primates that is consistent with the theory of evolution.

4 Conclusion

A brief introduction to randomization method has been given. Two toy examples have been studied to explain its logic. Example 1 represented a case in which the treatment had an effect on the studied characteristic, while Example 2 presented a case when chance was acting alone. Both examples aimed to set up the logic which is used in the analysis of a primate family dataset.

We analyzed a real dataset of five species of primates under a Bayesian statistical approach and used randomization to detect if this contained nonrandom information. The data were permuted to eliminate any kind of phylogenetic signal, and then the probability that these randomized data came from the tree model was calculated (marginal likelihood). This procedure was repeated several times, generating a distribution for the estimates. The probability for the original dataset was much higher than the maximum value of the same value of the randomized data. We would not expect such a probability if there was no tree signal. Therefore, we concluded that chance was not acting alone and these species have a tree-like relationship. The presence of a hierarchical structure provides evidence for descent from common ancestry.

The results given here are consistent with the theory of evolution and are added to the huge amount of evidence which supports it. For instance, 28 morphological datasets were analyzed and are in favor of the tree-like models [1]; in addition sequence data for 5 proteins from 11 species contains similar phylogenetic information [5]. In this line, we have shown that Bayesian inference provides the means to detect this phylogenetic signal through the marginal likelihood. In practice, it is unusual to find data that completely lack hierarchical structure [2] and the data analyzed here were not the exceptions.

All the analysis and plots have been produced in R-project [6].

Appendix 1

We analyze the dataset assuming a GTR + Γ_4 model and consider the following prior distributions on the parameters involved in the analysis:

- Branch lengths: $t_i|\mu \sim \text{Exp}(1/\mu)$, for $i = 1, \ldots, 8$, with $\mu \sim$ Inverse-Gamma (3,0.2).
- Relative rates: $q_i|\phi \sim \text{Exp}(\phi)$, for $i = 1, \ldots, 5$, with $\phi \sim \text{Exp}(1)$.
- Base frequencies: $\pi \sim$ Dirichlet(1,1,1,1).
- Gamma shape parameter: $\lambda \sim$ Gamma(0.5,1).

For more information about the parameters involved in the phylogenetic analysis, see [9].

Appendix 2

Nested sampling [8] is a Bayesian algorithm to estimate mainly the marginal likelihood. It requires a tunning parameter called *active points*. The precision of the estimate depends on the number of active points. The higher it is, the more accurate the estimate and the higher the computational cost are.

To estimate the observed marginal likelihood, we use 100 active points. This yields a standard deviation of 0.73 of the log-marginal likelihood estimate. For the 1000 randomized datasets, we use five active points in order to get a quick picture of their log-marginal likelihood distribution.

References

1. Archie, J.W.: A randomization test for phylogenetic information in systematic data. Syst. Zool. **38**, 239–252 (1989)
2. Baum, D.A., Smith, S.D.: Tree Thinking: An Introduction to Phylogenetic Biology. Roberts and Company Publishers, Greenwood Village (2012)
3. Darwin, C.: On the Origin of Species by Means of Natural Selection, or the Preservation of Favoured Races in the Struggle for Life. J. Murray, London (1859)
4. Maturana Russel, P., Brewer, B.J., Klaere, S., Bouckaert, R.: Model selection and parameter inference in phylogenetics using nested sampling. arXiv:1703.05471v3
5. Penny, D., Foulds, L.R., Hendy, M.D.: Testing the theory of evolution by comparing phylogenetic trees constructed from five different protein sequences. Nature **297**, 197–200 (1982)
6. R Core Team.: R: A Language and Environment for Statistical Computing. R Foundation for Statistical Computing, Vienna, Austria (2015)
7. Roos, C., Zinner, D., Kubatko, L.S., Schwarz, C., Yang, M., Meyer, D., Nash, S.D., Xing, J., Batzer, M.A., Brameier, M., Leendertz, F.H., Ziegler, T., Perwitasari-Farajallah, D., Nadler, T., Walter, L., Osterholz, M.: Nuclear versus mitochondrial DNA: evidence for hybridization in colobine monkeys. BMC Evol. Biol. **11**, 77 (2011)
8. Skilling, J.: Nested sampling for general Bayesian computation. Bayesian Anal. **1**, 833–860 (2006)
9. Yang, Z.: Molecular Evolution: A Statistical Approach. Oxford University Press, Oxford (2014)

A Comparison of Two Methods for Obtaining a Collective Posterior Distribution

Rafael Catoia Pulgrossi, Natalia Lombardi Oliveira, Adriano Polpo and Rafael Izbicki

Abstract Bayesian inference is a powerful method that allows individuals to update their knowledge about any phenomenon when more information about it becomes available. In this paradigm, before data is observed, an individual expresses his uncertainty about the phenomenon of interest through a prior probability distribution. Then, after data is observed, this distribution is updated using Bayes theorem. In many situations, however, one desires to evaluate the knowledge of a group rather than of a single individual. In this case, a way to combine information from different sources is by mixing their uncertainty. The mixture can be done in two ways: before or after the data is observed. Although in both cases, we achieve a collective posterior distribution, they can be substantially different. In this work, we present several comparisons between these two approaches with noninformative priors and use the Kullback–Leibler's divergence to quantify the amount of information that is gained by each collective distribution.

Keywords Collective posterior distributions · Mixing prior distributions · Mixing posterior distributions · Group decision making · Bayesian inference

---

R. C. Pulgrossi (✉) · N. L. Oliveira · A. Polpo · R. Izbicki
Federal University of São Carlos, Rod. Wasington Luiz, km235, São Carlos, Brazil
e-mail: rafael.catoia@gmail.com

N. L. Oliveira
e-mail: nat.nlo@gmail.com

A. Polpo
e-mail: polpo@ufscar.br

R. Izbicki
e-mail: rafaelizbicki@gmail.com

A. Polpo et al. (eds.), *Bayesian Inference and Maximum Entropy Methods in Science and Engineering*, Springer Proceedings in Mathematics & Statistics 239, https://doi.org/10.1007/978-3-319-91143-4_21

1 Introduction

Bayesian inference is an approach based on the updating knowledge about a phenomenon. This knowledge can always be represented at any moment by a probability function. As new information about the phenomenon becomes available, we learn about it and thus, the knowledge is updated, being Bayes' theorem the tool used for this update.

In more practical statistical terms, let's assume an individual is looking to learn about a parameter (phenomenon). His/her knowledge is updated when data becomes available, with the likelihood function being the function that carries all information the data has about the parameter. The probability density function that translates the individual's knowledge before data observation is called *prior distribution*, whereas after the update, it is called *posterior distribution*.

In many situations, there is more than a single individual who wants to learn about a phenomenon [1, 2, 5], and decisions must be taken at a group level. With that in mind, it is interesting to look at a group's knowledge and perhaps evaluate a distribution that expresses the uncertainty about this phenomenon in a collective way [6]. Considering that each individual expresses his/her knowledge through a probability distribution, a way to combine their uncertainty is by mixing these distributions [3]. There are two approaches for the update: (i) mixing each individual's prior distributions into a single prior distribution and updating it when the data becomes available; or (ii) obtaining the posterior distribution for each individual and then mixing them into a collective posterior.

Although in both cases, we achieve a collective posterior distribution, their results can be different. With that in mind, we present a comparison study for these two methods of combining information and use the Beta-Bernoulli family to compares these two approaches in different settings, highlighting the characteristics of each one.

2 Construction and Comparison of Collective Posteriors

In this section, we review mixture distributions and show how they can be used to combine information from a group of individuals.

2.1 *Mixture Distribution*

A mixture distribution is the probability distribution of a random variable derived by a convex combination of other independent random variables. Given a finite set of density functions $p_1(x), \ldots, p_k(x)$ and weights $w_1, \ldots, w_k$ such that $w_i \geq 0, i = 1, \ldots, k$ and $\sum_{i=1}^{k} w_i = 1$, the mixture distribution, $f(x)$, can be represented by: $f(x) = \Sigma_{i=1}^{k} w_i p_i(x)$.

Focusing on combining the knowledge of a group, it is possible to mix the individual probability functions either before or after data observation. We explore these in the sequence.

2.1.1 Collective Posterior Construction

Our focus is to express the uncertainty of a group through a distribution function obtained by mixing individual distribution functions that represent each individual's knowledge about the unknown parameter θ. There are two ways to mixture the individual distributions: before or after the update. In both approaches, consider $\pi_1(\theta), \ldots, \pi_k(\theta)$ prior distributions from k individuals and $w_1, \ldots, w_k$, with $0 \leq w_i \leq 1$ and $\sum_{i=1}^k w_i = 1$, their respective weights.

By mixing before the update, we construct a collective prior $\pi_c(\theta) = \sum_{i=1}^k w_i \pi_i(\theta)$ by mixing each individual's prior distribution and, when more information becomes available, it is updated through Bayes' Theorem, obtaining a collective posterior $\pi_c(\theta|x)$:

$$\pi_c(\theta|x) = \frac{\sum_{i=1}^k w_i \pi_i(\theta) L(\theta|x)}{\int_\Theta \sum_{i=1}^k w_i \pi_i(\theta) L(\theta|x) d\theta}. \tag{1}$$

Another way to obtain a collective posterior is by updating each prior with the available data, obtaining each individual posterior and after that, we achieve a collective posterior by mixing each individual posterior, obtaining:

$$\pi_c^*(\theta|x) = \sum_{i=1}^k w_i \pi_i(\theta|x). \tag{2}$$

Although in both cases, we achieve a collective posterior distribution, the results obtained by each approach can be substantially different. For instance, the posterior using the mixture of priors approach with weights w is equivalent to the mixture of posteriors with weights w'. Let $c_i = \int_\Theta \pi_i(\theta) L(\theta|x) d\theta$.

$$\begin{aligned} \pi_c(\theta|x) &= \frac{\sum_{i=1}^k w_i \pi_i(\theta) L(\theta|x) \frac{c_i}{c_i}}{\int_\Theta \sum_{i=1}^k w_i \pi_i(\theta) L(\theta|x) \frac{c_i}{c_i} d\theta} \\ &= \frac{\sum_{i=1}^k w_i \pi_i(\theta|x) c_i}{\int_\Theta \sum_{i=1}^k w_i \pi_i(\theta|x) c_i d\theta} \\ &= \sum_{i=1}^k \frac{w_i c_i}{\int_\Theta \sum_{i=1}^k w_i \pi_i(\theta|x) c_i d\theta} \pi_i(\theta|x) \\ &= \sum_{i=1}^k w_i' \pi_i(\theta|x). \end{aligned} \tag{3}$$

In order to compare the two approaches, we obtained the risk function and quantified the amount of information in each density with the Kullback–Leibler divergence.

3 The Beta-Bernoulli Case

In this section, we consider the Beta-Bernoulli conjugation, to construct and show how both methods perform. With that in mind, consider that θ represents a proportion, with $0 \leq \theta \leq 1$ and the knowledge about it a priori can be represented by a $Beta(\alpha, \beta)$ distribution, α, β being hyperparameters. We consider $Beta(\alpha_i, \beta_i)$, with $i = 1, \ldots, k$, to be the ith's individual's prior distribution of θ. Data comes from a Bernoulli distribution, that is, $X|\theta \sim Bernoulli(\theta)$, and the likelihood function for a sample size n is $L(\theta|x_1, \ldots, x_n) \propto \theta^S(1-\theta)^{n-S}$, in which $x_i = 0, 1$ and $S = \sum_{i=1}^n x_i$, $S \in \{0, 1, \ldots, n\}$.

Mixture before the update. Considering (1), first we construct the collective prior;

$$\pi_c(\theta) = \sum_{i=1}^{k} w_i \frac{\Gamma(\alpha_i, \beta_i)}{\Gamma(\alpha_i)\Gamma(\beta_i)} \theta^{\alpha_i - 1}(1-\theta)^{\beta_i - 1}.$$

When data becomes available, we update it, thus obtaining $\pi_c(\theta|x) = \sum_{i=1}^k w_i' \pi_i(\theta|x)$, in which

$$w_i' = \frac{w_i \frac{\Gamma(\alpha_i+\beta_i)}{\Gamma(\alpha_i)\Gamma(\beta_i)} \frac{\Gamma(\alpha_i+\beta_i+n)}{\Gamma(\alpha_i+S)\Gamma(\beta_i+n-S)}}{\left(\sum_{i=1}^k w_i \frac{\Gamma(\alpha_i+\beta_i)}{\Gamma(\alpha_i)\Gamma(\beta_i)} \frac{\Gamma(\alpha_i+S)\Gamma(\beta_i+n-S)}{\Gamma(\alpha_i+\beta_i+n)}\right)}. \tag{4}$$

Mixture after the update. In this approach, we update the individual's prior distribution separately and than mix the individual's posterior distributions, thus obtaining

$$\pi_c^*(\theta|x) = \sum_{i=1}^{k} w_i \frac{\Gamma(\alpha_i + \beta_i + n)}{\Gamma(\alpha_i + S)\Gamma(\beta_i + n - S)} \theta^{\alpha_i + S - 1}(1-\theta)^{\beta_i + n - S - 1} \tag{5}$$

4 Results

In this section, we present comparisons between both methods. The comparisons are not a simulation study; we consider all possible samples in each scenario. The scenarios are characterized by a choice of three prior distributions and we analyze the scenarios for different sample sizes. We consider three sources of prior information in each scenario, all of them representing the knowledge about an unknown proportion θ through a $Beta(\alpha, \beta)$ distribution:

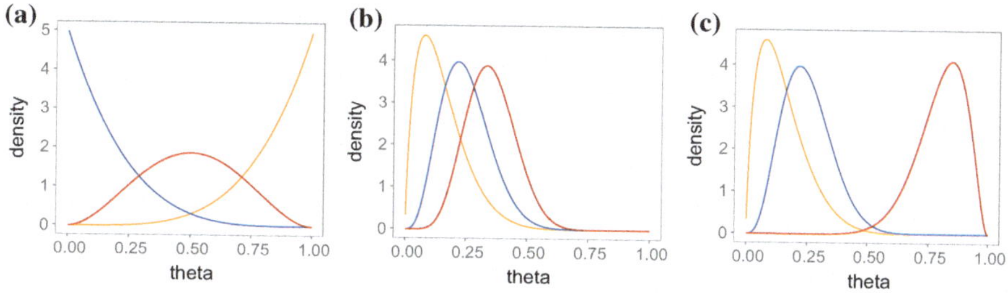

Fig. 1 Prior distributions. **a** is the symmetric scenario, **b** the concentrated scenario and **c** the discordant one

- **[Symmetrical]**. The prior distributions are $Beta(5, 1)$, $Beta(1, 5)$, $Beta(3, 3)$
- **[Concentrated]**. The priors are concentrated on the same region of the parameter space: we choose Beta priors with means 0.15, 0.25 and 0.35, all with 0.01 variance.
- **[Discordant]**. All priors have the same variance, 0.01, but the means are 0.15, 0.25, and 0.8.

All prior distributions are mixed with equal weights, that is, $w_i = 1/3, i = 1, 2, 3$. We use two different sample sizes: 15 and 30 and for each sample size, we evaluate all possible samples. The prior distributions used on each scenario (symmetric, concentrated and discordance scenarios) are graphically represented on Fig. 1.

In order to compare both methods of collective posterior construction, for each scenario and sample size, we present: (1) the posterior expected value and the variance of both methods; (2) the weights w' in order to understand how the weights in the posterior mixture behaves when using equal weights to mix prior distributions; (3) the squared risk function for θ of the estimators $E_{\pi_c}[\theta|x]$ and $E_{\pi_c^*}[\theta|x]$; (4) the Kullback–Leibler divergence between both methods, $\pi_c^*(\theta|x)$ and $\pi_c(\theta|x)$, and two noninformative priors ($Unif(0, 1)$ and Jeffrey's, that is a $Beta(0.5, 0.5)$) and their respective posteriors.

Looking at the first and second columns of Figs. 2, 3, 4 and 5, going from a size 15 to a size 30 sample does not change much of the comparison results. At Fig. 2, we observe that mixing before the update, represented by the blue circles, appears to be more flexible than mixing after the update. Moreover, we verify that when the sample size grows, the expected value of both methods tends to get closer to each other. For both sample sizes and for scenario 1 and 2, the triangles in Fig. 3 are always higher than the circles. In the third scenario (third column), the variance doesn't change much when using posterior mixture, whereas in prior mixture, when data information agreed with most prior distributions, the variance decreased, and when data information agreed with only one prior distribution, variance increased. Figure 4 shows that when using prior mixture with equal weights, the resulting posterior gives higher weights to the source whose information is more similar to the data.

Looking at Fig. 5, we verify that the risk in a sample size of 30 is lower than in a sample size of 15 in all scenarios. At the first column (Symmetric scenario), both methodologies have opposite behaviors: for extreme values of θ, mixture after update

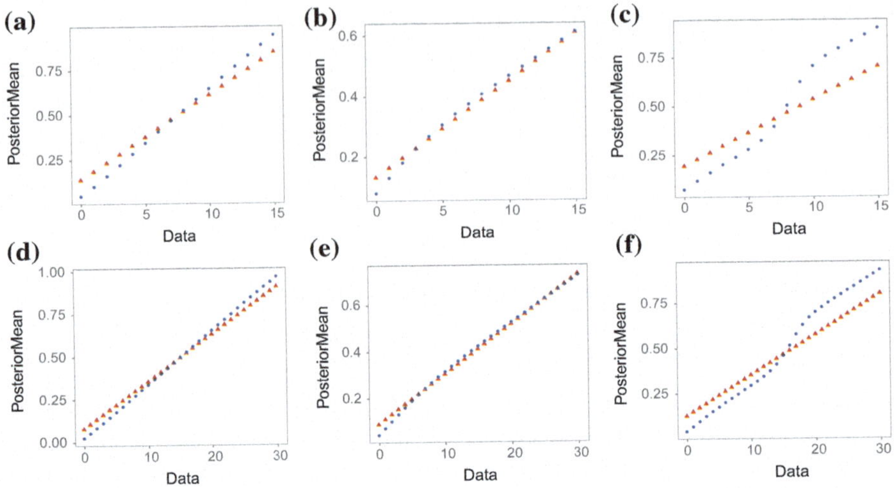

Fig. 2 Posterior mean versus S. Blue circles represent mixture before the update and red triangles, mixture after the update. Figures in the first row (of matrix of figures) consider sample size of 15, while the ones in the second row consider sample size of 30. Columns represent the scenarios: 1_{st} - Symmetric, 2_{nd} - Concentrated and 3_{rd} - Discordant

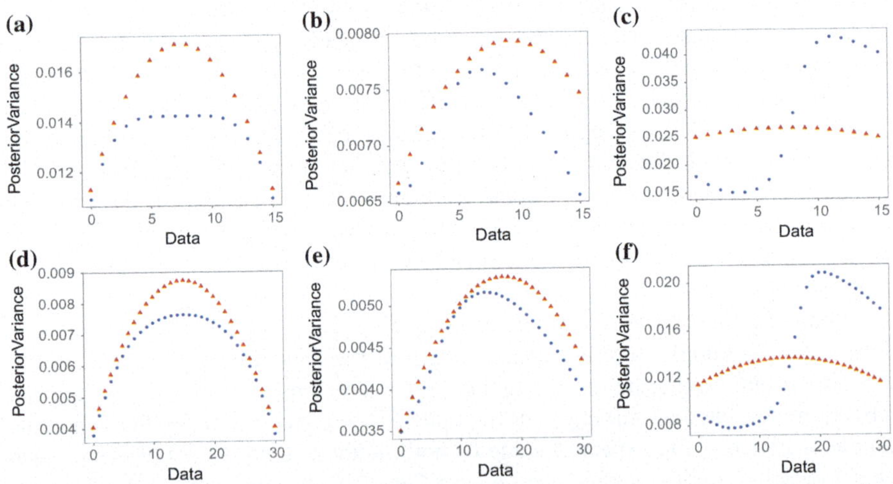

Fig. 3 Posterior variance versus S. Blue circles represent mixture before the update and red triangles, mixture after the update. Figures in the first row (of matrix of figures) consider sample size of 15, while the ones in the second row consider sample size of 30. Columns represent the scenarios: 1_{st} - Symmetric, 2_{nd} - Concentrated, and 3_{rd} - Discordant

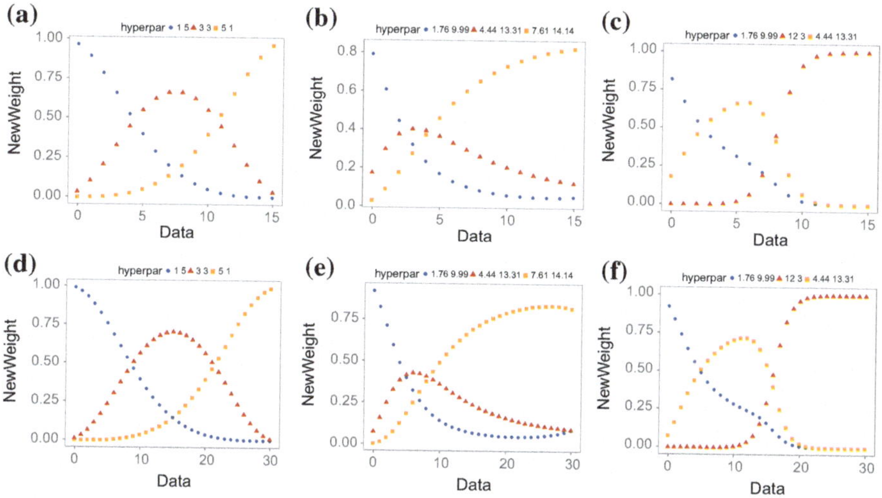

Fig. 4 W_i' versus S. The colors blue, red, and yellow differ the three prior sources. In the matrix of figures, the columns are the scenarios (symmetrical, concentrated, and discordant) and the lines represent sample sizes 15 and 30

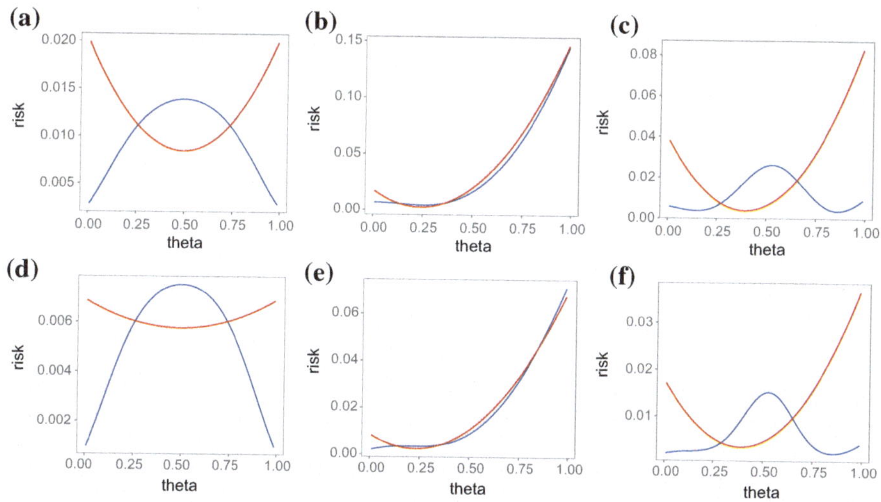

Fig. 5 Risk versus θ. Red line represents the risk of using $E_{\pi_{c*}}[\theta|x]$ and the blue one, the risk of using $E_{\pi_c}[\theta|X]$. The rows of the matrix of figures represent sample sizes (15 and 30) and columns represent the scenarios: 1_{st} - Symmetric, 2_{nd} - Concentrated, and 3_{rd} - Discordant

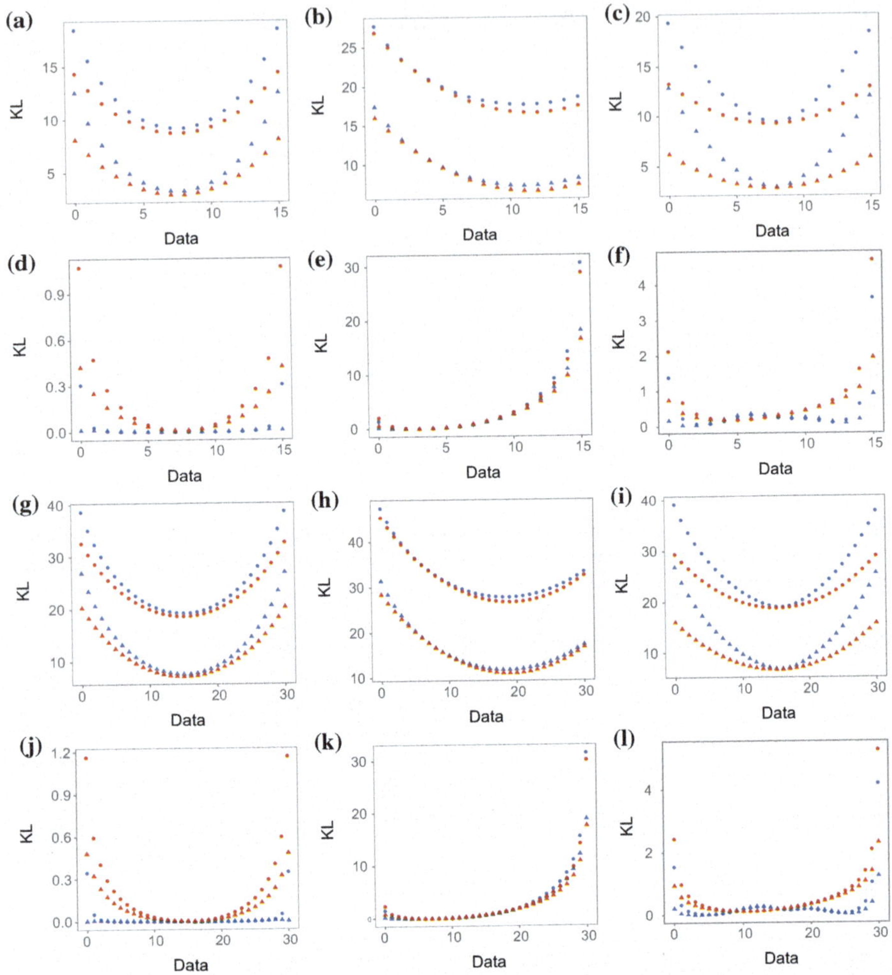

Fig. 6 Kullback Leibler divergence versus S for sample size of 15 (subfigures **a**–**f**) and sample size of 30 (subfigures **g**–**l**). The distributions that are used to compute the KL divergence depends on the color and shape of the points. For plots **a**, **b**, **c**, **g**, **h** and **i**, blue circles is the KL divergence between the posterior from a mixture of priors and the the Jeffrey's prior, blue triangle is the KL divergence between the posterior from a mixture of priors and the uniform prior, red circles is the KL divergence between the mixture of individual posteriors and the Jeffre's prior, red triangle is the KL divergence between the mixture of individual posteriors and the uniform prior. For plots **d**, **e**, **f**, **j**, **k**, and **l**, blue circles is the KL divergence between the posterior from a mixture of priors and the posterior obtained using Jeffrey's prior, blue triangle is the KL divergence between the posterior from a mixture of priors and the posterior obtained using a uniform prior, red circles is the KL divergence between the mixture of individual posteriors and the posterior obtained using Jeffre's prior, red triangle is the KL divergence between the mixture of individual posteriors and the posterior obtained using a uniform prior

(red) have higher risk than the other. For the concentrated scenario (second column), the two risks are quite similar but, for high probability values of θ a priori, mixture before the update presents a slightly lower risk than the other. At last, in the third scenario (third column), mixture before the update shows lower risks than the other for non-centered values of θ. This means that mixing very different priors before the update usually results in higher risk for mild values of theta than mixing after. At the same time, its maximum is lower than the maximum of the risk function when mixing these distributions after the update, meaning that mixing before the update is a more conservative approach. So, if one risk function has a lower maximum, but the other presents much lower risk on many parameter values, determining which approach works best is not a simple task.

In Fig. 6, we note that, for all scenarios, KL divergence is higher when sample becomes larger in all cases. In the first scenario, the KL divergence of mixture after the update is lower than the other for both noninformative priors. When we look at the posteriors from Jeffreys and Uniform, note that KL divergence is lower in the mixture before the update than after. That is, the posterior density from the mixture before the update results in a very similar density to a posterior from Jeffreys or Uniform prior. In the second scenario, (second column) both methods are very similar. At last, in the discordance scenario, mixing after the update presents lower KL divergence when compared to mixing before the update in both priors. When we look at posteriors from noninformative priors, the mixture before the update varies more than the other.

5 Final Comments

We presented a method to achieve a collective posterior distribution using mixture distributions in two different moments, generating different results. Mixing before the update is more flexible and gives higher weights to priors whose densities are concentrated near the likelihood function. Furthermore, when analyzing the KL divergence, we note that when priors cover the parameter space, each prior giving high probability to a different area, the collective posterior distribution from prior mixture is, in general, very similar when compared with a posterior distribution of a noninformative prior. Considering this, if we believe in the group's prior knowledge and we want to substantially use that information, mixing after the update seems like a better option.

The next step in this study is to understand how both methods perform in the Normal-Gamma conjugation family and in a general scenario using MCMC and Gibbs-Sampling [4]. Then, we will look into Bayes decision rule to find best approaches on a case-by-case basis based on the risk function. At last, we will apply both methodologies to a real dataset and compare them in that context.

Acknowledgements This work was partially supported by FAPESP grant 2017/03363-8.

References

1. Clemen, R.T., Winkler, R.L.: Combining probability distributions from experts in risk analysis. Risk Anal. **19**(2), 187–203 (1999)
2. Esteves, L.G., Izbicki, R., Stern, R.B.: Teaching decision theory proof strategies using a crowdsourcing problem. Am. Stat. (just-accepted) (2017)
3. Frühwirth-Schnatter, S.: Finite Mixture and Markov Switching Models. Springer Science & Business Media, New York (2006)
4. Gelman, A., Carlin, J.B., Stern, H.S., Rubin, D.B.: Bayesian Data Analysis, vol. 2. Chapman & Hall/CRC, Boca Raton (2014)
5. Izbicki, R., Stern, R.B.: Learning with many experts: model selection and sparsity. Stat. Anal. Data Min. ASA Data Sci. J. **6**(6), 565–577 (2013)
6. Kadane, J.B.: Principles of Uncertainty. CRC Press, Boca Raton (2011)

A Nonparametric Bayesian Approach for the Two-Sample Problem

Rafael de C. Ceregatti, Rafael Izbicki and Luis Ernesto B. Salasar

Abstract In this work, we propose a novel nonparametric Bayesian approach to the so-called two-sample problem. Let $X_1, \ldots, X_n$ and $Y_1, \ldots, Y_m$ be two independent i.i.d samples generated from P_1 and P_2, respectively. Using a nonparametric prior distribution for (P_1, P_2), we propose a new evidence index for the null hypothesis $H_0 : P_1 = P_2$ based on the posterior distribution of the distance $d(P_1, P_2)$ between P_1 and P_2. This evidence index is easy to compute, has an intuitive interpretation, and can also be justified from a Bayesian decision-theoretic framework. We provide a simulation study to show that our method achieves greater power than the Kolmogorov–Smirnov and the Wilcoxon tests in several settings. Finally, we apply the method to a dataset on Alzheimer's disease.

Keywords Bayesian inference · Hypothesis tests · Nonparametric inference

1 Introduction

One basic interest in Statistics is to test the difference between groups, e.g., testing the difference between a group of patients that received a drug and other that received placebo. This problem is known in the literature as the "two-sample problem" and

R. de C. Ceregatti (✉) · R. Izbicki (✉) · L. E. B. Salasar (✉)
Federal University of São Carlos, Rod. Washington Luís km 235, SP-310, São Carlos, São Paulo, Brazil
e-mail: rafaelceregattii@gmail.com

R. Izbicki
e-mail: rafaelizbicki@gmail.com

L. E. B. Salasar
e-mail: luis.salasar@gmail.com

A. Polpo et al. (eds.), *Bayesian Inference and Maximum Entropy Methods in Science and Engineering*, Springer Proceedings in Mathematics & Statistics 239, https://doi.org/10.1007/978-3-319-91143-4_22

consists in deciding whether two independent samples are drawn from the same population.

This problem has been extensively studied in the statistical literature. The classic Bayesian parametric formulation to this problem is in terms of the Bayes factor [5], however, how to define the hypothesis is a crucial question. [1] proposed a Bayesian two-sample test tailored to a specific problem. To general cases, [10] presented the Bayesian t-test. An alternative to Bayes factors for comparing groups is the Full Bayesian Significance Test [11, 12]. On the other hand, from a nonparametric perspective, there are the well-established Kolmogorov–Smirnov test [8] and the Wilcoxon test [7]. Unfortunately, very few attempts of attacking this problem from a Bayesian nonparametric perspective exist. The only exceptions we are aware is [2], which use Bayes factors for Dirichlet process; [14], which create a Bayesian nonparametric procedure for two-sample hypothesis test considering as prior the Polya tree process; and [6], which propose a Bayesian method to compare two-samples based on the Kolmogorov distance.

In this article, we propose to test the equality of the two populations by means of a nonparametric Bayesian evidence index, which is given by the posterior weighted mean of the distance $d(P_1, P_2)$ between P_1 and P_2, the probability distributions associated to each population. The remaining of the paper is organized as follows. In Sect. 2 we present our evidence index and a decision theoretical justification for it. In Sect. 3 we review the basic definition and properties of the Dirichlet process, which we use to model each of the distribution functions. In Sect. 4, we present a simulation study designed to compare our proposal with the usual Kolmogorov–Smirnov and Wilcoxon tests. In Sect. 5, we apply our method to a dataset of scale measurements for Alzheimer disease. Section 6 contains our final remarks.

2 The Nonparametric Bayesian Evidence Index

Assume that two independent samples $X_1, \ldots, X_n$, and $Y_1, \ldots, Y_m$ are drawn from P_1 and P_2, respectively. Our aim is to test the null hypothesis $H_0 : P_1 = P_2$ against the alternative $H_1 : P_1 \neq P_2$. Assuming, a suitable metric d between probability measures,[1] we can express the magnitude of the difference between the two populations P_1 and P_2 by $d(P_1, P_2)$. Using this metric, our problem can be reformulated as testing $H_0 : d(P_1, P_2) = 0$ against $H_1 : d(P_1, P_2) > 0$. In the following, we shall assume that the metric d is bounded above.

[1] Common choices for this metric are the Kolmogorov–Smirnov metric, the L2 metric and Lévy metric. For a survey of metrics between probability measures see [15].

2.1 Index Definition

Considering a given nonparametric prior for (P_1, P_2), let us assume that $\mathbb{P}^{x,y}$ is the posterior distribution for (P_1, P_2) given the observed samples $x = (x_1, \ldots, x_n)$ and $y = (y_1, \ldots, y_m)$. Our proposal is to measure the evidence against H_0 by the following index

$$I(H_0|x, y) = \int_0^M w(\varepsilon)\, \mathbb{P}^{x,y}\, (d(P_1, P_2) > \varepsilon)\, d\varepsilon, \tag{1}$$

where $w : [0, M] \longrightarrow (0, \infty]$ is a nonincreasing probability density function and M is the supremum of $d(P_1, P_2)$ when varying P_1 and P_2.

The idea behind this index is to express a discrepancy between the posterior distribution of $d(P_1, P_2)$ and 0. Next, we present a short explanation supporting the definition given in (1). Suppose that a positive ε value can be considered a "practical" significant distance between P_1 and P_2, that is, P_1 and P_2 with distance less than ε can be considered equal for practical purposes. Thus, the higher is the posterior probability of the event $[d(P_1, P_2) > \varepsilon]$, the higher is the evidence against H_0 (or in favor of H_1). But, in many instances, it might not be clear how to choose appropriately the ε value. In this situation, we propose to combine the different evidence values by taking a weighted average with respect to ε, which leads to expression (1) with w as the weight function. Notice that by choosing an uniform weight function, $w(\varepsilon) = 1/M$, the index (1) is proportional to the area below the survival curve of $d(P_1, P_2)$, which is known to be equal to the expected value of $d(P_1, P_2)$. But, since the evidence value obtained for smaller ε values are more relevant to measure the discrepancy between $d(P_1, P_2)$ and 0 than greater values, we advocate the use of a nonincreasing weight function. It is also worthwhile noting that, since w is a density function over $[0, M]$, the index varies in the $[0, 1]$ interval. Further, the index assumes the value 0 and 1 when $d(P1, P2)$ is almost surely equal to 0 and M, respectively.

The expression (1) of the index can be rewritten in a more convenient way for numerical computation as

$$I(H_0|x, y) = \mathbb{E}^{x,y}\big[W(d(P_1, P_2)\big], \tag{2}$$

where $\mathbb{E}^{x,y}$ stands for the expectation with respect to the probability measure $\mathbb{P}^{x,y}$. From (2), we see that a Monte Carlo approximation for our index is easily obtained based on the posterior simulations of (P_1, P_2).

In order to see that that (2) holds, first let us denote $D = d(P_1, P2)$ and $\mathbb{P}_D$ its distribution obtained when assuming that (P_1, P_2) is distributed according to $\mathbb{P}^{x,y}$. Thus,

$$I(H_0|x, y) = \int_0^M w(\varepsilon)\mathbb{P}_D\big((\varepsilon, M]\big)d\varepsilon = \int_0^M \int_0^M w(\varepsilon) I_{(\varepsilon,M]}(z) d\mathbb{P}_D(z) d\varepsilon,$$

which implies by the Fubini theorem, that

$$\begin{aligned}I(H_0|x, y) &= \int_0^M \int_0^M w(\varepsilon) I_{(\varepsilon, M]}(z) d\varepsilon d\mathbb{P}_D(z) = \int_0^M \int_0^z w(\varepsilon) d\varepsilon d\mathbb{P}_D(z) \\ &= \int_0^M W(z) d\mathbb{P}_D(z) = E[W(D)],\end{aligned}$$

where W is the cumulative distribution of the density w and $I_A(z)$ denotes the indicator function assuming 1 if $z \in A$ and 0 otherwise.

2.2 Decision-Theoretic Formulation

Our index proposal can also be motivated in the bayesian decision framework [4]. Let $\mathbb{D} = \{a, a^c\}$ be the decision space, where a stands for accepting H_0 and a^c for rejecting H_0. Let us consider, the following loss function for our decision problem:

$$L\big((P_1, P_2), d\big) = \begin{cases} c_0 W(d(P_1, P_2)), & \text{if } d = a, \\ c_1[1 - W(d(P_1, P_2))], & \text{if } d = a^c, \end{cases}$$

where c_0 and c_1 are positive real numbers representing the maximum loss when accepting and rejecting H_0, respectively. Observe that, if we decide to accept H_0, the loss function is zero if $d(P_1, P_2) = 0$ and increases with the value of $d(P_1, P_2)$. On the other hand, if we decide to reject H_0, then the function decreases with the value of the distance $d(P_1, P_2)$ and vanishes if $d(P_1, P_2) = M$.

For a decision $\delta(x, y) \in \mathbb{D}$, the posterior expected loss is given by

$$\mathbb{E}^{x,y}\big[L\big((P_1, P_2), \delta(x, y)\big)\big] = \begin{cases} c_0 \mathbb{E}^{x,y}[W(d(P_1, P_2))], & \text{if } \delta(x, y) = a, \\ c_1\big[1 - \mathbb{E}^{x,y}[W(d(P_1, P_2))]\big], & \text{if } \delta(x, y) = a^c. \end{cases}$$

Thus, the Bayes rule is given by rejecting H_0 if and only if

$$I(H_0|x, y) = \mathbb{E}^{x,y}\big[W(d(P_1, P_2))\big] > c, \tag{3}$$

where $c = c_1/(c_1 + c_0)$.

3 Dirichlet Process

Our approach to solve the two-sample problem is fairly general: if we can draw samples from the posterior, we can compute the index. Thus, the index can be applied to any prior distribution, such as Pólya trees and the Beta processes. In this paper, however, we focus on the one of the most used methods to perform Bayesian nonparametric inference, which is the Dirichlet process prior [9].

The Dirichlet process can be briefly described as follows. Consider $(\mathscr{X}, \mathscr{B})$ a measurable space related to an observable quantity, G a probability measure (base probability) on $(\mathscr{X}, \mathscr{B})$, and K a positive real number. A stochastic process $\{P(B), B \in \mathscr{B}\}$ is said to be a Dirichlet process if for every partition $B_1, \ldots, B_m$ the random vector $(P(B_1), \ldots, P(B_m))$ has Dirichlet distribution with parameter $(K\,G(B_1), \ldots, K\,G(B_m))$. Therefore, a Dirichlet process is a random probability measure on $(\mathscr{X}, \mathscr{B})$. It is useful to notice that, for a fixed B, $P(B)$ has a Beta distribution with parameters $KG(B)$ and $K(1 - G(B))$, which implies that $P(B)$ has mean $G(B)$ and variance $G(B)(1 - G(B))/(K + 1)$. Thus, we can interpret G as the mean of the Dirichlet process and K as a precision constant, that is, larger values of K implies less variability around the base probability measure G. Thus, the Dirichlet process can be regarded as a general and simple way to express uncertainty about a probability measure P on a given space.

The class of Dirichlet process priors possesses a very convenient conjugacy property: assuming $Z_1, \ldots, Z_n$ is a sample of size n from P and P follows a Dirichlet process prior with hyperparameters G and K, then the posterior for P is a Dirichlet process with hyperparameters $G' = \frac{K}{K+n}G + \frac{n}{K+n}F_n$ and $K' = K + n$, where F_n is the empirical distribution of the data. In particular, we see that the posterior mean G' of P is a weighted average between the prior belief G and the empirical distribution F_n obtained from the data. Therefore, in order to simulate from the posterior, we need only to know how to simulate from a general Dirichlet process.

A practical way to simulate from a Dirichlet process with hyperparameters G' and K' is applying the stick-breaking algorithm [16], which is summarized as follows:

1. Simulate $Y_1, Y_2, \ldots$ i.i.d values from the base measure G';
2. Simulate $\theta_1, \theta_2, \ldots$ i.i.d from the Beta$(1, K')$;
3. Define $p_1 = \theta_1$ and $p_i = \theta_i \prod_{j=1}^{i-1}(1 - \theta_j)$ for $i \geq 2$;
4. For $B \in \mathscr{B}$, define

$$P(B) = \sum_{i=1}^{\infty} p_i \delta_{Y_i}(B). \tag{4}$$

Consequently, from the above algorithm, we can easily sample a probability measure P from the Dirichlet process. But, in order to apply the algorithm, we need to specify the number of Y_i's and θ_i's draw from the base measure G' and the beta distribution Beta(1, K'). A reasonable choice for the number of simulated values l is to set l that makes the sum $\sum_{i=1}^{l} p_i$ approximately equal to 1. Since the samples from the Beta distribution depends only on the concentration parameter, for $K = 1$ we set the number l to be 300. The final step 4 gives a simulated probability measure P from the Dirichlet process. Expression (4) is known as Sethuraman's representation and makes clear that Dirichlet processes are discrete with probability one.

Table 1 Threshold c values (0.95 sample quantiles of the index) for different support distributions and λ values considering $n = m = 50$.

λ	Distributions		
	N(0,1)	LN(0,1)	U([0,1])
1	0,2848	0,2927	0,2826
2	0,4755	0,4875	0,4844
3	0,6218	0,6349	0,6241
4	0,7272	0,7232	0,7215

4 Prior Specification and Decision Procedure

In order to proceed to test the hypothesis $P_1 = P_2$ using our index, we need to specify the prior distribution for P_1 and P_2, choose a metric d and a weight function w. The prior for P_1 and P_2 is specified as two independent Dirichlet process with the same hyperparameters G and K. The concentration parameter K is set to 1 and G is chosen accordingly to the known support of the data: for observations taking values on the real line, we choose the standard gaussian distribution $N(0, 1)$; for observations taking values in the nonnegative real line, we choose the standard lognormal distribution $LN(0, 1)$ and for observations taking values in the $[0, 1]$ interval, we choose the uniform distribution $U(0, 1)$. The metric d considered is the Kolmogorov metric defined by $d(P_1, P_2) = \sup_x |P_1((-\infty, x]) - P_2((-\infty, x])|$ and, since the maximum of the Kolmogorov distance is 1, the weight function w is taken to be a Beta$(1, \lambda)$ density ($\lambda \geq 1$), which has cumulative weight function $W_\lambda(t) = 1 - (1 - t)^\lambda, t \in [0, 1]$.

Now, it only remains to decide how to choose the threshold value c for the decision criterion in (3). At this point, we follow the philosophical approach suggested in [13] and adopt a bayes/non-bayes compromise to select the threshold. The idea is to select the value c that controls the type I error, that is, given that the hypothesis H_0 is true, we declare it false with probability less than α, e.g., $\alpha = 0.05$. In order to do so, we simulate two samples (with same sizes as the original samples) from the same distribution[2] and calculate the evidence index for them. We repeat the latter procedure a large number of times and take c as the 0.95 sample quantile of the index values. Table 1 presents the obtained threshold c considering the three different settings for the population support and $\lambda = 1, 2, 3, 4$. Thus, the H_0 hypothesis should be rejected if the index calculated for a given value of λ exceeds the correspondent c value given in Table 1.

[2]This null distribution is defined to be N(0,1), LN(0,1), or U(0,1), in the same way as the base measure, accordingly to the known support of data.

5 Simulation Study

In this section, we present a simulation study to compare our decision criteria with the Kolmogorov–Smirnov and Wilcoxon tests. We consider eight scenarios representing different departures from the null:

(a) Normal mean shift: $\mathbf{X} \sim N(0, 1)$ and $\mathbf{Y} \sim N(\theta, 1), \theta = 0, \ldots, 1$
(b) Normal variance shift: $\mathbf{X} \sim N(0, 1)$ and $\mathbf{Y} \sim N(0, \theta), \theta = 1, \ldots, 3$
(c) Normal mixtures: $\mathbf{X} \sim N(0, 1)$ and $\mathbf{Y} \sim \frac{1}{2}N(-\theta, 1) + \frac{1}{2}N(\theta, 1), \theta = 0, \ldots, 2$
(d) Fat tails: $\mathbf{X} \sim N(0, 1)$ and $\mathbf{Y} \sim t(\theta^{-1}), \theta = 0.01, \ldots, 10$.
(e) Lognormal mean shift: $\log \mathbf{X} \sim N(0, 1)$ and $\log \mathbf{Y} \sim N(\theta, 1), \theta = 0, \ldots, 1.5$
(f) Lognormal variance shift: $\log \mathbf{X} \sim N(0, 1)$ and $\log \mathbf{Y} \sim N(0, \theta), \theta = 1, \ldots, 3$
(g) Normal skewness: $\mathbf{X} \sim N(0, 1)$ and $\mathbf{Y} \sim SN(0, 1, \theta, 0), \theta = 1, \ldots, 1.5$
(h) Beta symmetry: $\mathbf{X} \sim Beta(1, 1)$ and $\mathbf{Y} \sim Beta(\theta, \theta), \theta = 1, \ldots, 6$

The comparison is made in terms of the "power to detect the alternative." That is, we fix a threshold that controls the type I error in 5% and compute the power function for the respective tests. The power function is calculated in the following way. For each value of θ, we draw 50 observations from $\mathbf{X}$ and $\mathbf{Y}$. After that, we calculate the value of the index using $W_4(t) = 1 - (1 - t)^4$ as the cumulative weight function. We repeat these steps 1000 times and compute the proportion of times that we reject the null hypothesis.

Figure 1 indicates that the Wilcoxon test is able to detect changes in the location and skewness parameter (scenarios (a), (e) and (g)), but shows extremely low power in detecting the alternative for all other scenarios. The Kolmogorov–Smirnov test presents a medium power performance over all scenarios. On the other hand, the proposed index overperfomed its competitor in 5 scenarios ((b), (c), (d), (f) and (h)) and presented a similar performance to the best one in all other scenarios.

Additionally, we also investigate the consistency of the proposed method, that is, we study the power function under the alternative for increasing sample sizes. In order to do so, we fixed θ at 1, 3, 1, 1.25, 1, 3, 1, 3 for the scenarios (a) to (h) and simulated 1000 datasets. The results are reported in Fig. 2 for $n = m = 10, 20, 30 \ldots, 100$. Once again, the proposed index overperfomed its competitor in the same 5 scenarios ((b), (c), (d), (f), and (h)) and was quite similar to its competitors in the other scenarios.

6 Application

We apply our methods to a dataset of three groups of patients (HC: with healthy cognition, MCI: with mild cognitive decline and DA: with Alzheimer's disease) submitted to a questionnaire for Alzheimer's disease diagnostic (CAMCOG). More details on this dataset can be obtained in [3].

From Fig. 3, we see that CAMCOG scores have different behavior among the three groups. The group with Alzheimer's disease (DA) has the lowest scores, followed

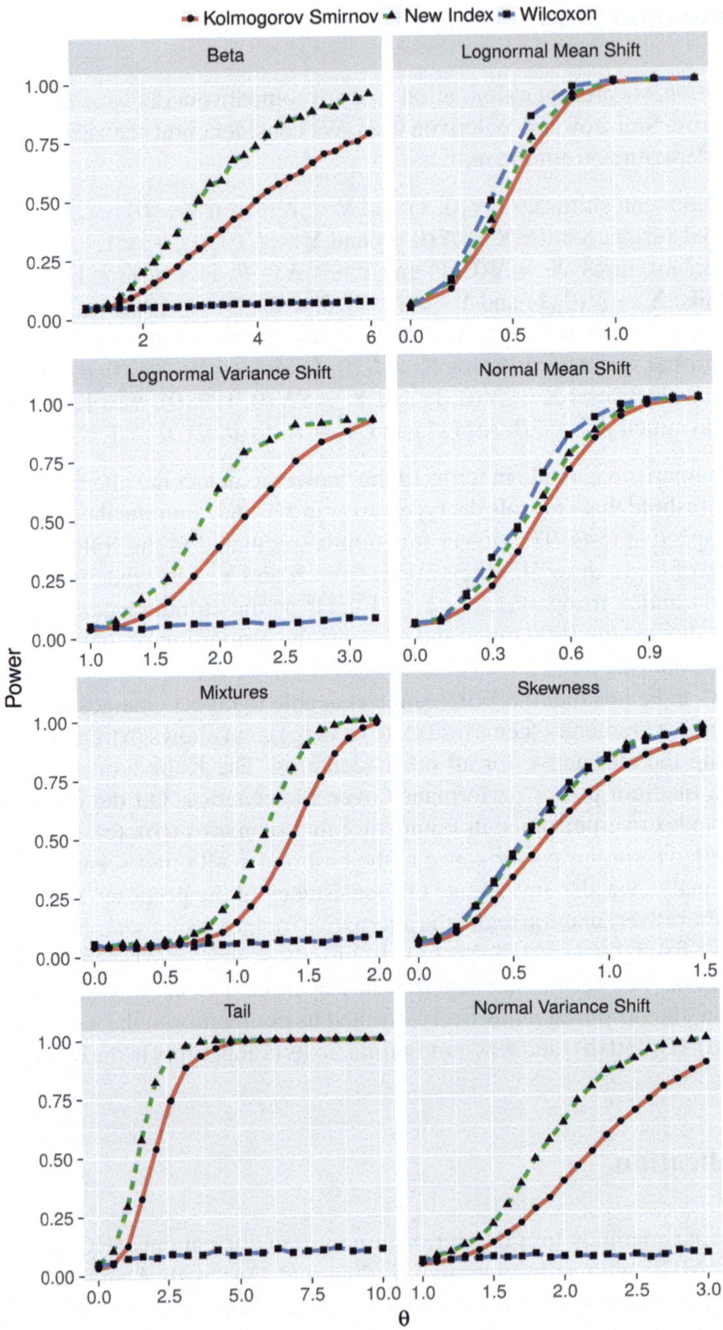

Fig. 1 Power function comparison for different settings

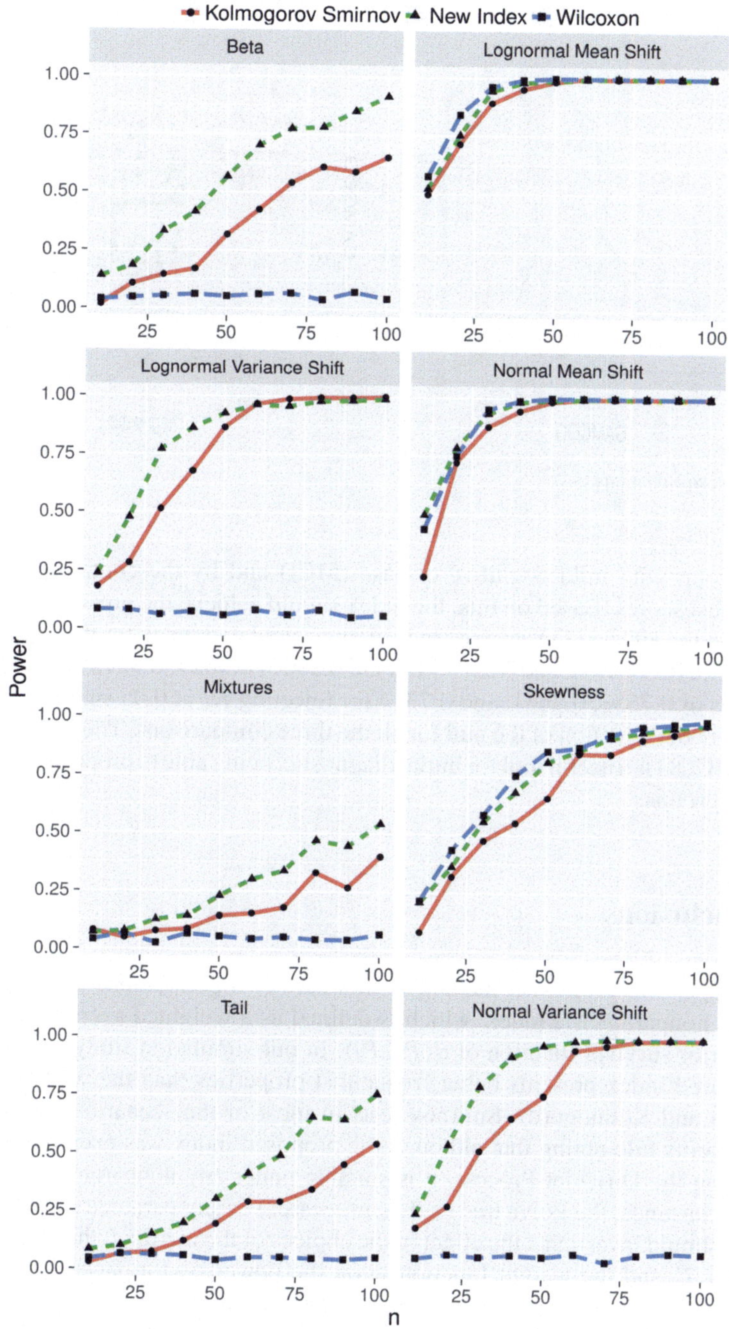

Fig. 2 Power for different settings changing the sample size

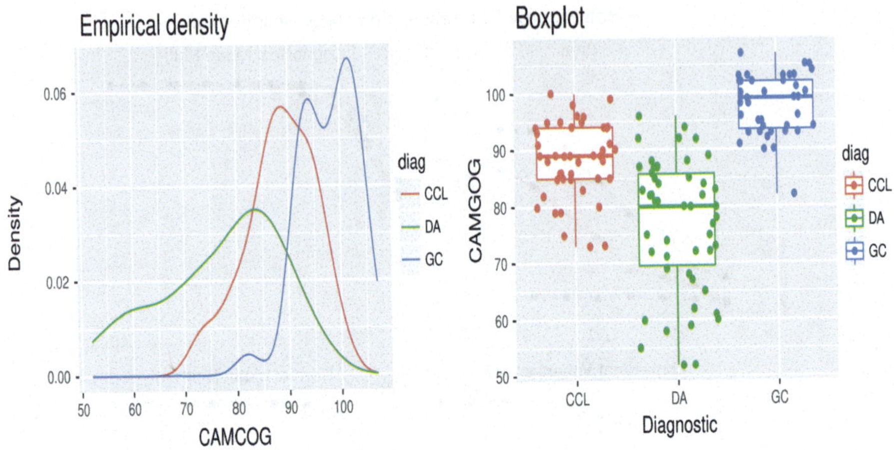

Fig. 3 Descriptive analysis

by the group with mild cognitive disease (MCI) and by the control group (HC) with highest scores. Based on this, the index should achieve the greatest value when comparing the HC and DA groups. Indeed, the calculated index for the comparisons HC vs DA, HC vs MCI, and MCI vs DA are 0.9993, 0.9629, 0.9312 with respective thresholds of 0.7558, 0.7681, and 0.7314 for rejecting H_0 at 0.05 significance level, implying in the rejection of the null for all the three comparisons. The latter indicates that CAMCOG is a useful tool for initial diagnostic, being able to properly distinguish the three groups.

7 Conclusions

In this paper, we propose a method to compare two populations that relies on a Bayesian nonparametric index, which is defined as a weighted average area below the posterior survival function of $d(P_1, P_2)$. In our simulation study, we show that the proposed index presents better frequentist properties than the well-established Wilcoxon and Kolmogorov–Smirnov tests in most of the scenarios considered. It is also worthwhile noting that although the proposed index was exemplified in this work using the Dirichlet Process, it is suitable under any nonparametric Bayesian prior (for instance, the Polya tree or Beta processes) or any parametric prior. Further research should investigate the effect of the choices of the metric d, the concentration parameter K and the weight function w on the power performance. Besides this, theoretical investigation of a consistency property can give greater support for the method.

Acknowledgements This work was partially supported by FAPESP grant 2017/03363-8.

References

1. Baldi, P., Long, A.D.: A Bayesian framework for the analysis of microarray expression data: regularized t-test and statistical inferences of gene changes. Bioinformatics **17**, 509–519 (2001)
2. Basu, S., Chib, S.: Marginal likelihood and Bayes factors for Dirichlet process mixture models. J. Am. Stat. Assoc. **98**, 224–235 (2003)
3. Cecato, J.F., Martinelli, J.E., Izbicki, R., Yassuda, M.S., Aprahamian, I.: A subtest analysis of The Montreal Cognitive Assessment (MoCA): which subtests can best discriminate between healthy controls, mild cognitive impairment and Alzheimer's disease. Int. Psychogeriatr. **29**, 825–832 (2016). https://doi.org/10.1017/S1041610215001982
4. DeGroot, M.H.: Optimal Statistical Decisions. Wiley, New York (2005)
5. Kass, R., Raftery, A.: Bayes factors. J. Am. Stat. Assoc. **90**, 773–795 (1995)
6. Labadi, L.A., Masuadi, E., Zarepour, M.: Two-sample Bayesian nonparametric goodness-of-fit test (2014). arXiv:1411.3427
7. Mann, H.B., Whitney, D.R.: On a test of whether one of two random variables is stochastically larger than the other. Ann. Math. Stat. **18**, 50–60 (1947)
8. Smirnov, N.: Table for estimating the goodness of fit of empirical distributions. Ann. Math. Stat. **19**, 279–281 (1948)
9. Ferguson, T.S.: A Bayesian analysis of some nonparametric problems. Ann. Stat. **1**, 209–230 (1973)
10. Gönen, M., Johnson, W.O., Lu, Y.W., Westfall, P.H.: The Bayesian two-sample t test. Am. Stat. **59**, 252–257 (2005)
11. de Bragança Pereira, C.A., Stern, J.M.: Evidence and credibility: full Bayesian significance test for precise hypotheses. Entropy **1**(4), 99–110 (1999)
12. Izbicki, R., Fossaluza, V., Hounie, A.G., Nakano, E.Y., de Bragança Pereira, C.A.: Testing allele homogeneity: the problem of nested hypotheses. BMC Genet. **13**(1), 1–11 (2012)
13. Good, I.J.: The Bayes/non-Bayes compromise: a brief review. J. Am. Stat. Assoc. **87**, 597–606 (1992)
14. Holmes, C.C., Caron, F., Griffin, J.E., Stephens, D.A.: Two-sample Bayesian nonparametric hypothesis testing. Bayesian Anal. **10**, 297–320 (2015)
15. Rachev, S.T., Klebanov, L.B., Stoyanov, S.V., Fabozzi, F.J.: The Methods of Distances in the Theory of Probability and Statistics. Springer, New York (2013)
16. Sethuraman, J.: A constructive definition of Dirichlet priors. Stat. Sin. **4**, 639–650 (1994)

Covariance Modeling for Multivariate Spatial Processes Based on Separable Approximations

Rafael S. Erbisti, Thais C. O. Fonseca and Mariane B. Alves

Abstract The computational treatment of high dimensionality problems is a challenge. In the context of geostatistics, analyzing multivariate data requires the specification of the cross-covariance function, which defines the dependence between the components of a response vector for all locations in the spatial domain. However, the computational cost to make inference and predictions can be prohibitive. As a result, the use of complex models might be unfeasible. In this paper, we consider a flexible nonseparable covariance model for multivariate spatiotemporal data and present a way to approximate the full covariance matrix from two separable matrices of minor dimensions. The method is applied only in the likelihood computation, keeping the interpretation of the original model. We present a simulation study comparing the inferential and predictive performance of our proposal and we see that the approximation provides important gains in computational efficiency without presenting substantial losses in predictive terms.

Keywords Nonseparable covariance · Likelihood approximation · Kronecker product · Predictive performance · Multivariate spatial process

R. S. Erbisti
Department of Statistics, Centro de Tecnologia, Federal University of Rio de Janeiro, Bloco I-044b - LSE, CEP, Rio de Janeiro 21941-909, Brazil
e-mail: rerbisti@dme.ufrj.br

T. C. O. Fonseca (✉) · M. B. Alves
Department of Statistics, Centro de Tecnologia, Federal University of Rio de Janeiro, Bloco C, CEP, Rio de Janeiro 21941-909, Brazil
e-mail: thais@im.ufrj.br

M. B. Alves
e-mail: mariane@im.ufrj.br

A. Polpo et al. (eds.), *Bayesian Inference and Maximum Entropy Methods in Science and Engineering*, Springer Proceedings in Mathematics & Statistics 239, https://doi.org/10.1007/978-3-319-91143-4_23

1 Introduction

With the increase of high-resolution geocoded data, the big n problem became crucial in the spatial and spatiotemporal setup. For instance, if Gaussianity is assumed, large covariance matrices need to be inverted in the inference procedure and computational effort is of cubic order on the number of locations. This limitation becomes even more important in the case of spatiotemporal or multivariate data. Even low-dimensional vectors observed over space may lead to huge covariance matrices, making the inference for unknown parameters not feasible. Thus, a compromise between complexity and parsimony is called for in this context.

In this paper, we work with multivariate spatial covariance functions to illustrate the high dimensionality problem. To treat the computational limitation we approximate the full covariance matrices using a decomposition based on the Kronecker product of two separable matrices of minor dimensions. These approximations have been applied to the likelihood function in order to obtain fast estimation of parameters but we still keep the interpretation and flexibility of the multivariate nonseparable model.

The remainder of the paper is organized as follows. Section 2 presents definitions and characteristics about multivariate process modeling. To allow for fast estimation of parameters, a fast algorithm is used to compute the likelihood function to allow for scalable modeling of large multivariate spatial data in Sect. 3. Section 4 presents a discussion on the proposed approach.

2 Multivariate Covariance Modeling

Consider a partial realization of a random function $Y(\mathbf{s})$, $\mathbf{s} \in D \subseteq \Re^d$, $d \geq 1$, where $\mathbf{s}$ denotes a spatial location. Usually, several quantities are measured for each location $\mathbf{s}$, resulting in a multivariate random vector $\mathbf{Y}(\mathbf{s}) = (Y_1(\mathbf{s}), \ldots, Y_p(\mathbf{s}))^T$, $\mathbf{s} \in D$. If Gaussianity is assumed, it suffices to define its mean and cross-covariance functions. Throughout this text, it is assumed that $\mathbf{Y}(\mathbf{s})$ is a spatially stationary process, that is

$$E[Y_i(\mathbf{s})] = m_i, \tag{1}$$

$$Cov[Y_i(\mathbf{s}), Y_j(\mathbf{s}+\mathbf{h})] = E[(Y_i(\mathbf{s}) - m_i)(Y_j(\mathbf{s}+\mathbf{h}) - m_j)] = C_{ij}(\mathbf{h}), \tag{2}$$

$\forall \mathbf{s}, \mathbf{s}+\mathbf{h} \in D; i, j = 1, \ldots, p$, with $\mathbf{h}$ the spatial separation vector and $C_{ij}(\mathbf{h})$ denoting the cross-covariance function. The covariance functions considered need to be valid, i.e., the resulting covariance matrix must be positive definite.

Separable covariance functions (see [1]) are defined by

$$C_{ij}(\mathbf{s}, \mathbf{s}') = a_{ij}\rho(\mathbf{s}, \mathbf{s}'), \tag{3}$$

with $A = \{a_{ij}\}$ a positive definite $p \times p$ matrix and $\rho(\cdot, \cdot)$ a valid correlation function. Let $\mathbf{Y}$ be a vectorized version of $Y_{ik} = Y_i(\mathbf{s}_k), k = 1, \ldots, n; i = 1, \ldots, p$. Then the covariance matrix is $\boldsymbol{\Sigma} = \mathbf{R} \otimes \mathbf{A}$, with $R_{kl} = \rho(\mathbf{s}_k, \mathbf{s}_l), k, l = 1, \ldots, n$. The condition of positive definiteness is respected if $\mathbf{R}$ and $\mathbf{A}$ are positive definite. This specification is computationally advantageous as inverses and determinants are obtained from smaller matrices, that is, $\boldsymbol{\Sigma}^{-1} = \mathbf{R}^{-1} \otimes \mathbf{A}^{-1}$ and $|\boldsymbol{\Sigma}| = |\mathbf{R}|^p |\mathbf{A}|^n$. However, this might not be a realistic assumption for different processes across space. The separable specification implies, for example, that when the spatial location varies, the covariance pattern for different components remains the same.

The nonseparable cross-covariance function based on mixing separable functions (see [2]) is given by

$$K_{ij}(\mathbf{s}, \xi_i, \xi_j) = \int \int C_1(\mathbf{s}; u) C_2(\xi_i, \xi_j; v) g_{ij}(u, v) du dv \tag{4}$$

with (U, V) a nonnegative bivariate random vector following a joint distribution $G_{ij}(u, v)$, independent of the process $\mathbf{Y}(\mathbf{s})$, ξ_i, ξ_j representing the components i, j, respectively, on a latent k-dimensional space as in [3] and $\mathbf{s}$ an arbitrary spatial location.

It is possible to analytically solve (4) defining $C_1(\mathbf{s}; u) = exp\{-\gamma_1(\mathbf{s})u\}$ and $C_2(\xi_i, \xi_j; v) = exp\{-\gamma_2(\xi_i, \xi_j)v\}$, where $\gamma_1(\mathbf{s})$ and $\gamma_2(\xi_i, \xi_j)$ are continuous functions on $\mathbf{s} \in \Re^d$ and $\xi_i, \xi_j \in \Re^k$, respectively.

The class of covariance functions generated in [2] is used in this work. The general model, based on (4) and described in [2] is given by

$$K_{ij}(\mathbf{s}, \xi_i, \xi_j) = C_{ij} C(h, \delta_{ij}) = \sigma_i \sigma_j \left(1 + \delta_{ij} + \frac{h}{b_{ij}}\right)^{-\alpha_0} \left(1 + \frac{h}{b_{ij}}\right)^{-\alpha_1} (1 + \delta_{ij})^{-\alpha_2} \tag{5}$$

where $h = \|\mathbf{s} - \mathbf{s}'\|$, $\mathbf{s}, \mathbf{s}' \in D$, δ_{ij} is the latent distance between the components i and j, $\delta_{ij} = \|\xi_i - \xi_j\|$, $\sigma_i, \sigma_j \in \Re$, b_{ij}'s are spatial range parameters, α_l are smoothness parameters, for $l = 1, 2$, and α_0 is a separability parameter. For more details about latent dimensions see [2–4].

If the same spatial range parameter for all components is considered then $b_{ij} = \phi$, $\forall i, j = 1, 2, \ldots, p$, and a particular case of the general function is obtained. Furthermore, if $\alpha_0 = 0$, the separable model is obtained and the resulting covariance function is in the Cauchy class.

3 Likelihood Computation for Nonseparable Covariance Models

We have presented a nonseparable covariance model which results in a full matrix $\boldsymbol{\Sigma}$ which might have high dimension and the computation of likelihoods in a Gaussian model require the inversion of this matrix. We investigate the use of separable approximations for the matrix $\boldsymbol{\Sigma}$ which will lead to fast computation of the likelihood function.

Let $(\mathbf{y}_t(\mathbf{s}_1), \ldots, \mathbf{y}_t(\mathbf{s}_n))$ be a matrix of multivariate data observed at spatial locations $\mathbf{s}_1, \ldots, \mathbf{s}_n \in D$ and at time t, where $\mathbf{y}_t(\mathbf{s}_i) = (y_{1t}(\mathbf{s}_i), \ldots, y_{pt}(\mathbf{s}_i))'$, $t = 1, \ldots, T$, is a $p-$dimensional vector. If the Gaussian assumption is made, the likelihood function with T independent replicates for the unknown parameters based on n spatial locations is given by

$$l(\mathbf{y}; \boldsymbol{\theta}) = (2\pi)^{\frac{-npT}{2}} |\boldsymbol{\Sigma}|^{-T/2} exp\left\{\frac{-1}{2}\sum_{t=1}^{T}(\mathbf{y}_t - \boldsymbol{\mu})'\boldsymbol{\Sigma}^{-1}(\mathbf{y}_t - \boldsymbol{\mu})\right\} \quad (6)$$

with $\mathbf{y}_t$ the vectorized version of $(\mathbf{y}_t(\mathbf{s}_1), \ldots, \mathbf{y}_t(\mathbf{s}_n))$ with np observations, $\boldsymbol{\mu} = \mathbf{X}\boldsymbol{\beta}$ the mean vector, $\boldsymbol{\Sigma}$ the covariance matrix with dimension $np \times np$, and $\boldsymbol{\theta}$ the parameter vector.

3.1 Separable Approximations

Reference [5] investigates the use of singular decompositions of a full matrix [6] in the context of nonseparable spatiotemporal covariance matrices. The work considers a decomposition based on the separable matrices which allow for fast inversions and determinant computations. Thus, instead of $np \times np$ matrices, the approximation uses only $n \times n$ and $p \times p$ matrices. We consider the same separable approximation in order to compute likelihoods for the nonseparable multivariate spatial models presented in Sect. 2. The aim is to obtain matrices $\mathbf{R} \in \Re^{n\times n}$ and $\mathbf{A} \in \Re^{p\times p}$ such that the Frobenius norm[1] of $\|\boldsymbol{\Sigma} - \mathbf{R} \otimes \mathbf{A}\|_F$ is minimized, for a given full covariance matrix $\boldsymbol{\Sigma}$. The author shows that the solution to this problem is given by the singular value decomposition of a permuted version of $\boldsymbol{\Sigma} \in \Re^{np\times np}$.

The idea is to rearrange $\boldsymbol{\Sigma}$ obtaining another matrix $\Im(\boldsymbol{\Sigma}) \in \Re^{n^2\times p^2}$, such that the sum of squares that arises in $\|\boldsymbol{\Sigma} - \mathbf{R} \otimes \mathbf{A}\|_F$ equals the sum of squares in $\|\Im(\boldsymbol{\Sigma}) - vec(\mathbf{R}) \otimes vec(\mathbf{A})^T\|_F$. It is showed in [6] that $\|\boldsymbol{\Sigma} - \mathbf{R} \otimes \mathbf{A}\|_F = \|\Im(\boldsymbol{\Sigma}) - vec(\mathbf{R}) \otimes vec(\mathbf{A})^T\|_F$ and $\|\boldsymbol{\Sigma}\|_F = \|\Im(\boldsymbol{\Sigma})\|_F$.

The problem then reduces to finding the rank of the rectangular matrix $\Im(\boldsymbol{\Sigma}) \in \Re^{n^2\times p^2}$. The solution is based on the singular value decomposition of $\Im(\boldsymbol{\Sigma})$, where $\mathbf{U}^T\Im(\boldsymbol{\Sigma})\mathbf{V} = diag(w_1, \ldots, w_r)$, $\mathbf{U} \in \Re^{n^2\times n^2}$ and $\mathbf{V} \in \Re^{p^2\times p^2}$ are orthogonal matrices, $w_1 \geq w_2 \geq \ldots \geq w_r \geq 0$ and $r = rank(\Im(\boldsymbol{\Sigma})) = min\{n^2, p^2\}$. The solution can also be found in [6] and is given by:

$$vec(\mathbf{R}) = \sqrt{w_1}\boldsymbol{u}_1 \qquad vec(\mathbf{A}) = \sqrt{w_1}\boldsymbol{v}_1 \quad (7)$$

[1]The Frobenius norm of a $n \times n$ matrix $\mathbf{B}$ ($\|\mathbf{B}\|_F$) is given by $\|\mathbf{B}\|_F = \left(\sum_{i=1}^{n}\sum_{j=1}^{n} b_{ij}^2\right)^{1/2}$.

with $\boldsymbol{u}_1$ denoting the first column of the matrix $\mathbf{U} \in \Re^{n^2 \times n^2}$ and $\boldsymbol{v}_1$, the first column of $\mathbf{V} \in \Re^{p^2 \times p^2}$.

In order to measure the quality of the approximation, [5] defines an approximation error, denoted by $\kappa_{\boldsymbol{\Sigma}}(\mathbf{R}, \mathbf{A})$, as follows:

$$\kappa_{\boldsymbol{\Sigma}}(\mathbf{R}, \mathbf{A}) = \frac{\|\boldsymbol{\Sigma} - \mathbf{R} \otimes \mathbf{A}\|_F}{\|\boldsymbol{\Sigma}\|_F}. \tag{8}$$

$\kappa_{\boldsymbol{\Sigma}}(\mathbf{R}, \mathbf{A})$ varies between zero (if $\boldsymbol{\Sigma}$ is separable) and $\sqrt{1 - \frac{1}{r}}$, and is minimized by $\mathbf{R}$ and $\mathbf{A}$ given above. A standardized error index, varying between zero and one is given by:

$$\kappa^*_{\boldsymbol{\Sigma}}(\mathbf{R}, \mathbf{A}) = \frac{\kappa_{\boldsymbol{\Sigma}}(\mathbf{R}, \mathbf{A})}{\sqrt{1 - \frac{1}{r}}}. \tag{9}$$

From the covariance structure proposed in Eq. (5) with $\alpha_1 = \alpha_2 = 1$, we investigate the sensitivity of the separability approximation error index as a function of α_0, for $p = 2$ and $p = 3$. Note that we use the idea previously applied to the context of nonseparable spatiotemporal covariance matrices in the context of nonseparable multivariate spatial covariance matrices. In Fig. 1, we can see that the separability approximation error index is not larger than 5% for a covariance structure in which all the components have the same spatial range. From Fig. 1a note that there is no error when α_0 is zero, which reduces to the separable case. If different spatial ranges are considered, it is possible to see in Fig. 1b that the error index does not start at zero because if $\alpha_0 = 0$ the separable case is not obtained.

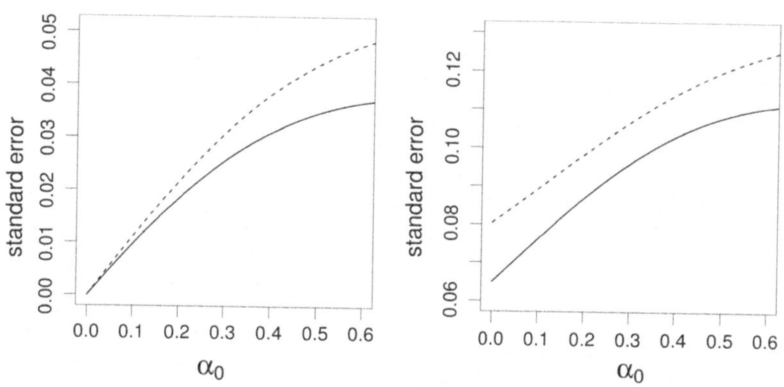

(a) Equal spatial ranges.

(b) Different spatial ranges.

Fig. 1 Separability approximation error index as a function of α_0. Full line: $p = 2$; dashed line: $p = 3$

3.2 Sensitivity Study

We present a sensitivity study of the approximation structure investigated in the spatiotemporal context by [5], however, used here for the multivariate spatial case. We consider different scenarios and measure the errors obtained in the likelihood approximation. Moreover, we compare the inferential and predictive results obtained as we apply the full nonseparable model with and without separable approximation for the covariance matrix in the likelihood computation, as well as considering a separable model.

Consider a bivariate dataset of 200 spatial locations in the $[0, 1] \times [0, 1]$ square. The observations were generated from the model $\mathbf{y} = \boldsymbol{\mu} + \boldsymbol{\varepsilon}$. We consider a Gaussian process, so $\mathbf{y} \sim N_{np}(\boldsymbol{\mu}, \boldsymbol{\Sigma})$, where $\boldsymbol{\Sigma}_{ij}$ is obtained from the covariance function defined in (5) with $\alpha_1 = \alpha_2 = 1$ and $b_{ij} = \phi$, $i, j = 1, 2$. Therefore, consider the following parameter specification: $\boldsymbol{\Theta} = (\boldsymbol{\mu}, \delta_{12}, \phi, \alpha_0, \sigma_1, \sigma_2)$ with $\boldsymbol{\mu}_1 = \boldsymbol{\mu}_2 = 0$, $\delta_{12} = 2, \phi = 0.1, \alpha_0 = 0.5$ and $\sigma_1 = \sigma_2 = 1$. In this example, we generate only one dataset in the region of interest. We plot the likelihood contour with both structures, using the separable approximation for the covariance matrix and its full original structure. From Fig. 2 it can be seen that the approximate structure is very similar to the full structure. Note that in some cases the approximate likelihood and the exact one are almost coincident. It seems that the approximations are satisfactory.

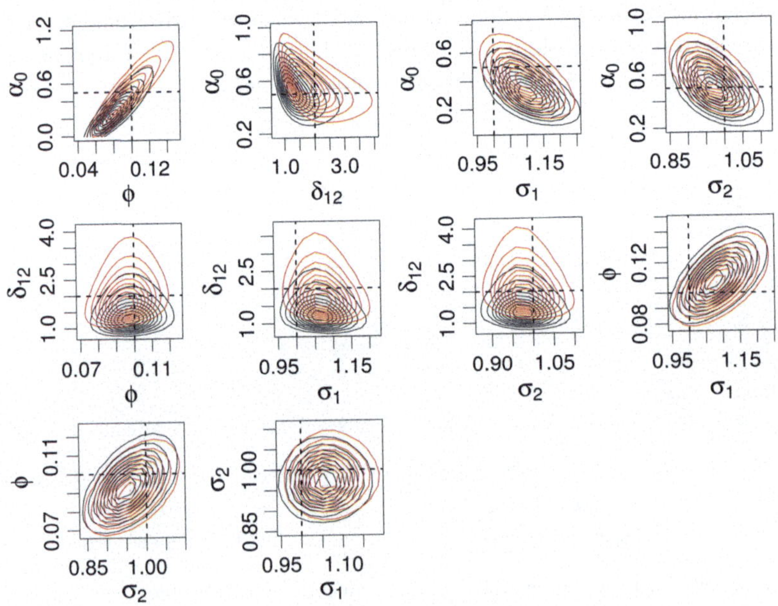

Fig. 2 Likelihood contour plots. Black line: full structure. Red line: approximate structure. Dashed black line: true value of parameters

Table 1 Necessary time (in seconds) to calculate the likelihood function based on a full covariance matrix and an approximate structure. (Intel(R) Core(TM) i7-3630QM, 2.40 GHz, 6 GB RAM)

n	$p = 2$		$p = 3$		$p = 5$		$p = 8$	
	Full	Approx.	Full	Approx.	Full	Approx.	Full	Approx.
100	2.3	0.8	3.6	0.6	8.9	0.9	28.0	2.1
200	12.3	3.1	13.9	1.6	44.8	4.2	187.1	11.1
500	74.0	13.4	143.1	12.1	618.6	30.5	2409.5	86.9
700	148.2	20.9	388.4	29.4	1649.7	65.8	6520.7	182.2
1000	374.6	52.6	1020.7	66.2	4673.9	133.5	19180.3	446.2

We analyzed the necessary time to calculate the likelihood function based on a full covariance matrix and a covariance matrix with approximate structure. We generated $p = 2, 3, 5$, and 8 variables in a dataset with $n = 100, 200, 500, 700$, and 1000 spatial locations in the $[0, 1] \times [0, 1]$ square. In this example, 200 replicates were generated in the region of interest.

Table 1 shows that the separable approach provides important gains in computational efficiency. Note that the time to calculate the likelihood function is substantially lower when we use the approximate structure.

We also analyzed the time reduction using the separable approximations. Figure 3 shows that the time to calculate the likelihood function decreases as the size of the covariance matrix increases. Indeed, if we increase the variable numbers or spatial locations or both, the greater will be the computational gain with the approximate structure.

Finally, we compare the predictive results obtained by the separable model, the separable approximation for the covariance matrix in the likelihood of the nonseparable model and the results obtained with the nonseparable original covariance structure, without any approximations for the likelihood. For that purpose, we generated five datasets from the nonseparable structure described in Sect. 2. We use a less general function than proposed in Eq. (5). For each dataset, we generated $p = 2$ variables in $n = 110$ spatial locations in the $[0, 1] \times [0, 1]$ square considering the following parameter specification $\boldsymbol{\Theta} = (\boldsymbol{\beta}, \delta_{12}, \phi, \alpha_0, \sigma_1, \sigma_2)$ with $\boldsymbol{\beta} = (1, -0.2, -0.8, 0.5, 1.5, 0.6, -0.5, -0.8)$, $\delta_{12} = 2$, $\phi = 0.2$, $\alpha_0 = 1$, $\sigma_1 = 1.5$ and $\sigma_2 = 1$. The covariance functions used for the separable and nonseparable models are respectively shown in Eqs. (10) and (11):

$$C_{ij}(\mathbf{s}) = \begin{cases} a_{11}\left(1 + \left(\frac{h}{\phi}\right)^2\right)^{-1} & (i = j = 1) \\ a_{22}\left(1 + \left(\frac{h}{\phi}\right)^2\right)^{-1} & (i = j = 2) \\ a_{12}\left(1 + \left(\frac{h}{\phi}\right)^2\right)^{-1} & (i \neq j), \end{cases} \tag{10}$$

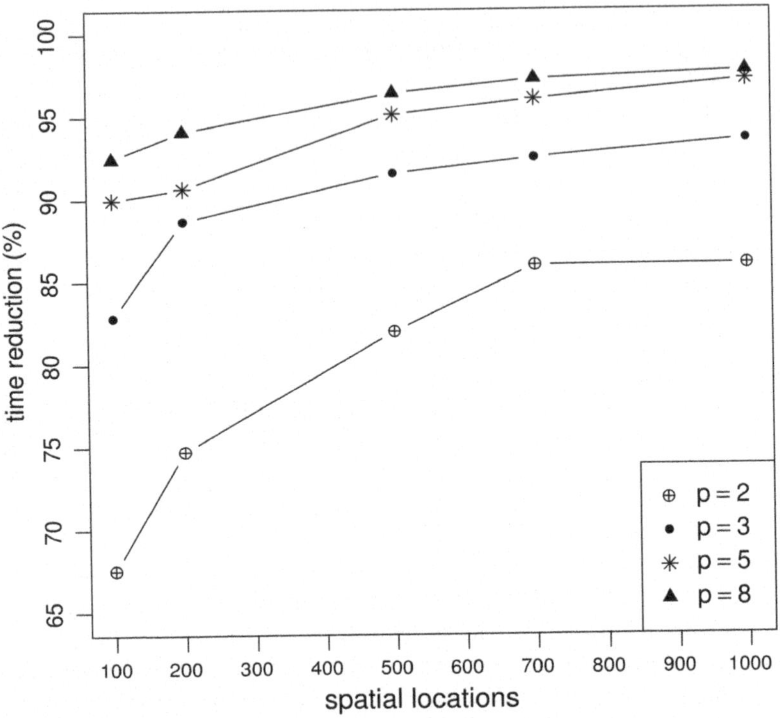

Fig. 3 Computational time reduction (in percent) in calculation of the likelihood function using approximate structure

$$C(h,\delta_{ij}) = \begin{cases} \sigma_1^2\left(1+\frac{h}{\phi}\right)^{-(\alpha_0+1)} & (i=j=1) \\ \sigma_2^2\left(1+\frac{h}{\phi}\right)^{-(\alpha_0+1)} & (i=j=2) \\ \sigma_1\sigma_2\left(1+\delta_{12}+\frac{h}{\phi}\right)^{-\alpha_0}\left(1+\frac{h}{\phi}\right)^{-1}(1+\delta_{12})^{-1} & (i \neq j), \end{cases} \quad (11)$$

with $h = \|\mathbf{s}-\mathbf{s}\|$ and $\delta_{12} = \|\xi_1 - \xi_2\|$. We adopted $T = 30$ independent replicates. The observations were generated from the model $\mathbf{y} = \boldsymbol{X\beta} + \boldsymbol{\varepsilon}$. We consider a Gaussian process, so $\mathbf{y}_t \sim N_{np}(\boldsymbol{X\beta}, \boldsymbol{\Sigma})$, $t = 1, \ldots, T$, where $\boldsymbol{\Sigma}$ is $np \times np$ covariance matrix and $\boldsymbol{X}$ are independent variables (latitude, longitude, and altitude). For the nonseparable models, we use as follows priors: $\sigma_i \sim N(0, 100)$, $i = 1, 2$, $\delta_{12} \sim Ga(1, 0.5)$, $\phi \sim Ga(0.1 \times med(d_s), 0.1)$, with $med(d_s) = 0.502$, $\boldsymbol{\beta} \sim N_8(\mathbf{0}, 1000\mathbf{I}_8)$ and $\alpha_0 \sim Ga(1, 0.25)$. For the separable models, we use as follows priors: $\mathbf{A} \sim InverseWishart(\mathbf{I}_2, 3)$, $\phi \sim Ga(0.25 \times med(d_s), 0.25)$, with $med(d_s) = 0.502$ and $\boldsymbol{\beta} \sim N_8(\mathbf{0}, 1000\mathbf{I}_8)$. The simulation method used was MCMC. For convergence monitoring we use the algorithms present in the `Coda` package in `R` (see [7]).

Table 2 Predictive model comparison. SEP: separable model. NSEP APP: nonseparable approximate model. NSEP: nonseparable model

Data	Average IS			LPML		
	SEP	NSEP APP.	NSEP	SEP	NSEP APP.	NSEP
1	153.37	126.73	126.60	−12389.77	−11654.74	−11571.50
2	152.29	127.01	126.42	−12339.71	−11644.76	−11593.16
3	151.58	126.98	126.80	−12330.21	−11713.76	−11657.84
4	153.47	128.50	128.71	−12383.98	−11706.85	−11790.66
5	152.34	125.93	125.74	−12388.03	−11673.80	−11645.60
Mean	152.61	127.03	126.85	−12366.34	−11678.78	−11651.75

Furthermore, the data of five spatial locations were removed from the training data and used for prediction validation. Therefore, we estimate the model using information about $n = 105$ spatial locations. After the estimation of the models for each dataset, we were able to compute measures of predictive performance for each model. The IS (Interval Score) and LPML (Logarithm of the Pseudo Marginal Likelihood) comparison measures are described in [8, 9], respectively. Table 2 presents the comparison of the models in predictive terms. We can see that in predictive terms the approximation leads to very similar results to the full case. Note that the separable model presents the worst results as it is not able to accommodate the nonseparable structure. Furthermore, the nonseparable approximation has performance very similar to the original full nonseparable model.

4 Discussion

In this work, we have investigated the performance of an approximation for the full nonseparable covariance model using the decomposition based on the Kronecker product of two separable matrices of minor dimensions. A sensitivity study was performed showing that the approximate approach provides important gains in computational efficiency while keeping the predictive power. Although taking advantage of approximations to compute the likelihood, our proposal keeps interpretation and flexibility.

In terms of prediction, the nonseparable model presents better results than the separable model, even when the approximation is considered. We conclude that it is better to consider a separable approximation of the nonseparable described model than to consider the separable structure. The nonseparable approximation reduces considerably the computational cost and keeps predictive power, which is usually the main focus of spatiotemporal data analysis.

References

1. Banerjee, S., Carlin, B.P., Gelfand, A.E.: Hierarchical Modeling and Analysis for Spatial Data. Monographs on Statistics and Applied Probability, 1st edn. Chapman & Hall/CRC, London (2004)
2. Erbisti, R., Fonseca, T., Alves, M.: Bayesian covariance modeling of multivariate spatial random fields. Version 1, 20 July (2017). arXiv:1707.06697
3. Apanasovich, T.V., Genton, M.G.: Cross-covariance functions for multivariate random fields based on latent dimensions. Biometrika **97**(1), 15–30 (2010)
4. Cox, T.F., Cox, M.A.A.: Multidimensional Scaling. Chapman & Hall/CRC, London (2000)
5. Genton, M.G.: Separable Approximations Of Space-time Covariance Matrices. Environmetrics **18**, 681–695 (2007)
6. Golub, G.H., Van Loan, C.F.: Matrix Computations. The Johns Hopkings University Press, Baltimore (1996)
7. Plummer, M., Best, N., Cowles, K., Vines, K.: CODA: convergence diagnosis and output analysis for MCMC. R News **6**, 7–11 (2006)
8. Gneiting, T., Raftery, A.E.: Strictly proper scoring rules, prediction and estimation. J. Am. Stat. Assoc. **102**(477), 360–378 (2007)
9. Ibrahim, J.G., Chen, M.-H., Sinha, D.: Bayesian Survival Analysis, 1st edn. Springer, Berlin (2001)

Uncertainty Quantification and Cumulative Distribution Function: How are they Related?

Roberta Lima and Rubens Sampaio

Abstract Uncertainty is described by the cumulative distribution function (CDF). Using, the CDF one describes all the main cases: the discrete case, the case when a absolutely continuous probability density exists, and the singular case, when it does not, or combinations of the three preceding cases. The reason one does not see any mention of uncertainty quantification in classical books, as Feller's and Chung's, is that they found no reason to call a CDF by another name. However, one has to acknowledge that to use a CDF to describe uncertainty is clumsy. The comparison of CDF to see which is more uncertain is not evident. One feels that there must be a simpler way. Why not to use some small set of statistics to reduce a CDF to a simpler measure, easier to grasp? This seems a great idea and, indeed, one finds it in the literature. Several books deal with the problem. We focus the discussion on three main cases: (1) to use mean and standard deviation to construct an envelope with them; (2) to use coefficient of variation; (3) to use Shannon entropy, a number, that could allow an ordering for the uncertainties of all CDF that have entropy, a most desirable thing. The reductions (to replace the CDF for a small set of statistics) may indeed work in some cases. But they do not always work and, moreover, the different measures they define may not be compatible. That is, the ordering of uncertainty may vary depending what set one chooses. So the great idea does not work so far, but they are happily used in the literature. One of the objectives of this paper is to show, with examples, that the three reductions used to "measure" uncertainties are not compatible. The reason it took so long to find out the mistake is that these reductions

R. Lima (✉) · R. Sampaio
PUC-Rio, Department of Mechanical Engineering, Rua Marquês de São Vicente, 225, Gávea 22451-900, Brazil
e-mail: robertalima@puc-rio.br

R. Sampaio
e-mail: rsampaio@puc-rio.br

A. Polpo et al. (eds.), *Bayesian Inference and Maximum Entropy Methods in Science and Engineering*, Springer Proceedings in Mathematics & Statistics 239, https://doi.org/10.1007/978-3-319-91143-4_24

methods are applied to very complex problem that hide well the unsuitability of the reductions. Once one tests them with simpler examples one clearly sees their inadequacy. So, let us safely continue to use the CDF while a good reduction is not found!

Keywords Uncertainty quantification · Cumulative distribution function Measures of uncertainty · Statistics · Entropy · Variance · Coefficient of variation

1 Introduction

Uncertainty is, certainly, described by the cumulative distribution function (CDF). Since all random variables have a CDF, one associates an uncertainty to each of them. Using the CDF, one describes the three main cases and their combinations: when there is a probability density function, the discrete case, and the singular case. That is, of course, the reason why one does not see any mention to uncertainty quantification in classical books as [1, 2]. The authors saw no reason to call a CDF by another name. CDF is uncertainty. The prescription of a CDF is its quantification.

However, one has to acknowledge that to use a CDF to describe uncertainty is clumsy. One feels that there must be a simpler measure, easier to grasp. Why not use some small set of statistics to reduce a CDF to a simpler and more easy to grasp measure? This seems a great idea and, indeed, one finds it in the literature. It is common to find papers using statistics as measures. Variance, coefficient of variation, and Shannon entropy are the statistics most used. The idea is to associate a number to the CDF.

We focus the discussion on three main cases:

1. to use the Shannon entropy;
2. to use mean and standard deviation to construct an envelope with them to make a nice graph;
3. to use mean and coefficient and variation.

In the probabilistic context, the Shannon entropy, S, [3, 4] of a random variable is viewed as a measure of the information carried by the associated probabilistic distribution. Sometimes, the Shannon entropy is used as a synonym of uncertainty. In this paper, uncertainty is the CDF, entropy is a statistics computed from a CDF. It reflects some properties of the CDF, but not all. In the case of discrete random variables, S is defined using the mass function. In the case of continuous random variables with a derivative, using the probability density function.

For some probabilities, one can associate a mean μ, variance σ^2, and coefficient of variation $\delta = \sigma/\mu$ (ratio between the standard deviation and mean). The mean of a probability mass, or distribution, is the best approximation of it by a number, and the absolute error of this approximation, in the mean square sense, is the variance. The coefficient of variation is a measure of the relative error.

It is common to find papers using statistics as measures of uncertainty, as we can see in [5–10]. Please note that Statements as *given a random variable with fixed mean, when the variance grows, the level of uncertainties also grows* or *for two random variables with different mean and variance, the random variable with the higher coefficient of variation is always the more uncertain* are meaningless.

The reductions—to replace the CDF for a small set of statistics—may indeed work in some cases. But they do not always work and, moreover, the different measures they define may not be compatible. That is, the ordering of uncertainty, in the measure defined by the statistics, may vary depending on what set of statistics one chooses. So the great idea does not work so far, but it is happily used in the literature.

This paper is organized as follows. In Sect. 2, we discuss briefly the meaning of the mean, variance, coefficient of variation, and the Shannon entropy for continuous random variables. Then, in Sect. 3, we show some examples of density functions of continuous random variables for which an increase of variance or coefficient of variation may not cause an increase of the Shannon entropy. The addressed density functions are well known and are frequently found as a probabilistic model in several applications. We analyze a bimodal family and, Gaussian and Gamma density functions.

2 Mean, Variance, Coefficient of Variation, and Shannon Entropy

2.1 Mean, Variance, and Coefficient of Variation

Given a probability space $(\Omega, \mathtt{F}, Pr)$, where Ω is a sample space, $\mathtt{F}$ is an event space, and Pr a probability measure on $(\Omega, \mathtt{F})$. If X is a continuous random variable on $(\Omega, \mathtt{F}, Pr)$ with density function p, then the expectation or mean of X is defined by

$$\mu = E[X] = \int_{-\infty}^{\infty} x p(x)\, dx\,, \tag{1}$$

whenever this integral converges absolutely (in that $\int_{-\infty}^{\infty} |xp(x)|\, dx < \infty$) [11]. The variance of X is defined by

$$\sigma^2 = E[(X-\mu)^2] = \int_{-\infty}^{\infty} (x-\mu)^2 p(x)\, dx\,, \tag{2}$$

whenever this integral converges absolutely.

Random variables possess a Hilbertian structure [12, 13]. Let $L^2(\Omega)$ be a Hilbert space of random variables. The norm of an element $X \in L^2(\Omega)$ is:

$$\|X\| = \sqrt{E[X^2]} \,. \tag{3}$$

If we would like to compute the best approximation of a random variable X by a constant, we may determine the value $m \in \mathbb{R}$ such that

$$m = \arg\min \{\|X - \lambda\| : \lambda \in \mathbb{R}\} \,. \tag{4}$$

We look for the orthogonal projection of X onto a linear subspace having dimension 1, which is given by

$$V = \left\{Z \in L^2(\Omega) : Z \text{ is constant: } Z(\omega) = \lambda \in \mathbb{R},\ \forall \omega \in \Omega\right\} \tag{5}$$

Using the definition of the norm given in Eq. (3)

$$\begin{aligned} m &= \arg\min E[(X - \lambda)^2] \\ &= \arg\min \left\{E[X^2] - 2E[X]\lambda + \lambda^2\right\} \end{aligned} \tag{6}$$

Computing the derivative and making it equal to zero, it is possible to find the solution

$$m = E[X] \tag{7}$$

Thus, the mean is the best approximation of a random variable X by a constant, as shown in Fig. 1.

The norm of the absolute error of the approximation is

$$\|X - m\| = \sqrt{E[(X - E[X])^2]} = \sqrt{\sigma^2} \,, \tag{8}$$

i.e., the error is the square root of the variance of X. The variance and coefficient of variation are related to errors, respectively the absolute and relative errors, of the approximation of X by a constant. These statistics are not measures of the uncertainty of X, in the sense, they cannot give the CDF.

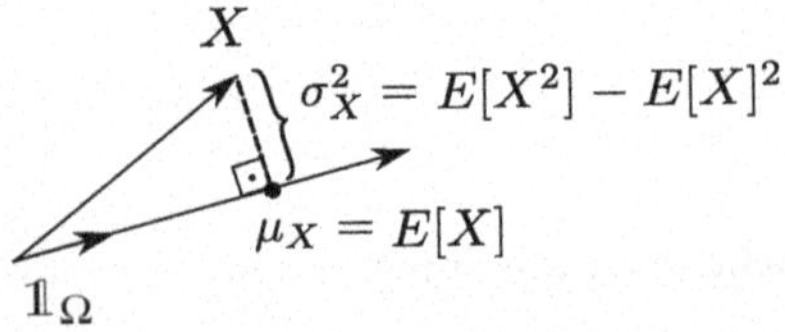

Fig. 1 Orthogonal projection of X onto a linear subspace having dimension 1

2.2 Shannon Entropy

The Shannon entropy of a continuous random variable is [14]

$$S(X) = -\int_{-\infty}^{\infty} p(x) \ln p(x)\, dx\,. \tag{9}$$

We set $0 \ln 0 = 0$. The entropy of a continuous random variable can take negative values. For example: if X is uniformly distributed in $[0, 2^n]$, then $S = -\int_0^{2^n} \frac{1}{2^n} \ln\left(\frac{1}{2^n}\right) dx = \ln 2^n$.

3 Examples

3.1 Bimodal Density Function

Consider, a family of continuous random variables X_d, parameterized by d, with a bimodal density function p_d symmetrically distributed around its mean μ. The function p_d, sketched in Fig. 2, is

$$p_d(x) = \begin{cases} 1, & x \in [\mu - (d/2) - (1/2)\,,\, \mu - (d/2)]\,, \\ 1, & x \in [\mu + (d/2)\,,\, \mu + (d/2) + (1/2)]\,, \\ 0, & \text{in all others cases.} \end{cases} \tag{10}$$

The variance of this family of random variables X_d is

$$\sigma_d^2 = E[(X_d - \mu)^2] = \frac{1}{12}(3d^2 + 3d + 1)\,. \tag{11}$$

The entropy is

$$S_d = -\int_{-\infty}^{\infty} p_d(x) \ln p_d(x)\, dx = 0\,. \tag{12}$$

Fig. 2 Family of bimodal density functions p_d

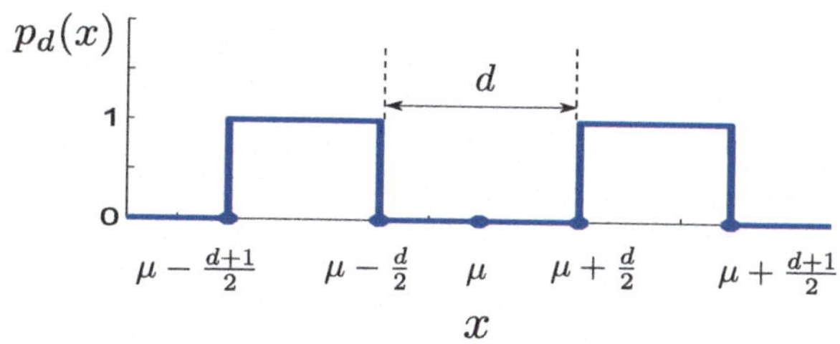

Observing Eqs. (2), and (3), we verify that for a given value of μ, as d (distance between peaks of the bimodal density function) increases, the variance, σ_d^2, and the coefficient of variation, $\delta_d = \sigma_d/\mu$ (for $\mu \neq 0$), increase. However, the entropy S_d remains constant and equal to zero.

This example of the bimodal family shows quite well that entropy, mean, variance, and coefficient of variation are different things. One can vary $\mu, \sigma^2, \delta = \sigma/\mu$ independently with fixed entropy.

3.2 Gaussian Density Function

The next example deals with a continuous random variable. Consider a family of continuous random variables X_μ, parametrized by μ, with a Gaussian density function p_μ

$$p(x)_\mu = \frac{1}{\sqrt{2\sigma^2\pi}} e^{-\frac{(x-\mu)^2}{2\sigma^2}}, \tag{13}$$

where μ is the mean and σ^2 the variance. The entropy of the Gaussian density function is

$$S_\mu = -\int_{-\infty}^{\infty} p(x) \ln p(x)\, dx = \frac{1}{2} \ln (2\sigma^2\pi\, e). \tag{14}$$

Observing Eq. (14), we verify that the mean μ does not enter the final formula of the entropy. This means that all Gaussian functions with a common variance σ^2 have the same entropy. A translation changes μ and $\delta = \sigma/\mu$, but does not change the value of S_μ. Increasing the mean, δ decreases, and decreasing the mean, δ increases.

3.3 Gamma Density Function

The next example also shows that it is possible that an increase of the variance corresponds to a decrease of entropy. Consider, a family of continuous random variables X with a Gamma density function p

$$p(x) = \frac{1}{\Gamma(k)\theta^k} x^{k-1} e^{-\frac{x}{\theta}}. \tag{15}$$

written as a function of the parameters $k > 0$ and $\theta > 0$, a shape parameter and a scale parameter, respectively. The mean of X is $\mu = k\theta$ and the variance $\sigma^2 = k\theta^2$. The entropy is [15]

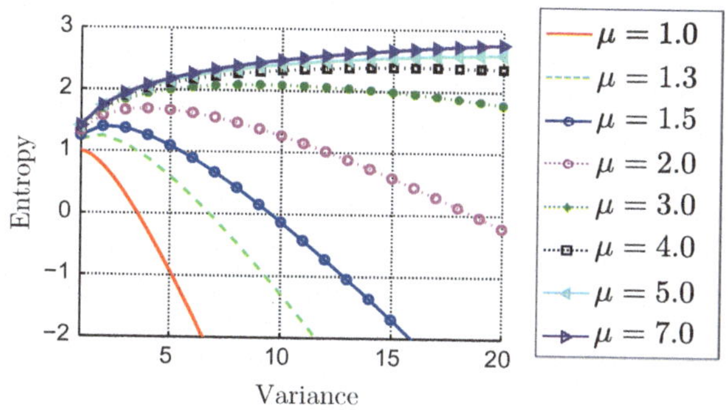

Fig. 3 Entropy as function of the variance of a continuous random variable with a Gamma density function

$$S = -\int_{-\infty}^{\infty} p(x) \ln p(x)\, dx = k + \ln \theta + \ln[\Gamma(k)] + (1 - k)\psi(k). \quad (16)$$

where ψ is the digamma function. In Fig. 3 it is shown the graph of the entropy S as function of the variance σ^2 for different values of the mean μ of X. We verify that for the smaller value of the mean $\mu = 1.0$, the value of S decreases as the variance increases. When $\mu = 3.0$, an interesting behavior can be observed. For values of variance lower than 9.0, S increases as the variance increases. However, for values of variance above 9.0, S decreases as the variance increases. This means that entropy and σ^2 may not vary in the same sense. Thus, if the Shannon entropy S is considered a measure of the level of uncertainty, the variance cannot be taken as a measure of the level of uncertainty either.

4 Conclusions

In this paper, we show that the use of statistics such as variance, σ^2, coefficient of variation, $\delta = \sigma/\mu$, and Shannon entropy, S, as measures of uncertainty of a random variable is not a good practice.

The use of these statistics as measures of uncertainties may lead to contradiction. In this paper, we constructed simple examples for which the increase of variance, or coefficient of variation, does not correspond to an increase of the Shannon entropy. In particular, we showed that for the Gamma density function, it is possible to have a decrease of entropy with an increase of the variance.

Besides not always giving right results, the use of statistics as a measure of uncertainty presents others inconsistencies. In relation to the Shannon entropy, an inconsistency appears, for example, if we compare the entropy of discrete and continuous random variables. While the entropy of discrete random variables is always a positive value, the entropy of continuous random variables can assume values in $\mathbb{R}$. In relation to the use of moments as measures of uncertainties, there are two inconsistencies. The first one appears due to the fact that there are random variables that do not have any moments, for example the random variables with Cauchy density function. The second is that it is not possible to use variance and coefficient of variation for random vectors in $\mathbb{R}^n$. In more than one dimension, the correspondent to the variance is a covariance matrix in $\mathbb{R}^{n\times n}$. It is worth to remark that in the case of multivariate distributions, where the probability is distributed in a subset of $\mathbb{R}^n$, for $n \geq 2$, the mean is a vector and the variance is a matrix.

The quest for a set of statistics to measure uncertainty has not yet an answer. The solutions found in the literature do not solve the problem.

Acknowledgements The authors acknowledge the support given by FAPERJ, CNPq, and CAPES.

References

1. Chung, K.: A Course in Probability Theory. Academic Press, London (1974)
2. Feller, W.: An Introduction to Probability Theory and its Applications, vol. I and II. Wiley, New York (1957)
3. Jaynes, E.: Information theory and statistical mechanics. Phys. Rev. **106**(4), 620–630 (1957)
4. Shannon, C.: A mathematical theory of communication. Bell Syst. Tech. **27**(379–423), 623–659 (1948)
5. Chen, J., Eeden, C., Zidek, J.: Uncertainty and the conditional variance. Stat. Probab. Lett. **80**, 1764–1770 (2010)
6. Conrad, K.: Probability distributions and maximum entropy, pp. 1–27 (2016). http://www.math.uconn.edu/~kconrad/blurbs/analysis/entropypost.pdf. Accessed 24 July 2016
7. Khosravi, A., Nahavandi, S.: An optimized mean variance estimation method for uncertainty quantification of wind power forecasts. Electr. Power Energy Syst. **61**, 446–454 (2014)
8. Motra, H., Hildebrand, J., Wuttke, F.: The Monte Carlo method for evaluating measurement uncertainty: application for determining the properties of materials. Probab. Eng. Mech. 1–9 (2016) (in press)
9. Nordström, J., Wahlsten, M.: Variance reduction through robust design of boundary conditions for stochastic hyperbolic systems of equations. J. Comput. Phys. **282**, 1–22 (2015)
10. Zidek, J., Eeden, C.: Uncertainty, Entropy, Variance and the Effect of Partial Information. Lecture Notes-Monograph Series, vol. 42. Institute of Mathematical Statistics (2003)
11. Grimmett, G., Welsh, D.: Probability an Introduction. Oxford Science Publications, New York (1986)
12. Souza de Cursi, E., Sampaio, R.: Uncertainty Quantification and Stochastic Modeling with Matlab. Elsevier, ISTE Press (2015)
13. Sampaio, R., Lima, R.: Modelagem Estocástica e Geração de Amostras de Variáveis e Vetores Aleatórios. Notas de Matemática Aplicada, vol. 70. SBMAC (2012). http://www.sbmac.org.br/arquivos/notas/livro_70.pdf
14. Ebrahimi, N., Maasoumi, E., Soofi, E.: Ordering univariate distributions by entropy and variance. J. Econ. **90**, 317–336 (1999)
15. Khodabin, M., Ahmadabadi, A.: Some properties of generalized gamma distribution. Math. Sci. **4**(1), 9–28 (2010)

Maximum Entropy Analysis of Flow Networks with Structural Uncertainty (Graph Ensembles)

Robert K. Niven, Michael Schlegel, Markus Abel, Steven H. Waldrip and Roger Guimera

Abstract This study examines MaxEnt methods for probabilistic inference of the state of flow networks, including pipe flow, electrical and transport networks, subject to physical laws and observed moments. While these typically assume networks of invariant graph structure, we here consider higher-level MaxEnt schemes, in which the network structure constitutes part of the uncertainty in the problem specification. In physics, most studies on the statistical mechanics of graphs invoke the Shannon entropy $H_G^{Sh} = -\sum_{\Omega_G} P(G) \ln P(G)$, where G is the graph and Ω_G is the graph ensemble. We argue that these should adopt the relative entropy $H_G = -\sum_{\Omega_G} P(G) \ln P(G)/Q(G)$, where $Q(G)$ is the graph prior associated with the graph macrostate G. By this method, the user is able to employ a simplified accounting over graph macrostates rather than need to count individual graphs. Using

---

R. K. Niven (✉) · S. H. Waldrip
School of Engineering and Information Technology, The University of New South Wales, Canberra, NSW 2600, Australia
e-mail: r.niven@adfa.edu.au

S. H. Waldrip
e-mail: Steven.Waldrip@student.adfa.edu.au

M. Schlegel
Technische Universität Berlin, Berlin, Germany
e-mail: Michael.Schlegel@tu-berlin.de

M. Abel
Ambrosys GmbH/University of Potsdam, Potsdam, Germany
e-mail: markus.abel@ambrosys.de

R. Guimera
Rovira i Virgili University, Tarragona, Spain
e-mail: roger.guimera@urv.cat

A. Polpo et al. (eds.), *Bayesian Inference and Maximum Entropy Methods in Science and Engineering*, Springer Proceedings in Mathematices & Statistics 239, https://doi.org/10.1007/978-3-319-91143-4_25

combinatorial methods, we here derive a variety of graph priors for different graph ensembles, using different macrostate partitioning schemes based on the node or edge counts. A variety of such priors are listed herein, for ensembles of undirected or directed graphs.

Keywords Maximum entropy · Graphs · Networks · Graph priors · Graph ensemble

1 Introduction

Over the past few years, we have developed a MaxEnt framework to infer the state of flow on all types of flow networks, for example, pipe flow, electrical, communications and transport networks [1–4]. In this approach, the user adopts the relative entropy:

$$H_{\mathbf{X}} = -\int_{\Omega_{\mathbf{X}}} P(\mathbf{X}) \ln \frac{P(\mathbf{X})}{Q(\mathbf{X})} d\mathbf{X} \tag{1}$$

in which $\mathbf{X}$ are the unknown network parameters (such as flow rates and potentials), $P(\mathbf{X})$ is a joint probability density function (pdf) over $\mathbf{X}$, $Q(\mathbf{X})$ is the prior pdf, and $\Omega_{\mathbf{X}}$ is the domain of $\mathbf{X}$. The entropy (1) is maximized, subject to the constraints on the network, to infer the state of the network. The constraints necessarily include all relevant physical laws (such as Kirchhoff's node and loop laws), as well as any physical observations measured at particular nodes, edges or over components of the network. The resulting inference is expressed in terms of the pdf $P(\mathbf{X})$, which can either be used directly, or from which the moments or other statistical features of the flow (e.g., mean, mode, variances) can be extracted. The MaxEnt method, therefore, provides one approach to extend previous deterministic methods for flow network analysis, applicable only to fully determined networks, to a probabilistic framework which can handle incomplete information.

In the past decade, there has been a tremendous surge of interest in the structural properties of networks in statistical physics (and other fields), especially the emergent scaling features of the Internet and human social networks [5–8]. Such studies generally consider the probability $P(G)$ of a graph G within a graph ensemble Ω_G, almost always inferred by maximizing the Shannon entropy [9]:

$$H_G^{Sh} = -\sum_{G \in \Omega_G} P(G) \ln P(G) \tag{2}$$

While correct, this formulation does not exploit the fundamental advantage of statistical mechanics, based on the separate counting of observable macrostates and their underlying microstates. Instead of (2), network analysts and graph theorists would be better advised to adopt the discrete relative entropy (negative Kullback–Leibler) function [10]

$$H_G = -\sum_{G \in \Omega_G} P(G) \ln \frac{P(G)}{Q(G)} \tag{3}$$

now based on the graph macrostates G, defined as equivalence classes (sets) of graphs which partition the ensemble Ω_G. $P(G)$ and $Q(G)$ now represent the posterior and prior probabilities of the macrostate G within the graph ensemble Ω_G. Eq. (3) ensures that maximizing (3), subject only to normalization, gives the inferred state $P^*(G) = Q(G)$ [11]. Further constraints will then restrict the ensemble, either by removing (microcanonical ensemble) or weighting (canonical ensemble) its constituent graphs, causing $P^*(G)$ to deviate from $Q(G)$ consistent with these constraints. In contrast, the Shannon form (2) requires the counting of each individual graph in an ensemble, which may be quite onerous for large ensembles, and does not provide the user with any insights from the network structure.

We can indeed unite the above fields, to present a MaxEnt framework for probabilistic inference of flows on a network, subject to uncertainty in the flow parameters and *in the network structure itself*. This entails use of the relative entropy function:

$$H_{G,\mathbf{X}(G)} = -\sum\nolimits_{\Omega_G} \int_{\Omega_{\mathbf{X}(G)}} P(\mathbf{X}(G), G) \ln \frac{P(\mathbf{X}(G), G)}{Q(\mathbf{X}(G), G)} d\mathbf{X} \tag{4}$$

where $P(\mathbf{X}(G), G)$ and $Q(\mathbf{X}(G), G)$ are the joint posterior and prior pdfs, defined over parameters $\mathbf{X}$ and graph macrostates G. As a first step, analyzes using (4) will generally invoke the Bayesian separation:

$$Q(\mathbf{X}(G), G) = Q(G) Q(\mathbf{X}(G)|G) \tag{5}$$

based on the distinct graphical and flow priors $Q(G)$ and $Q(\mathbf{X}(G)|G)$. For some flow networks, complete separability $Q(\mathbf{X}(G), G) = Q(G)Q(\mathbf{X})$ may be possible.

The aim of this study is to formally derive the priors $Q(G)$ for graph macrostates in a variety of graph ensembles, as a prelude to later studies on the joint graph and flow parameter priors $Q(\mathbf{X}(G), G)$. We here note that graph priors will not only depend on the graph ensemble selected, but also on the rule (equivalence relation) used to partition the ensemble into graph macrostates. Different partitioning rules will obviously give rise to different priors —there are many ways to count cats in a collection of user-selected baskets of cats. The choice of ensemble and partitioning scheme must, therefore, be made by the user, and so will depend on his/her purpose, although some approaches will be more mathematically tractable or fruitful.

2 Derivation of Graph Priors

In statistical physics, the *degeneracy* $g(G)$ of a discrete macrostate G can be defined as its statistical weight or number of occurrences in the ensemble [12], counting each component graph (or microstate) once each. If the entire ensemble Ω_G is also countable and finite, then the prior can be calculated by

$$Q(G) = \frac{g(G)}{|\Omega_G|} \tag{6}$$

where $|\Omega_G|$ is the cardinal number of Ω_G. If the macrostate and ensemble are both countably infinite or uncountable, it may be possible to define the prior by a limiting process applied to (6), although many such priors will be found to vanish asymptotically.

We here calculate priors for various graph macrostates G in a range of graph ensembles Ω_G, using partitioning rules based on the numbers of nodes and/or edges. These include both undirected and directed graph ensembles, each discussed in turn. The complete sets of results are summarized, respectively, in Tables 1 and 2. In the following, all nodes and edges are considered distinguishable (are labeled), and so are counted according to their multiplicities in each macrostate and the ensemble. Where present, self-edges are each counted only once.

2.1 Undirected Graph Priors

In turn, we discuss various partitioning schemes for undirected graph ensembles, which are appropriate for the analysis of potential-driven flows, such as electricity, pipe flow and chemical networks. All results are tabulated in Table 1.

(1) We first consider an ensemble of simple graphs with N nodes, which can be partitioned into macrostates based on the number of edges M. We disallow self-loops. By a little consideration, it will be seen that the degeneracy of each such macrostate can be derived by the allocation of M indistinguishable digits (such as 1s) to the upper triangle of elements A_{ij} of the adjacency matrix $\mathcal{A}$, with all occupancies restricted to $\{0, 1\}$. This gives degeneracy $g = \binom{T_{N-1}}{M}$, based on the $(N-1)$th triangle number $T_{N-1} = \frac{1}{2}N(N-1)$ of independent matrix elements. By summation or direct allocation, the ensemble itself can be shown to have cardinal number $2^{T_{N-1}}$, hence, the prior is obtained as $Q(G) = \binom{T_{N-1}}{M}/2^{T_{N-1}}$.

(2) We then embed the above 'microcanonical' ensemble into a 'canonical' ensemble of all simple undirected graphs with $n \leq N$ nodes, for fixed N. We consider three different partitioning schemes:

(a) A partitioning scheme based on the macrostates with n nodes and M edges. By construction from (1), we directly obtain the degeneracy $g = \binom{T_{n-1}}{M}$ and ensemble dimension $\sum_{n=1}^{N} 2^{T_{n-1}}$, hence giving the corresponding prior.

(b) A partitioning scheme based on the macrostates with M edges, regardless of the number of nodes. From (a), the degeneracy is then $g = \sum_{n=1}^{N} \binom{T_{n-1}}{M}$, while the ensemble dimension is unchanged, giving the corresponding prior.

(c) A partitioning scheme based on the macrostates with n nodes, regardless of the number of edges. The degeneracy is then given by the subset ensemble with n nodes, of dimension $2^{T_{n-1}}$; using the known ensemble dimension then gives the prior.

(3) We now consider the ensemble of single-edge undirected graphs with N nodes, now allowing for self-loops. We consider two partitioning schemes:

(a) In the first scheme, we partition the ensemble into macrostates of graphs with N nodes, L self-edges and m non-self-edges. The number of graphs in each such macrostate is given by the number of ways to allocate m elements in the upper triangle of the adjacency matrix, without self-loops $\binom{T_{N-1}}{m}$, multiplied by the number of ways to allocate L self-loops amongst the N diagonal elements $\binom{N}{L}$. The ensemble now has dimension 2^{T_N}, based on the Nth triangle number $T_N = \frac{1}{2}(N+1)N$, since it includes the diagonal adjacency elements A_{ii}. The prior is thus $Q(G) = \binom{T_{N-1}}{m}\binom{N}{L}/2^{T_N}$.

(b) In the second scheme, we form graph macrostates with N nodes and M edges, regardless of the number of self-loops. We now use a simpler allocation scheme of elements to the upper triangle of the adjacency matrix, including diagonal elements, giving the degeneracy $\binom{T_N}{M}$, and corresponding prior. For $m = M - L$, it can be verified that $\sum_{L=0}^{N} \binom{T_{N-1}}{m}\binom{N}{L} = \binom{T_N}{M}$, so the two partitions give the same results (although the first requires more information).

(4) We again embed the above 'microcanonical' ensembles into a 'canonical' ensemble of all undirected graphs with $n \le N$ nodes, allowing self-loops. We again consider several partitioning schemes:

(a) Graphs with n nodes, m non-self-edges and L self-edges;
(b) Graphs with any nodes, m non-self-edges and L self-edges;
(c) Graphs with any nodes, M total edges including self-edges;
(d) Graphs with n nodes and any edges including self-edges.

The resulting degeneracies, ensemble dimension, and priors follow by construction from those in (3), and are set out in Table 1.

(5) We now consider the ensemble of undirected multigraphs — i.e., with the possibility of parallel edges including self-loops —and with N nodes. To keep the ensemble finite, we restrict the total number of edges to C. We wish to examine graph macrostates with N nodes and $M \le C$ edges. Following the previous logic, we must now consider the allocation of M edges to T_N adjacency matrix

elements, without restriction on occupancy, giving the degeneracy $g = \binom{T_N+M-1}{M}$ (n.b., similar to the allocation scheme for Bose–Einstein statistics [12–14]). By summation of this result over $M = 0...C$, it can be shown the ensemble has dimension $\binom{T_N+C}{C}$, leading to the corresponding prior.

(6) We can further embed the above ensemble (5) in a larger 'canonical ensemble' of undirected multigraphs with $n \leq N$ nodes, again with the restriction of maximum C edges. We again consider several partitions:

(a) Graphs with n nodes, M total edges;
(b) Graphs with any nodes, M total edges
(c) Graphs with n nodes and any total edges.

The degeneracies, ensemble dimension, and priors follow by construction from (5), and are set out in Table 1.

2.2 Directed Graph Priors

We now replicate the above ensembles and partitioning schemes, but this time for directed graph structures, generally required for the analysis of transportation networks. These results are set out in the same pattern in Table 2 as for the undirected graph ensembles, and mostly exhibit the same features, but with the following distinctions:

(i) The macrostates of simple or single-edge digraphs, based on the allocation of edges to the $N \times N$ adjacency matrix, now must account for $2T_{N-1}$ independent elements if there are no self-loops, or N^2 elements with self-loops.

(ii) The macrostates of multidigraphs are now based on the allocation of M edges to N^2 elements, without restriction on occupancies, giving the degeneracy $g = \binom{N^2+M-1}{M}$ and a corresponding ensemble dimension of $\binom{N^2+C}{C}$.

2.3 Asymptotic Limits

From Tables 1 and 2, most of the calculated priors for the canonical ensembles (those with $n \leq N$) vanish in the asymptotic limit $N \to \infty$. Interestingly, some do not appear to do so. One such prior is that for multigraph macrostates identified by $U_N^{\text{multi},M}$ in the undirected multigraph ensemble (Table 1). While, we do not have a mathematical proof, numerical analyses suggest the following limits:

Table 1 Dimensions, degeneracies and prior probabilities for various partitions of undirected graph ensembles

Ensemble			Macrostate				
Symbol Ω_G	Description	Dimension $\lvert\Omega_G\rvert$	Symbol G	Description	Degeneracy $g(G)$	Prior Prob. $Q(G)$	Label
Ω_{U_N}	All simple undirected graphs with N nodes	$2^{T_{N-1}}$	U_N^M	Graphs with N nodes, M edges	$\binom{T_{N-1}}{M}$	$\dfrac{\binom{T_{N-1}}{M}}{2^{T_{N-1}}}$	(U1)
$\Omega_{U_{n\le N}}$	All simple undirected graphs with $n \le N$ nodes	$\sum\limits_{n=1}^{N} 2^{T_{n-1}}$	U_n^M	Graphs with n nodes, M edges	$\binom{T_{n-1}}{M}$	$\dfrac{\binom{T_{n-1}}{M}}{\sum\limits_{n=1}^{N} 2^{T_{n-1}}}$	(U2)
			U^M	Graphs with any nodes, M edges	$\sum\limits_{n=1}^{N} \binom{T_{n-1}}{M}$	$\dfrac{\sum\limits_{n=1}^{N} \binom{T_{n-1}}{M}}{\sum\limits_{n=1}^{N} 2^{T_{n-1}}}$	(U3)
			U_n	Graphs with n nodes, any edges	$2^{T_{n-1}}$	$\dfrac{2^{T_{n-1}}}{\sum\limits_{n=1}^{N} 2^{T_{n-1}}}$	(U4)

(continued)

Table 1 (continued)

Ensemble			Macrostate				
Symbol Ω_G	Description	Dimension $\|\Omega_G\|$	Symbol G	Description	Degeneracy $g(G)$	Prior Prob. $Q(G)$	Label
$\Omega_{U_N^{\circlearrowleft}}$	All single-edge undirected graphs with N nodes, incl. self-edges	2^{T_N}	$U_N^{L\circlearrowleft,m}$	Graphs with N nodes, m non-self-edges, L self-edges	$\binom{T_{N-1}}{m}\binom{N}{L}$	$\frac{\binom{T_{N-1}}{m}\binom{N}{L}}{2^{T_N}}$	(U5)
			$U_N^{\circlearrowleft,M}$	Graphs with N nodes, M total edges incl. self-edges	$\binom{T_N}{M}$	$\frac{\binom{T_N}{M}}{2^{T_N}}$	(U6)
$\Omega_{U_{n\le N}^{\circlearrowleft}}$	All single-edge undirected graphs with $n \le N$ nodes, incl. self-edges	$\sum_{n=1}^{N} 2^{T_n}$	$U_n^{L\circlearrowleft,m}$	Graphs with n nodes, m non-self-edges, L self-edges	$\binom{T_{n-1}}{m}\binom{n}{L}$	$\frac{\binom{T_{n-1}}{m}\binom{n}{L}}{\sum_{n=1}^{N} 2^{T_n}}$	(U7)
			$U^{L\circlearrowleft,m}$	Graphs with any nodes, m non-self-edges, L self-edges	$\sum_{n=1}^{N}\binom{T_{n-1}}{m}\binom{n}{L}$	$\frac{\sum_{n=1}^{N}\binom{T_{n-1}}{m}\binom{n}{L}}{\sum_{n=1}^{N} 2^{T_n}}$	(U8)
			$U^{\circlearrowleft,M}$	Graphs with any nodes, M total edges incl. self-edges	$\sum_{n=1}^{N}\binom{T_n}{M}$	$\frac{\sum_{n=1}^{N}\binom{T_n}{M}}{\sum_{n=1}^{N} 2^{T_n}}$	(U9)

(continued)

Table 1 (continued)

Ensemble			Macrostate				
Symbol Ω_G	Description	Dimension $\lvert\Omega_G\rvert$	Symbol G	Description	Degeneracy $g(G)$	Prior Prob. $Q(G)$	Label
			$U_n^{\circlearrowleft}$	Graphs with n nodes, any edges incl. self-edges	2^{T_n}	$\frac{2^{T_n}}{\sum_{n=1}^{N} 2^{T_n}}$	(U10)
$\Omega_{U_N^{\text{multi},M\le C}}$	All undirected multigraphs with N nodes, up to C edges incl. self-edges	$\binom{T_N+C}{C}$	$U_N^{\text{multi},M}$	Multigraphs with N nodes, M total edges	$\binom{T_N+M-1}{M}$	$\frac{\binom{T_N+M-1}{M}}{\binom{T_N+C}{C}}$	(U11)
$\Omega_{U_{n\le N}^{\text{multi},M\le C}}$	All undirected multigraphs with $n \le N$ nodes, up to C edges incl. self-edges	$\sum_{n=1}^{N} \binom{T_n+C}{C}$	$U_n^{\text{multi},M}$	Multigraphs with n nodes, M total edges	$\binom{T_n+M-1}{M}$	$\frac{\binom{T_n+M-1}{M}}{\sum_{n=1}^{N} \binom{T_n+C}{C}}$	(U12)
			$U^{\text{multi},M}$	Multigraphs with any nodes, M total edges	$\sum_{n=1}^{N} \binom{T_n+M-1}{M}$	$\frac{\sum_{n=1}^{N} \binom{T_n+M-1}{M}}{\sum_{n=1}^{N} \binom{T_n+C}{C}}$	(U13)
			U_n^{multi}	Multigraphs with n nodes, any total edges	$\binom{T_n+C}{C}$	$\frac{\binom{T_n+C}{C}}{\sum_{n=1}^{N} \binom{T_n+C}{C}}$	(U14)

Table 2 Dimensions, degeneracies and prior probabilities for various partitions of directed graph ensembles

Ensemble			Macrostate				
Symbol Ω_G	Description	Dimension $\lvert\Omega_G\rvert$	Symbol G	Description	Degeneracy $g(G)$	Prior Prob. $Q(G)$	Label
Ω_{D_N}	All simple digraphs with N nodes	$4^{T_{N-1}}$	D_N^M	Digraphs with N nodes, M edges	$\binom{2T_{N-1}}{M}$	$\dfrac{\binom{2T_{N-1}}{M}}{4^{T_{N-1}}}$	(D1)
$\Omega_{D_{n\le N}}$	All simple digraphs with $n \le N$ nodes	$\sum\limits_{n=1}^{N} 4^{T_{n-1}}$	D_n^M	Digraphs with n nodes, M edges	$\binom{2T_{n-1}}{M}$	$\dfrac{\binom{2T_{n-1}}{M}}{\sum\limits_{n=1}^{N} 4^{T_{n-1}}}$	(D2)
			D^M	Digraphs with any nodes, M edges	$\sum\limits_{n=1}^{N} \binom{2T_{n-1}}{M}$	$\dfrac{\sum\limits_{n=1}^{N} \binom{2T_{n-1}}{M}}{\sum\limits_{n=1}^{N} 4^{T_{n-1}}}$	(D3)
			D_n	Digraphs with n nodes, any edges	$4^{T_{n-1}}$	$\dfrac{4^{T_{n-1}}}{\sum\limits_{n=1}^{N} 4^{T_{n-1}}}$	(D4)

(continued)

Table 2 (continued)

Ensemble			Macrostate				
Symbol Ω_G	Description	Dimension $\lvert\Omega_G\rvert$	Symbol G	Description	Degeneracy $g(G)$	Prior Prob. $Q(G)$	Label
$\Omega_{D_N^{\circlearrowleft}}$	All non-multiedge digraphs with N nodes, incl. self-edges	2^{N^2}	$D_N^{L\circlearrowleft,m}$	Digraphs with N nodes, m non-self-edges, L self-edges	$\binom{2T_{N-1}}{m}\binom{N}{L}$	$\dfrac{\binom{2T_{N-1}}{m}\binom{N}{L}}{2^{N^2}}$	(D5)
			$D_N^{\circlearrowleft,M}$	Digraphs with N nodes, M total edges incl. self-edges	$\binom{N^2}{M}$	$\dfrac{\binom{N^2}{M}}{2^{N^2}}$	(D6)
$\Omega_{D_{n\le N}^{\circlearrowleft}}$	All non-multiedge digraphs with $n \le N$ nodes, incl. self-edges	$\sum\limits_{n=1}^{N} 2^{n^2}$	$D_n^{L\circlearrowleft,m}$	Digraphs with n nodes, m non-self-edges, L self-edges	$\binom{2T_{n-1}}{m}\binom{n}{L}$	$\dfrac{\binom{2T_{n-1}}{m}\binom{n}{L}}{\sum\limits_{n=1}^{N} 2^{n^2}}$	(D7)
			$D^{L\circlearrowleft,m}$	Digraphs with any nodes, m non-self-edges, L self-edges	$\sum\limits_{n=1}^{N}\binom{2T_{n-1}}{m}\binom{n}{L}$	$\dfrac{\sum\limits_{n=1}^{N}\binom{2T_{n-1}}{m}\binom{n}{L}}{\sum\limits_{n=1}^{N} 2^{n^2}}$	(D8)
			$D^{\circlearrowleft,M}$	Digraphs with any nodes, M total edges incl. self-edges	$\sum\limits_{n=1}^{N}\binom{n^2}{M}$	$\dfrac{\sum\limits_{n=1}^{N}\binom{n^2}{M}}{\sum\limits_{n=1}^{N} 2^{n^2}}$	(D9)

(continued)

Table 2 (continued)

Ensemble			Macrostate				
Symbol Ω_G	Description	Dimension $\lvert\Omega_G\rvert$	Symbol G	Description	Degeneracy $g(G)$	Prior Prob. $Q(G)$	Label
			$D_n^{\circlearrowleft}$	Digraphs with n nodes, any edges incl. self-edges	2^{n^2}	$\frac{2^{n^2}}{\sum_{n=1}^{N} 2^{n^2}}$	(D10)
$\Omega_{D_N^{\text{multi},M\le C}}$	All multidigraphs with N nodes, up to C edges	$\binom{N^2+C}{C}$	$D_N^{\text{multi},M}$	Multidigraphs with N nodes, M total edges	$\binom{N^2+M-1}{M}$	$\frac{\binom{N^2+M-1}{M}}{\binom{N^2+C}{C}}$	(D11)
$\Omega_{D_{n\le N}^{\text{multi},M\le C}}$	All multidigraphs with $n \le N$ nodes, up to C edges	$\sum_{n=1}^{N} \binom{n^2+C}{C}$	$D_n^{\text{multi},M}$	Multidigraphs with n nodes, M total edges	$\binom{n^2+M-1}{M}$	$\frac{\binom{n^2+M-1}{M}}{\sum_{n=1}^{N} \binom{n^2+C}{C}}$	(D12)
			$D^{\text{multi},M}$	Multidigraphs with any nodes, M total edges	$\sum_{n=1}^{N} \binom{n^2+M-1}{M}$	$\frac{\sum_{n=1}^{N} \binom{n^2+M-1}{M}}{\sum_{n=1}^{N} \binom{n^2+C}{C}}$	(D13)
			D_n^{multi}	Multidigraphs with n nodes, any total edges	$\binom{n^2+C}{C}$	$\frac{\binom{n^2+C}{C}}{\sum_{n=1}^{N} \binom{n^2+C}{C}}$	(D14)

$$Q(U_N^{\text{multi},M}) = \frac{\binom{T_N+M-1}{M}}{\binom{T_N+C}{C}} \xrightarrow{N\to\infty} \begin{cases} 0 & \text{if } M < C \\ & \text{or } M = C > O(N^2) \\ \frac{1}{2\alpha-1} & \text{if } M = C = O(\alpha N^2) \\ 1 & \text{if } M = C < O(N^2) \end{cases} \tag{7}$$

For the analogous multidigraph macrostate identified by $D_N^{\text{multi},M}$ in the multidigraph ensemble (Table 2), the above limits appear to be repeated, but with limit $\frac{1}{\alpha-1}$ for $M = C = \alpha N^2$. In both cases, the prior appears to vanish asymptotically for $C \to \infty$.

If these asymptotic limits (and others) can be established more rigorously, they provide the means to derive graph priors for macrostates in countably infinite graph ensembles, for which it is not possible to conduct statistical mechanics based on the counting of individual graphs.

3 Conclusions

We consider graph priors for various ensembles of undirected and directed graphs, to simplify the analysis of flow networks with uncertainty in the network structure. By combinatorial reasoning, we formally derive a collection of graph priors for various choices of graph macrostates in graph ensembles, partitioned according to the numbers of nodes and/or edges of graphs in the macrostate. The results are discussed and listed in tabular form. For simple graphs (no self-edges), single-edge graphs (allowing self-edges) or multigraphs, the 'microcanonical ensemble' constructed with a fixed number of nodes N can be embedded in a higher order 'canonical ensemble' with up to N nodes, allowing construction of more and more complicated ensembles. While most calculated priors appear to vanish asymptotically for countably infinite ensembles, some asymptotic limits have been identified numerically, for multigraphs and multidigraphs macrostates in certain ensembles. Such asymptotic results suggest a method to derive graph priors for macrostates in countably infinite graph ensembles, which cannot be handled by the counting of individual graphs.

Acknowledgements This project acknowledges funding support from the Australian Research Council Discovery Projects Grant DP140104402, Go8/DAAD Australia-Germany Joint Research Cooperation Scheme RG123832 and the French Agence Nationale de la Recherche Chair of Excellence (TUCOROM) and the Institute Prime, Poitiers, France.

References

1. Waldrip, S.H., Niven, R.K.: Maximum entropy derivation of quasi-Newton methods. SIAM J. Optim. **26**(4), 2495–2511 (2016)
2. Waldrip, S.H., Niven, R.K.: Comparison between Bayesian and maximum entropy analyses of flow networks. Entropy **19**(2), 58 (2017)

3. Waldrip, S.H., Niven, R.K., Abel, M., Schlegel, M.: Maximum entropy analysis of hydraulic pipe flow networks. J. Hydraul. Eng. ASCE **142**(9), 04016028 (2016)
4. Waldrip, S.H., Niven, R.K., Abel, M., Schlegel, M.: Reduced-parameter method for maximum entropy analysis of hydraulic pipe flow networks. J. Hydraul. Eng. ASCE **30** (2017). (accepted)
5. Albert, A., Barabási, A.-L.: Statistical mechanics of complex networks. Rev. Modern Phys. **74**, 47–97 (2001)
6. Park, J., Newman, M.E.J.: Statistical mechanics of networks. Phys. Rev. E **70**, 066117 (2004)
7. Bianconi, G.: Entropy of network ensembles. Phys. Rev. E **79**, 036114 (2009)
8. Anand, K., Bianconi, G.: Entropy measures for networks: toward an information theory of complex topologies. Phys. Rev. E **80**, 045102(R) (2009)
9. Shannon, C.E.: A mathematical theory of communication. Bell Sys. Tech. J. **27**, 379, 623 (1948)
10. Kullback, S., Leibler, R.A.: On information and sufficiency. Ann. Math. Stat. **22**, 79–86 (1951)
11. Kapur, J.N., Kesevan, H.K.: Entropy Optimisation Principles with Applications. Academic Press Inc., Boston (1992)
12. Brillouin, L.: Les Statistiques Quantiques et Leurs Applications. Les Presses Universitaires de France, Paris (1930)
13. Niven, R.K., Grendar, M.: Generalized classical, quantum and intermediate statistics and the Polya urn model. Phys. Lett. A **373**, 621–626 (2009)
14. Niven, R.K.: Combinatorial entropies and statistics. Eur. Phys. J. B **70**, 49–63 (2009)

Optimization Employing Gaussian Process-Based Surrogates

R. Preuss and U. von Toussaint

Abstract The optimization of complex plasma-wall interaction and material science models is tantamount with long-running and expensive computer simulations. This indicates the use of surrogate-based methods in the optimization process. A Gaussian process (GP)-based Bayesian adaptive exploration method has been developed and validated on mock examples. The self-consistent adjustment of hyperparameters according to the information present in the data turns out to be the main benefit from the Bayesian approach. While the overall properties and performance is favorable (in terms of expensive function evaluations), the optimal balance between local and global exploitation still mandates further research for strongly multimodal optimization problems.

Keywords Global optimization · Gaussian process · Parametric studies Bayesian inference

1 Introduction

The modeling of particle transport and plasma-wall interaction in the scrape-off layer in fusion plasmas is obtained numerically by the interplay of two extensive codes either describing the plasma solving a fluid equation or the transport of neutrals by a Monte Carlo method. Each code part produces data sets the other part of the code needs to proceed—a circumstance which leads to running times in the order of weeks. Still, after years of computer runs for multiple parameter settings, quite a large database has been gathered with over 1500 entries. With this at one's disposal one is tempted to employ some surrogate modeling in order to explore the dataset for modes in certain data ranges motivated by physics, or simply to give advice for

R. Preuss (✉) · U. von Toussaint
Max-Planck-Institut für Plasmaphysik, 85748 Garching, Germany
e-mail: preuss@ipp.mpg.de

U. von Toussaint
e-mail: udt@ipp.mpg.de

A. Polpo et al. (eds.), *Bayesian Inference and Maximum Entropy Methods in Science and Engineering*, Springer Proceedings in Mathematics & Statistics 239, https://doi.org/10.1007/978-3-319-91143-4_26

which parameter setting the next computer run has to be done in order to increase the information content of the database most effectively.

A long established method in global optimization of complex multimodal models is the construction of a response surface via fast surrogate models [20]. The numerically easy accessible surrogate is employed to find the maximum of the response surface which coordinates are fed back to the original function. With the outcome obtained the surrogate model gets reparameterized and the whole procedure is iterated till success. Unfortunately, a lot of pitfalls are out to spoil the result by pretending delusive maxima of the surrogate model without reference to the real ones of the complex model [6, 9]. We propose to employ the prediction of function values by the Gaussian process (GP) method for the surrogate model and to profit from the capabilities of a Bayesian approach to self-consistently adjust hyperparameters according to the information present in the data. The later turns out to be the main contribution to let the surrogate model describe the unknown model behind the data as close as possible.

2 The Gaussian Process Method

The GP method has been appreciated much in the fields of neural networks and machine learning [3–5, 13, 16]. Residing on this, further work showed the applicability of active data selection via variance-based criterions [7, 21]. In general, for unknown functions costly to evaluate Bayesian optimization [15] was deployed either with sequential [11, 20] or batch design [1], and recently, in combination of both [2, 8]. Very first efforts in geosciences [10] tackling the problem above with so-called kriging [14] can be subordinated to the realm of Gaussian process methods as well. The presentation of the GP method in this the paper was already introduced at [18], and follows in notation—and apart from small amendments —the very instructive book of Rasmussen & Williams [19].

The problem of predicting function values in a multidimensional space supported by given data is a regression problem for a nontrivial function of unknown shape. Given n input data vectors $\boldsymbol{x}_i$ of dimension N_{dim} (with matrix $\boldsymbol{X} = (\boldsymbol{x}_1, \boldsymbol{x}_2, \ldots, \boldsymbol{x}_n)$) and corresponding target data $\boldsymbol{y} = (y_1, \ldots, y_n)^T$ blurred by Gaussian noise of variance σ_d^2 the sought quantity is the target value f_* at test input vector $\boldsymbol{x}_*$. The latter would be generated by a function $f(\boldsymbol{x})$

$$y = f(\boldsymbol{x}) + \varepsilon, \tag{1}$$

where $\langle\varepsilon\rangle = 0$ and $\langle\varepsilon^2\rangle = \sigma_d^2$. Since we are completely ignorant about the (complex) model describing function our approach is to employ the Gaussian process method, with which any uniformly continuous function may be represented. As a statistical process it is fully defined by its covariance function and called Gaussian, because any collection of random variables produced by this process has a Gaussian distribution.

The Gaussian process method defines a distribution over functions. One can think of the analysis as taking place in a space of functions (function-space view) which is conceptually different to the familiar view of solving the regression problem of, for instance, the standard linear model (SLM)

$$f^{\mathrm{SLM}}(\boldsymbol{x}) = \boldsymbol{x}^T \boldsymbol{w}, \tag{2}$$

in the space of the weights $\boldsymbol{w}$ (weight-space view). At this point it is instructive to restate the results for the latter: the predictive distribution depending on mean $\bar{f}_*$ and variance for a test input data point $\boldsymbol{x}_*$ is given by

$$p(f_*^{\mathrm{SLM}}|\boldsymbol{X}, \boldsymbol{y}, \boldsymbol{x}_*) \propto \mathcal{N}\left(\bar{f}_*^{\mathrm{SLM}}, \mathrm{var}(f_*^{\mathrm{SLM}})\right), \tag{3}$$

with

$$\bar{f}_*^{\mathrm{SLM}} = \frac{1}{\sigma_d^2}\boldsymbol{x}_*^T \left[\sigma_d^{-2}\boldsymbol{X}\boldsymbol{X}^T + \Sigma_p^{-1}\right]^{-1} \boldsymbol{X}\boldsymbol{y}, \tag{4}$$

$$\mathrm{var}(f_*^{\mathrm{SLM}}) = \boldsymbol{x}_*^T \left[\sigma_d^{-2}\boldsymbol{X}\boldsymbol{X}^T + \Sigma_p^{-1}\right]^{-1} \boldsymbol{x}_*. \tag{5}$$

Σ_p is the covariance in a Gaussian prior on the weights. Now, we transform these results to the function-space view of the Gaussian process method. As stated above the defining quantity of the Gaussian process method is the covariance function. Its choice is decisive for the inference we want to apply. It is the place where we incorporate all the properties which we would like our (hidden) function describing our problem to have in order to influence the result. For example, the neighborhood of two input data vectors $\boldsymbol{x}_p$ and $\boldsymbol{x}_q$ should be of relevance for the smoothness of the result. This shall be expressed by a length scale λ which represents the long-range dependence of the two vectors. For the covariance function itself, we employ a Gaussian type exponent with the negative squared value of the distance between two vectors $\boldsymbol{x}_p$ and $\boldsymbol{x}_q$

$$k(\boldsymbol{x}_p, \boldsymbol{x}_q) = \sigma_f^2 \exp\left\{-\frac{1}{2}\left|\frac{\boldsymbol{x}_p - \boldsymbol{x}_q}{\lambda}\right|^2\right\}. \tag{6}$$

σ_f^2 is the signal variance. If one is ignorant about this value, literature proposes to set it to one as default value (Chaps. 2.3 and 5.4 in [19]). However, in probability theory, we consider it as an hyperparameter to be marginalized over (see next chapter). To avoid lengthy formulae, we abbreviate the covariance matrix of the input data as $(\boldsymbol{K})_{ij} = k(\boldsymbol{x}_i, \boldsymbol{x}_j)$ and the vector of covariances between test point and input data as $(\boldsymbol{k}_*)_i = k(\boldsymbol{x}_*, \boldsymbol{x}_i)$.

Moreover, we consider the degree of information which the data contain by a term $\sigma_n^2\boldsymbol{\Delta}$ to be composed of an overall variance σ_n^2 accounting that the data are noisy and the matrix $\boldsymbol{\Delta}$ with the variances $\boldsymbol{\sigma}_d^2$ of the given input data on its diagonal and zero otherwise. While σ_n^2 is a hyperparameter, the matrix entry $(\sigma_d)_i$ is the relative uncertainty estimation of a single data point y_i and provided by the experimentalist. If no uncertainties of the input data are given, $\boldsymbol{\Delta}$ is set to the identity matrix. It can be shown (Chap. 2.2 in [19]) that in analogy to Eq. (3) for given λ, σ_f and σ_n the probability distribution for a single function value f_* at test input $\boldsymbol{x}_*$ is

$$p(f_*|\boldsymbol{X}, \boldsymbol{y}, \boldsymbol{x}_*) \propto \mathcal{N}\left(\bar{f}_*, \mathrm{var}(f_*)\right), \tag{7}$$

with mean

$$\bar{f}_* = \boldsymbol{k}_*^T \left(\boldsymbol{K} + \sigma_n^2 \boldsymbol{\Delta}\right)^{-1} \boldsymbol{y}, \tag{8}$$

and variance

$$\mathrm{var}(f_*) = \boldsymbol{k}(\boldsymbol{x}_*, \boldsymbol{x}_*) - \boldsymbol{k}_*^T \left(\boldsymbol{K} + \sigma_n^2 \boldsymbol{\Delta}\right)^{-1} \boldsymbol{k}_*. \tag{9}$$

3 Marginalizing the HyperParameters

The hyperparameters $\boldsymbol{\theta} = (\lambda, \sigma_f, \sigma_n)^T$ determine the result of the Gaussian process method. Since we do not know a priori, which setting is useful, we marginalize over them later on in order to get the target values $\boldsymbol{f}_*$ for test inputs $\boldsymbol{X}_*$. Their moments are

$$\langle \boldsymbol{\theta}^m \rangle = \frac{1}{Z} \int \mathrm{d}\boldsymbol{\theta}\, \boldsymbol{\theta}^m p(\boldsymbol{\theta}|\boldsymbol{y}) = \frac{1}{Z} \int \mathrm{d}\boldsymbol{\theta}\, \boldsymbol{\theta}^m p(\boldsymbol{y}|\boldsymbol{\theta}) p(\boldsymbol{\theta}), \qquad Z = \int \mathrm{d}\boldsymbol{\theta}\; p(\boldsymbol{\theta}|\boldsymbol{y}), \tag{10}$$

where our special interest is the first (expectation value) and second central (variance or rather square root thereof, i.e., standard deviation) moment listed in all subsequent tables.

For the choice of the prior not much is to be expected. A sensible choice would be to assume them in the order of one with a variance of the same size, but confined to be positive

$$p(\theta_i) \propto \mathcal{N}(1, 1) \;\; \forall \;\; \theta_i \geq 0 \quad \text{and} \quad p(\theta_i) = 0 \;\; \text{otherwise}. \tag{11}$$

Depending on the application one should check on these assumptions and be cautious that the prior of the hyperparameters should not influence the result.

The marginal likelihood $p(\boldsymbol{y}|\boldsymbol{\theta})$ is obtained by

$$p(\boldsymbol{y}|\boldsymbol{\theta}) = \int \mathrm{d}\boldsymbol{f}\; p(\boldsymbol{y}|\boldsymbol{f}, \boldsymbol{\theta}) p(\boldsymbol{f}|\boldsymbol{\theta}). \tag{12}$$

As we deal with the Gaussian process the probability functions are of Gaussian type, with the likelihood as $p(\boldsymbol{y}|\boldsymbol{f}, \boldsymbol{\theta}) \propto \mathcal{N}(\boldsymbol{f}, \sigma_n \boldsymbol{\Delta})$ and the prior for $\boldsymbol{f}$ as $p(\boldsymbol{f}|\boldsymbol{\theta}) \propto \mathcal{N}(\boldsymbol{0}, \boldsymbol{K})$ (end of Chap. 2.2, page 19 in [19]). Thus, the integration in Eq. (12) yields

$$\log p(\boldsymbol{y}|\boldsymbol{\theta}) = \mathrm{const} - \frac{1}{2} \boldsymbol{y}^T \left[\boldsymbol{K}(\boldsymbol{\theta}) + \sigma_n^2 \boldsymbol{\Delta}\right]^{-1} \boldsymbol{y} - \frac{1}{2} \log \left|\boldsymbol{K}(\boldsymbol{\theta}) + \sigma_n^2 \boldsymbol{\Delta}\right|. \tag{13}$$

The expectation value for the targets $\boldsymbol{f}_*$ at test inputs $\boldsymbol{X}_*$ employs the marginal likelihood and priors for the hyperparameters from above

$$\langle \boldsymbol{f}_* \rangle = \int \mathrm{d}\boldsymbol{\theta}\, \bar{\boldsymbol{f}}_* \frac{p(\boldsymbol{y}|\boldsymbol{\theta})p(\boldsymbol{\theta})}{\int \mathrm{d}\boldsymbol{\theta}'\, p(\boldsymbol{y}|\boldsymbol{\theta}')p(\boldsymbol{\theta}')}, \tag{14}$$

where the fraction contains the sampling density in Markov chain Monte Carlo.

Rescaling of the input data and whitening of the output is performed in order to do the analysis not hampered by large scales or biased from a linear trend. All data has been back-transformed for display.

4 Global Optimization Scheme

The task is to find the maximum of a complex model function costly to evaluate with respect to certain input variables. The number of input variables is the dimension of the problem. We start with a set of data points obtained from the complex simulations within a region of interest (ROI) which advantageously covers the maximum we are looking for (surrogate modeling fails outside of data supported region). In order to avoid any bias the input variables are chosen from a multidimensional Sobol sequence. Next, the Gaussian process method is applied to determine a surrogate model which establishes a response surface to be easily accessible for numerical optimizers. To circumvent pitfalls resulting from delusive maxima in the surrogate model without relation to actual maxima in the original model the response surface is adjusted to represent an expected improvement of the data set in finding the final maximum. This is achieved by defining the response surface to result from the sum of the variance and the expectation value of the surrogate at certain input values. It was shown before that the calculation of the model at a point of the largest variance within the response surface coincides with largest reduction in entropy, i.e., maximal information gain [12]. This means that if we want to infer from the response surface of the surrogate to the maximum of the original model, we have to consider likewise those data points which sum up with their variances to be larger than the maximal expectation value. The routine used for finding the maximum employs just (inverse) line minimization in multiple dimensions as is performed by Powell's routine found, e.g., in Chap. 10.5 of [17]. Since this routine only finds the next local maximum starting at some initial point one has to be cautious about the choice of the initial point. For this, we chose the maximal value of the respective sum of variance and expectation value of the surrogate for all possible points being in the middle between all possible pairs of points in the dataset. Due to the properties of the Gaussian process, these are the locations where the variances will be largest. Eventually, the optimization routine returns the position of a maximum found on the response surface and an additional data point gets simulated with the complex model. The whole procedure is iterated till a newly found maximum differs from the previous one only within computationally accuracy.

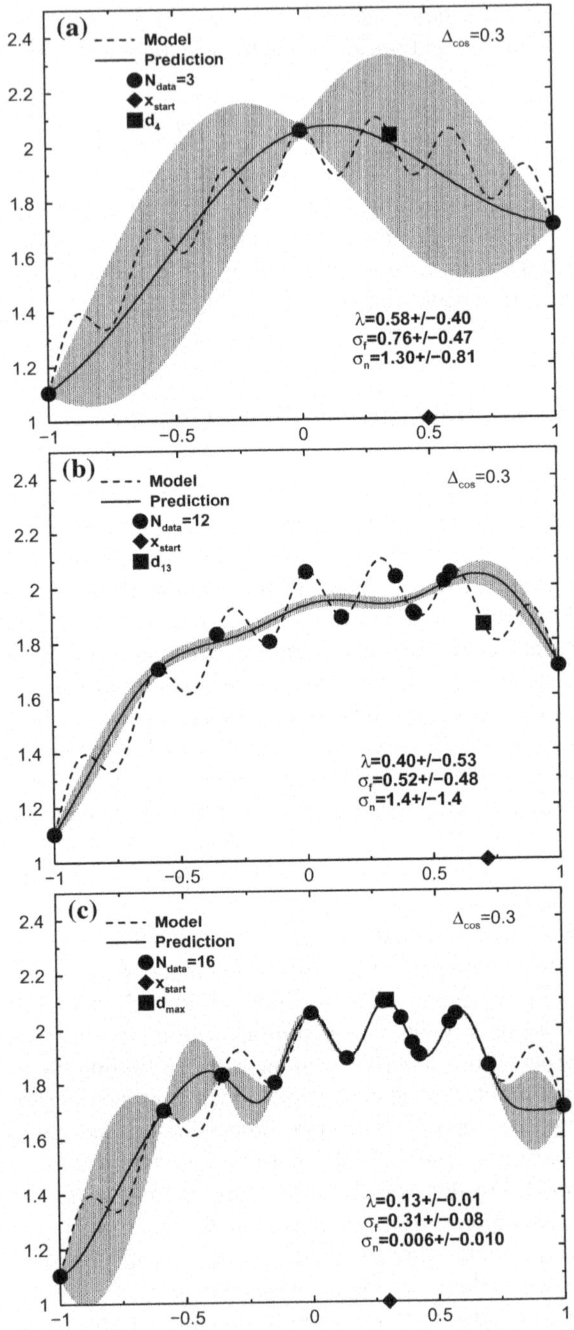

Fig. 1 One-dimensional case for a model with a cosine fine structure of $\Delta_{\cos} = 0.3$ on top of a broad parabola around x = 0.3 (dashed line, see Eq. (15)). The gray-shaded area is the uncertainty range of the prediction, i.e., surrogate model (thin line), obtained as a Gaussian process with $N_{\rm data}$ data (filled circles). The position of the highest sum of variance and expectation value in the middle of any two points from the dataset is shown on the baseline (filled diamond). The position of the maximum found for the sum of variance and expectation value is input to the model and gives an additional data point (filled square). **a** Initial condition with $N_{\rm data} = 3$. **b** After nine iterations. **c** The procedure succeeds for a total of $N_{\rm data} = 16$ data points

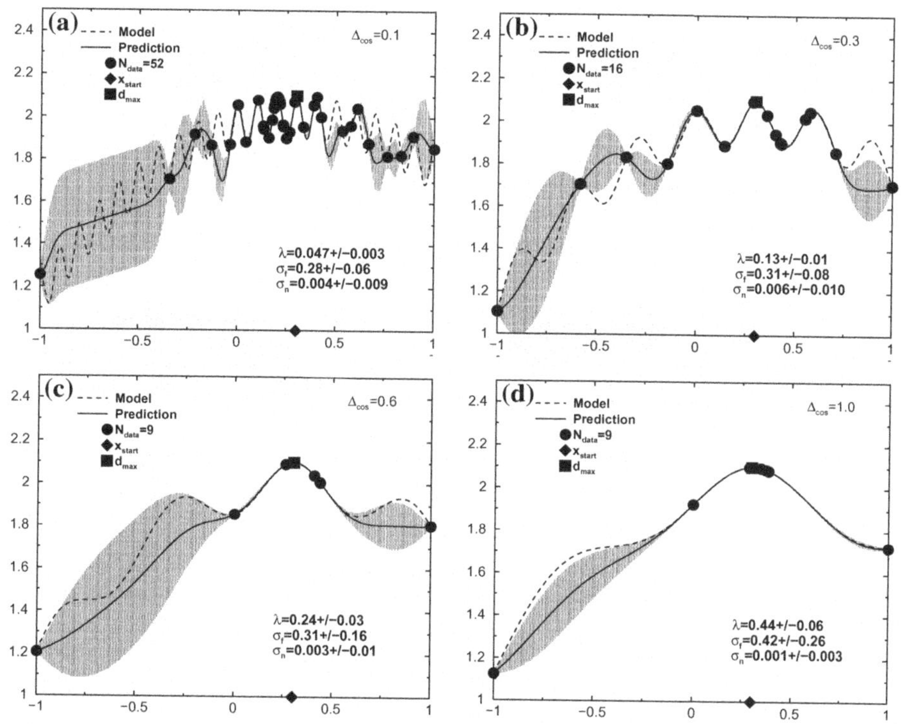

Fig. 2 One-dimensional case: final surrogate response for various cosine fine structures $\Delta_{\cos} \in \{(a)\ 0.1, (b)\ 0.3, (c)\ 0.6, (d)\ 1.0\}$ of the complex model

5 Results in One and Two Dimensions

In order to demonstrate proof of principle for our algorithm above we examine a function with a broad maximum and a cosine structure on top of it:

$$y = 2 - \sum_{i=1}^{N_{\text{dim}}} \left\{ \frac{1}{2}(x_i - 0.3)^2 - \frac{1}{10} \cos\left[\frac{2\pi(x_i - 0.3)}{\Delta_{\cos}}\right]\right\}. \tag{15}$$

The global maximum is set for an input vector with 0.3, while the variability of the function within $[-1:1]$ as ROI is given by the factor $\Delta_{\cos}$ which will be chosen in between 0.1 and 1.

Figure 1 shows the result for $\Delta_{\cos} = 0.3$. Though this still seems to be a moderate variability with respect to the ROI, the model shows a lot of local extrema which constitute pitfalls for the search of the global maximum. From the initial data set of $N_{\text{data}} = 3$ already the data point for $x = 0$ represents a local maximum which is hard to overtop for methods ignoring the uncertainty of the surrogate model (see [9]). Still with $N_{\text{data}} = 12$ data points the surrogate model shows a delusive maximum close

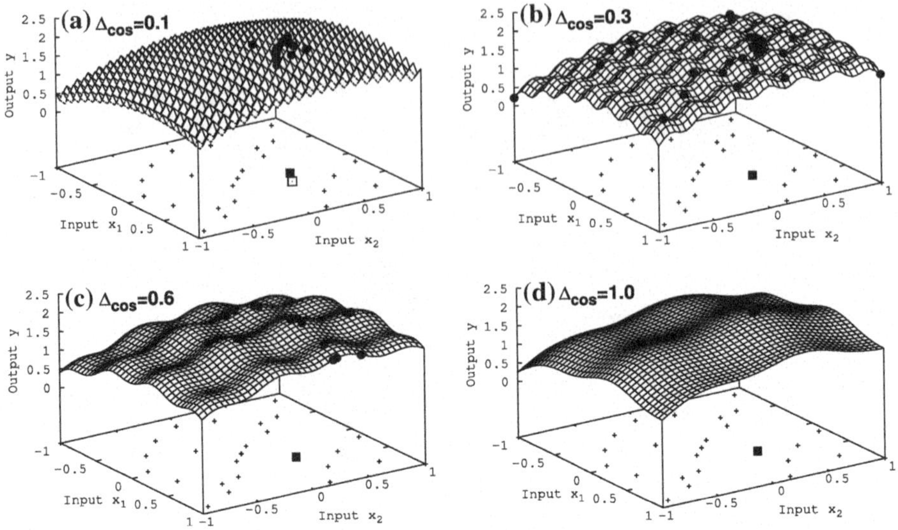

Fig. 3 Two-dimensional case: surface of the complex model for various cosine fine structures $\Delta_{\cos} \in \{$(a) 0.1, (b) 0.3, (c) 0.6, (d) 1.0$\}$ The initial 25 input data are shown in the basement (plus signs), and on top of the surface the additional data points which were acquired during the iteration of the procedure (filled circles). All maxima found (filled square) are the true ones, apart from the case of $\Delta_{\cos} = 0.1$ in **a**, where the true maximum is represented by the open square

to the right border of the ROI (Fig. 1b). The values of the hyperparameters (λ, σ_f, σ_n) induce a broadly structured surrogate ignoring the local extrema of the hidden model function (their expectation values are shown at the lower right position of each figure). This changes in Fig. 1c as the procedure succeedingly pins the correct extrema, acquires additional data points up to the final number of $N_{\text{data}} = 16$. Here, the algorithm benefits from the self-adjusting skills of the Bayesian approach to adjust the hyperparameters in the covariance of the Gaussian process to adapt the surrogate model to fine structures on top of broader extrema of the complex model. In the end, the proper functional behavior as well as the true maximum are reproduced.

The final surrogate responses for various cosine fine structures are depicted in Fig. 2a. Though for the case with the highest variability $\Delta_{\cos} = 0.1$ nearly 50 calls to the complex model are needed, the algorithm finally succeeds in finding the correct maximum. In all four cases, the true maximum is found. Again the region of the most eligible local maxima gets highly resolved.

Eventually, we turn in Fig. 3 to the two-dimensional case. Again the true maximum of the complex is found, apart from the case of $\Delta_{\cos} = 0.1$ in Fig. 2a, where the algorithm gets erroneously stuck in a local maximum close to the true one. Since with smaller $\Delta_{\cos}$ the extensions of the local extrema shrink in size compared to the broader maximum in Eq. (15), it gets more and more ambitious to converge to the correct result, i.e., the correct local extrema with neighboring almost equal choices. It is the task for further examinations to classify the relationship of the expected improvement in changing from one local extrema to another with respect to underlying larger extremal structures.

6 Summary and Conclusion

An algorithm for global optimization was demonstrated to autonomously converge to qualified maxima. The fully Bayesian approach benefits from the self-adjusting skills with hyperparameters in the Gaussian process, which enables the surrogate model to adapt to fine structures on top of broader extrema of the complex model. To be instructive, the procedure was characterized for two low-dimensional examples. It is left to ongoing research to show the feasibility of the proposed method for identifying extrema in higher dimensions.

References

1. Azimi, J., Fern, A., Fern, X.Z.: Batch Bayesian optimization via simulation matching. In: Lafferty, J.D., Williams, C.K.I., Shawe-Taylor, J., Zemel, R.S., Culotta, A. (eds.) Advances in Neural Information Processing Systems, vol. 23, pp. 109–117. Curran Associates Inc, New York (2010)
2. Azimi, J., Jalali, A., Fern, X.Z.: Hybrid batch Bayesian optimization. In: Proceedings of the 29th International Conference on Machine Learning. Edinburgh, Scotland, UK (2012)
3. Barber, D.: Bayesian Reasoning and Machine Learning. Cambridge University Press, Cambridge (2012)
4. Bishop, C.: Neural Networks for Pattern Recognition. Oxford University Press, Oxford (1996)
5. Cohn, D.: Neural network exploration using optimal experiment design. Neural Netw. **9**, 1071–1083 (1996)
6. Duvigneau, R.: Optimization using surrogate models. Talk given on 27 May 2015. https://team.inria.fr/acumes/files/2015/05/cours_meta.pdf. Accessed 30 June 2017
7. Gramacy, R.B., Lee, H.K.H.: Adaptive design and analysis of supercomputer experiments. Technometrics **51**, 130–145 (2009)
8. Gonzalez, J., Osborne, M., Lawrence, N.D.: GLASSES: relieving the Myopia of Bayesian optimisation. J. Mach. Learn. Res. **51**, 790–799 (2016)
9. Jones, D.: A taxonomy of global optimization methods based on response surfaces. J. Global Optim. **21**, 345–383 (2001)
10. Krige, D.G.: A statistical approach to some basic mine valuation problems on the Witwatersrand. J. Chem. Metal. Min. Soc. S. Afr. **52**, 119–139 (1951)
11. Locatelli, M.: Bayesian algorithms for one-dimensional global optimization. J. Global Optim. **10**, 57–76 (1997)
12. MacKay, D.J.C.: Information-based objective functions for active data selection. Neural Comput. **4**, 590 (1992)
13. MacKay, D.J.C.: Information Theory, Inference, and Learning Algorithms. Cambridge University Press, Cambridge (2003)
14. Matheron, G.: Principles of geostatistics. Econ. Geol. **58**, 1246–1266 (1963)
15. Mockus, J.: Bayesian Approach to Global Optimization. Springer, Berlin (1989)
16. Neal, R.M.: Monte Carlo implementation of Gaussian process models for Bayesian regression and classification. Technical report No. 9702, pp. 1–24. Department of Statistics, University of Toronto (1997)

17. Press, W.H., Teukolsky, S.A., Vetterling, W.T., Flannery, B.P.: Numerical Recipes: The Art of Scientific Computing. Cambridge University Press, Cambridge (2007)
18. Preuss, R., von Toussaint, U.: Prediction of plasma simulation data with the Gaussian process method. In: Niven, R. (ed.) Bayesian Inference and Maximum Entropy Methods in Science and Engineering, AIP Conference Proeedings, vol. 1636, pp. 118–123. Melville, New York (2014)
19. Rasmussen, C., Williams, C.: Gaussian Processes for Machine Learning. MIT Press, Cambridge (2006)
20. Sacks, J., Welch, W.J., Mitchell, T.J., Wynn, H.P.: Design and analysis of computer experiments. Stat. Sci. **4**, 409–423 (1989)
21. Seo, S., Wallat, M., Graepel, T., Obermayer, K.: Gaussian process regression: active data selection and test point rejection. In: Proceedings of the International Joint Conference on Neural Networks, pp. 241–246. IEEE, New York (2000)

Bayesian and Maximum Entropy Analyses of Flow Networks with Non-Gaussian Priors and Soft Constraints

Steven H. Waldrip and Robert K. Niven

Abstract We have recently developed new maximum entropy (MaxEnt) and Bayesian methods for the analysis of flow networks, including pipe flow, electrical and transportation networks. Both methods of inference update a prior probability density function (pdf) with new information, in the form of data or constraints, to obtain a posterior pdf for the system. We here examine the effects of non-Gaussian prior pdfs, including truncated normal and beta distributions, both analytically and by the use of numerical examples, to explore the differences and similarities between the MaxEnt and Bayesian formulations. We also examine 'soft constraints' imposed within the prior.

Keywords Maximum entropy · Bayes' theorm · Flow networks

1 Introduction

Fluid and energy flows on networks are an important problem. Traditionally, these systems have been analysed using deterministic methods, which do not consider uncertainty. To account for uncertainty, a probabilistic framework is required. Two methods for probabilistic inference are applied here: Bayesian inference using Bayes' rule, and maximum entropy (MaxEnt) analysis.

Bayes' theorem can be derived from the product rule of probabilities, whereas the MaxEnt method for inference can be derived from an axiomatic approach

S. H. Waldrip · R. K. Niven (✉)
School of Engineering and Information Technology, The University of New South Wales, Canberra, NSW 2600, Australia
e-mail: r.niven@adfa.edu.au

S. H. Waldrip
e-mail: Steven.Waldrip@student.adfa.edu.au

A. Polpo et al. (eds.), *Bayesian Inference and Maximum Entropy Methods in Science and Engineering*, Springer Proceedings in Mathematices & Statistics 239, https://doi.org/10.1007/978-3-319-91143-4_27

[2, 15, 17] or by a combinatorial method [1, 9, 14, 16]. The maximum relative entropy method (MaxEnt), equivalent to the minimum Kullback–Leibler divergence [12], is a method of inference used to infer or update a probability distribution describing an under-determined system, which respects all constraints imposed on the system and is closest to the prior distribution [11].

There have been many studies on the connection between Bayes' theorem and the MaxEnt method, with some authors suggesting that one can be obtained from the other (in either direction) [3–5, 21]. In one example, the current authors have compared the probability distributions for quasi-Newton rules obtained by inferring the Jacobian or Hessian using Bayesian inference [6, 7] or the MaxEnt method [20], In both cases with the same Gaussian prior. It was found that both methods obtained the same posterior means, but the covariance matrices were different.

In this study, we extend the work of [18] to consider non-Gaussian prior distributions. Firstly, in Sect. 2 we develop a Bayesian method to analyse flow networks. In Sect. 3, we present a MaxEnt theory using soft constraints implemented using the prior pdf. In Sect. 4, we compare the distributions obtained by the two methods by a case study.

2 Bayesian Analysis

Consider, a flow network with N flow rates assembled into the vector $\boldsymbol{\Psi}$. To avoid inconsistencies due to different network representations, we consider a basis set $\boldsymbol{X}$ of n flow rates selected from $\boldsymbol{\Psi}$ as parameters of the joint pdf used to represent the uncertainty. The indices of the basis set $\boldsymbol{X}$ in $\boldsymbol{\Psi}$ are given by the set $\mathcal{B}$, while the indices of the complementary non-basis set of flow rates in $\boldsymbol{\Psi}$ are given by set $\mathcal{N}$. The derivation of the Bayesian method requires a prior belief of the state of the system, represented as a prior pdf $q(\boldsymbol{X})$, which is updated using observed data to a posterior pdf according to Bayes' rule:

$$p(\boldsymbol{X}|\boldsymbol{y}) = \frac{p(\boldsymbol{y}|\boldsymbol{X})q(\boldsymbol{X})}{\underset{\Omega}{\int \cdots \int} p(\boldsymbol{y}|\boldsymbol{X})q(\boldsymbol{X})d\boldsymbol{X}} \tag{1}$$

where $p(\boldsymbol{y}|\boldsymbol{X})$ is the likelihood function, the denominator allows for normalisation, $\boldsymbol{X}$ is the basis set of flow rates, $\boldsymbol{y}$ is the vector of observed data and Ω is the domain of $\boldsymbol{X}$. The flow rates $\bar{\boldsymbol{X}}$ not included in the basis set are taken as functions of the model parameters $\boldsymbol{X}$, using:

$$\bar{\boldsymbol{X}} = \boldsymbol{V}\boldsymbol{X} \tag{2}$$

$$\boldsymbol{V} = -\boldsymbol{A}^{-1}_{i\in\mathcal{V},j\in\mathcal{N}}\boldsymbol{A}_{i\in\mathcal{V},j\in\mathcal{B}}, \tag{3}$$

$$\boldsymbol{A} = \left[\boldsymbol{C},\ \boldsymbol{W}\text{diag}(\boldsymbol{K}),\ \boldsymbol{F},\ \boldsymbol{T}\text{diag}(\boldsymbol{K})\right]^\top \tag{4}$$

in which the same nomenclature is used as [18], as follows:

- diag() places the elements of a vector on the diagonal of a square matrix;
- the set $\mathcal{V}$ contains the $N - n$ indices of the equations required to uniquely define $\bar{\boldsymbol{X}}$ from $\boldsymbol{X}$;
- the matrix $\boldsymbol{C}$ is a $c \times N$ connectivity matrix where c is the number of nodes, it containing elements $\{-1, 0, 1\}$. Its entries indicate membership of edge to the node, given by 0 if the edge is not connected to the node, 1 if the assumed direction of Ψ is entering the node and -1 otherwise;
- the vector $\boldsymbol{K}$ is a $N \times 1$ vector of flow resistances;
- the matrix $\boldsymbol{W}$ is a $w \times N$ loop matrix containing elements $\{-1, 0, 1\}$, where w is the number of independent cycles (loops) within the network. Its entries indicate membership of edges within a loop, given by 0 if the edge is not in the loop, 1 if the assumed direction of Q_m is in a clockwise direction around the loop and -1 otherwise;
- the matrix $\boldsymbol{F}$ is a $N_{\hat{\Psi}} \times N$ matrix containing either 0 or 1 in each of its elements. Each row will have a single 1 on the index corresponding to the dimension of the observed link;
- $N_{\hat{\Psi}}$ is the number of flow rate observation locations;
- the matrix $\boldsymbol{T}$ is a $h_c \times N$ pseudo-loop matrix containing $\{-1, 0, 1\}$, where h_c is the number of potential difference constraints applied. The pseudo-loop matrix contains paths between nodes of known pressure or potential values. The entries indicate membership of edges within the potential difference constraint, given by 0 if the edge is not in the constraint, 1 if the assumed direction of Psi_m is defined as in the direction from node 0 to node j, and -1 otherwise; and
- $\langle \boldsymbol{Y_T} \rangle$ is the $h_c \times 1$ vector of mean potential differences between a chosen location H_0 and H_j, for all nodes with potential observations.

The prior is chosen to represent one's belief of the system state before incorporating any measured data. Although any distribution which represents what is believed about the system state could be chosen, in this study, we select from the following, defined over a subset of the real domain:

- The truncated normal distribution, given by

$$q(\boldsymbol{X}) = \frac{\exp\left(-\frac{1}{2}(\boldsymbol{X} - \boldsymbol{m})^\top \boldsymbol{\Sigma}^{-1} (\boldsymbol{X} - \boldsymbol{m})\right)}{\dot{\kappa}} \tag{5}$$

  where $\boldsymbol{m}$ is the $n \times 1$ vector of location parameters, $\boldsymbol{\Sigma}$ is the $n \times n$ matrix of prior scale parameters and $\dot{\kappa}$ is a constant for normalisation.
- The beta distribution, given by

$$q(\boldsymbol{X}) = \prod_{i=1}^{n} \frac{(X_i - l_i)^{a_i - 1}(u_i - X_i)^{b_i - 1}}{B(a_i, b_i)(u_i - l_i)^{a_i + b_i - 1}} \tag{6}$$

where a_i and b_i are the distribution parameters and l and u are, respectively, the lower and upper edges of the domain.

In Bayes' method, likelihood functions are used to incorporate the physics of the system as well as any observed data, as follows:

- The likelihood function to incorporate conservation of mass at each node or Kirchhoff's first law (or the flow rate for incompressible systems) is given by delta functions defined by the limit of a Gaussian distribution

$$-2\ln(p(\mathbf{0}|\boldsymbol{X})) \propto \lim_{\boldsymbol{\Sigma}_C\to\mathbf{0}} \left(\mathbf{0}-\left(\boldsymbol{C}_X+\boldsymbol{C}_{\bar{X}}\right)\boldsymbol{X}\right)^{\top} \boldsymbol{\Sigma}_C{}^{-1}\left(\mathbf{0}-\left(\boldsymbol{C}_X+\boldsymbol{C}_{\bar{X}}\right)\boldsymbol{X}\right). \tag{7}$$

$$\boldsymbol{C}_X=\boldsymbol{C}_{i\notin\mathcal{V},j\in\mathcal{B}}, \qquad \boldsymbol{C}_{\bar{X}}=\boldsymbol{C}_{i\notin\mathcal{V},j\in\mathcal{N}}\boldsymbol{V}. \tag{8}$$

- The likelihood function to incorporate the loop laws for each loop, Kirchhoff's second law, is given by delta functions defined by the limit of a Gaussian distribution

$$-2\ln(p(\mathbf{0}|\boldsymbol{X})) \propto \lim_{\boldsymbol{\Sigma}_W\to\mathbf{0}} \left(\mathbf{0}-\left(\boldsymbol{W}_X+\boldsymbol{W}_{\bar{X}}\right)\boldsymbol{X}\right)^{\top} \boldsymbol{\Sigma}_W{}^{-1}\left(\mathbf{0}-\left(\boldsymbol{W}_X+\boldsymbol{W}_{\bar{X}}\right)\boldsymbol{X}\right). \tag{9}$$

$$\boldsymbol{W}_X=\boldsymbol{W}_{i\notin\mathcal{V},j\in\mathcal{B}}\mathrm{diag}\left(\boldsymbol{K}_{i\in\mathcal{B}}\right), \qquad \boldsymbol{W}_{\bar{X}}=\boldsymbol{W}_{i\notin\mathcal{V},j\in\mathcal{N}}\mathrm{diag}\left(\boldsymbol{K}_{i\in\mathcal{N}}\right)\boldsymbol{V}. \tag{10}$$

- Observed flow rates can be constrained using the following likelihood functions:
  - for a truncated normal case

$$-2\ln(p(\boldsymbol{Y}_F|\boldsymbol{X})) \propto \left(\boldsymbol{Y}_F-\left(\boldsymbol{F}_X+\boldsymbol{F}_{\bar{X}}\right)\boldsymbol{X}\right)^{\top} \boldsymbol{\Sigma}_F{}^{-1}\left(\boldsymbol{Y}_F-\left(\boldsymbol{F}_X+\boldsymbol{F}_{\bar{X}}\right)\boldsymbol{X}\right), \tag{11}$$

$$\boldsymbol{F}_X=\boldsymbol{F}_{i\notin\mathcal{V},j\in\mathcal{B}}, \qquad \boldsymbol{F}_{\bar{X}}=\boldsymbol{F}_{i\notin\mathcal{V},j\in\mathcal{N}}\boldsymbol{V}. \tag{12}$$

  where $\boldsymbol{Y}_F$ is a $N_{\hat{\psi}}\times 1$ vector that has the mode flow rate of each observation for a link in its elements, $\boldsymbol{\Sigma}_F$ is the $N_{\hat{\psi}}\times N_{\hat{\psi}}$ matrix of scale parameters of the observations.
  - for a beta distribution case

$$\ln(p(\boldsymbol{Y}_F|\boldsymbol{X}))=(\boldsymbol{a}_F-1)\odot\ln(\left(\boldsymbol{F}_X+\boldsymbol{F}_{\bar{X}}\right)(\boldsymbol{X}-\boldsymbol{l}))+(\boldsymbol{b}_F-1)\odot\ln(\left(\boldsymbol{F}_X+\boldsymbol{F}_{\bar{X}}\right)(\boldsymbol{u}-\boldsymbol{X})) \tag{13}$$

  where $\odot$ is an element wise multiplication and $\boldsymbol{a}_F$ and $\boldsymbol{b}_F$ are $N_{\hat{\psi}}\times 1$ vectors in which each of the j elements can be found respectively from

$$a_{Fj}=-\frac{((l_j-Y_{Fj})(\Sigma_{Fjj}-l_jY_{Fj}+l_ju_j-Y_{Fj}u_j+Y_{Fj}^2))}{(l_j\Sigma_{Fjj}-u_j\Sigma_{Fjj})} \tag{14}$$

$$b_{Fj}=-\frac{(Y_{Fj}^3-2Y_{Fj}^2u_j-l_jY_{Fj}^2+Y_{Fj}u_j^2+2l_jY_{Fj}u_j+\Sigma_{Fjj}Y_{Fj}-l_ju_j^2-\Sigma_{Fjj}u_j)}{(l_j\Sigma_{Fjj}-u_j\Sigma_{Fjj})} \tag{15}$$

- Observed potential differences can be constrained using the following likelihood functions:
  - for a truncated normal

$$-2\ln(p(\boldsymbol{Y_T}|\boldsymbol{X})) \propto \left(\boldsymbol{Y_T} - \left(\boldsymbol{T_X} + \boldsymbol{T_{\bar{X}}}\right)\boldsymbol{X}\right)^\top \boldsymbol{\Sigma_T}^{-1}\left(\boldsymbol{Y_T} - \left(\boldsymbol{T_X} + \boldsymbol{T_{\bar{X}}}\right)\boldsymbol{X}\right), \tag{16}$$

  where $\boldsymbol{Y_T}$ is a $h_c \times 1$ vector that has the mode potential difference of an observation between two points in each element, $\boldsymbol{\Sigma_T}$ is the $h_c \times h_c$ scale parameter matrix of the observations and

$$\boldsymbol{T_X} = \boldsymbol{T}_{i\notin\mathcal{V},j\in\mathcal{B}}\text{diag}\left(\boldsymbol{K}_{i\in\mathcal{B}}\right), \qquad \boldsymbol{T_{\bar{X}}} = \boldsymbol{T}_{i\notin\mathcal{V},j\in\mathcal{N}}\text{diag}\left(\boldsymbol{K}_{i\in\mathcal{N}}\right)\boldsymbol{V}. \tag{17}$$

  - for a beta distribution

$$\ln(p(\boldsymbol{Y_T}|\boldsymbol{X})) = (\boldsymbol{a_T} - 1) \odot \ln((\boldsymbol{T_X} + \boldsymbol{T_{\bar{X}}})(\boldsymbol{X} - \boldsymbol{l})) + (\boldsymbol{b_T} - 1) \odot \ln((\boldsymbol{T_X} + \boldsymbol{T_{\bar{X}}})(\boldsymbol{u} - \boldsymbol{X})) \tag{18}$$

  where $\boldsymbol{a_T}$ and $\boldsymbol{b_t}$ are $h_c \times 1$ vector in which each of the j elements can be found respectively from (14) and (15) replacing a_{Fj} Y_{Fj} Σ_{Fjj} b_{Fj} with a_{Tj} Y_{Tj} Σ_{Tjj} b_{Tj}

Bayes' rule is applied to update each of the prior functions with the likelihood functions of the same type for observed flows and potential differences but normal or delta probability distributions for conservation laws in all cases. The normal distribution prior (5) is updated multiplying it with (7), (9), (11) and (16). Its properties can be obtained by expanding and dropping all terms which are not functions of $\boldsymbol{X}$, combining like factors and completing the square giving the posterior in the form

$$-2\ln(p(\boldsymbol{X}|\boldsymbol{y})) \propto (\boldsymbol{X} - \boldsymbol{\mu})^\top \boldsymbol{\Sigma_p}^{-1}(\boldsymbol{X} - \boldsymbol{\mu}), \tag{19}$$

where the location and scale parameters are given respectively by

$$\boldsymbol{\mu} = \boldsymbol{m} + \boldsymbol{\Sigma}\boldsymbol{O}^\top\left(\boldsymbol{S} + \boldsymbol{O}\boldsymbol{\Sigma}\boldsymbol{O}^\top\right)^{-1}(\boldsymbol{y} - \boldsymbol{O}\boldsymbol{m}). \tag{20}$$

$$\boldsymbol{\Sigma_p} = \boldsymbol{\Sigma} - \boldsymbol{\Sigma}\boldsymbol{O}^\top\left(\boldsymbol{S} + \boldsymbol{O}\boldsymbol{\Sigma}\boldsymbol{O}^\top\right)^{-1}\boldsymbol{O}\boldsymbol{\Sigma}. \tag{21}$$

in which

$$\boldsymbol{O} = \left[\boldsymbol{C_X} + \boldsymbol{C_{\bar{X}}}\ \ \boldsymbol{W_X} + \boldsymbol{W_{\bar{X}}}\ \ \boldsymbol{F_X} + \boldsymbol{F_{\bar{X}}}\ \ \boldsymbol{T_X} + \boldsymbol{T_{\bar{X}}}\right]^\top \tag{22}$$

$$\boldsymbol{S}^{-1} = \begin{bmatrix} \boldsymbol{\Sigma_C}^{-1} & \boldsymbol{0} & \boldsymbol{0} & \boldsymbol{0} \\ \boldsymbol{0} & \boldsymbol{\Sigma_W}^{-1} & \boldsymbol{0} & \boldsymbol{0} \\ \boldsymbol{0} & \boldsymbol{0} & \boldsymbol{\Sigma_F}^{-1} & \boldsymbol{0} \\ \boldsymbol{0} & \boldsymbol{0} & \boldsymbol{0} & \boldsymbol{\Sigma_T}^{-1} \end{bmatrix} \tag{23}$$

$$y = \begin{bmatrix} \mathbf{0} \; \mathbf{0} \; Y_F \; Y_T \end{bmatrix}^\top \tag{24}$$

The posterior mean flow rates can then be obtained from

$$\langle X \rangle = \int_{l_1}^{u_1} \ldots \int_{l_n}^{u_n} X p(X|y) dX = \int_{l_1}^{u_1} \ldots \int_{l_n}^{u_n} X \frac{\exp\left(-\frac{1}{2}(X-\mu)^\top \Sigma_p{}^{-1}(X-\mu)\right)}{\hat{\kappa}} dX \tag{25}$$

where $\hat{\kappa}$ is a normalisation constant. The integral is evaluated using numerical methods, in this study using the R package 'tmvtnorm' based on the method of [13]. The covariance matrix is obtained from

$$\langle XX^\top \rangle - \langle X \rangle \langle X \rangle^\top = \int_{l_1}^{u_1} \ldots \int_{l_n}^{u_n} XX^\top p(X|y) dX - \langle X \rangle \langle X \rangle^\top \tag{26}$$

The beta distribution prior (6) is updated by multiplying it with (7), (9), (13) and (18).

$$\begin{aligned} \ln(p(X|y)) \propto -\frac{1}{2} X^\top \begin{bmatrix} C_X + C_{\bar{X}} \\ W_X + W_{\bar{X}} \end{bmatrix}^\top \begin{bmatrix} \Sigma_C & \mathbf{0} \\ \mathbf{0} & \Sigma_W \end{bmatrix}^{-1} \begin{bmatrix} C_X + C_{\bar{X}} \\ W_X + W_{\bar{X}} \end{bmatrix} X \\ (a-1) \odot \ln(X - l) + (b-1) \odot \ln(u - X) \quad (27) \\ +(a_T - 1) \odot \ln(\left(T_X + T_{\bar{X}}\right)(X - l)) + (b_T - 1) \odot \ln(\left(T_X + T_{\bar{X}}\right)(u - X)) \\ +(a_T - 1) \odot \ln(\left(T_X + T_{\bar{X}}\right)(X - l)) + (b_T - 1) \odot \ln(\left(T_X + T_{\bar{X}}\right)(u - X)) \end{aligned}$$

These posteriors are examined numerically in Sect. 4.

3 MaxEnt Analysis with Soft Constraints

The maximum entropy method follows the algorithm of Jaynes [8, 10]. For this, we define a pdf which expresses the uncertainty in the parameter set X and in the parameter observations Y_F and Y_T. The joint probability is defined to be:

$$p(X)dX = \text{Prob}(X \le \Upsilon_X \le X + dX, Y_F \le \Upsilon_{Y_F} \le Y_F + dY_F, Y_T \le \Upsilon_{Y_T} \le Y_T + dY_T), \tag{28}$$

where Υ_X, Υ_{Y_F} and Υ_{Y_T} are the vectors of the random variables for X, Y_F and Y_T, respectively. We also assume that each of the flow rate and potential difference constraints are applied as soft constraints. This choice of pdf gives the following relative entropy or negative Kullback–Leibler function [12], over the space of uncertainties:

$$\mathfrak{H} = -\int_{l_1}^{u_1} \cdots \int_{l_{n+n_O}}^{u_{n+n_O}} p(X, Y_F, Y_T) \ln \frac{p(X, Y_F, Y_T)}{q(X, Y_F, Y_T)} dX dY_F dY_T, \tag{29}$$

where $n_o = N_{\hat{\psi}} + h_c$, the number of data observations, $q(\boldsymbol{X}, \boldsymbol{Y_F}, \boldsymbol{Y_T})$ is the prior pdf, and l_i and u_i are the lower and upper bounds of the ith flow rate. The relative entropy is then maximised subject to the constraints on the system. The following constraints are always required:

- Normalisation of probability:

$$1 = \int_{l_1}^{u_1} \cdots \int_{l_{n+n_O}}^{u_{n+n_O}} p(\boldsymbol{X}, \boldsymbol{Y_F}, \boldsymbol{Y_T}) d\boldsymbol{X} d\boldsymbol{Y_F} d\boldsymbol{Y_T}. \tag{30}$$

- Kirchhoff's first law, for the conservation of flow rates at each internal node, here imposed in the mean:

$$\boldsymbol{0} = \left(\boldsymbol{C_X} + \boldsymbol{C_{\bar{X}}}\right) \left(\int_{l_1}^{u_1} \cdots \int_{l_{n+n_O}}^{u_{n+n_O}} \boldsymbol{X} p(\boldsymbol{X}, \boldsymbol{Y_F}, \boldsymbol{Y_T}) d\boldsymbol{X} d\boldsymbol{Y_F} d\boldsymbol{Y_T} \right). \tag{31}$$

- Kirchhoff's second law, which requires the potential difference to vanish around each enclosed loop, again imposed in the mean:

$$\boldsymbol{0} = \left(\boldsymbol{W_X} + \boldsymbol{W_{\bar{X}}}\right) \left(\int_{l_1}^{u_1} \cdots \int_{l_{n+n_O}}^{u_{n+n_O}} \boldsymbol{X} p(\boldsymbol{X}, \boldsymbol{Y_F}, \boldsymbol{Y_T}) d\boldsymbol{X} d\boldsymbol{Y_F} d\boldsymbol{Y_T} \right). \tag{32}$$

We also allow for any of the following constraints:

- A set of specified inflow/outflow and internal flow rate constraints:

$$\boldsymbol{0} = \int_{l_1}^{u_1} \cdots \int_{l_{n+n_O}}^{u_{n+n_O}} \left(\left(\boldsymbol{F_X} + \boldsymbol{F_{\bar{X}}}\right) \boldsymbol{X} - \boldsymbol{Y_F} \right) p(\boldsymbol{X}, \boldsymbol{Y_F}, \boldsymbol{Y_T}) d\boldsymbol{X} d\boldsymbol{Y_F} d\boldsymbol{Y_T}. \tag{33}$$

- Potential difference constraints between pairs of nodes:

$$\boldsymbol{0} = \int_{l_1}^{u_1} \cdots \int_{l_{n+n_O}}^{u_{n+n_O}} \left(\left(\boldsymbol{T_X} + \boldsymbol{T_{\bar{X}}}\right) \boldsymbol{X} - \boldsymbol{Y_T} \right) p(\boldsymbol{X}, \boldsymbol{Y_F}, \boldsymbol{Y_T}) d\boldsymbol{X} d\boldsymbol{Y_F} d\boldsymbol{Y_T}. \tag{34}$$

Second or higher order constraints were not applied to the MaxEnt analysis in this study, see [19] for potential higher order constraints. After identifying the constraints, the entropy (29) is then maximised subject to (30)–(32) and whichever of (33) and (34) apply. Applying the calculus of variations, we form the Lagrangian and maximise, giving as the final result:

$$p^*(\boldsymbol{X}, \boldsymbol{Y_F}, \boldsymbol{Y_T}) = q(\boldsymbol{X}, \boldsymbol{Y_F}, \boldsymbol{Y_T}) e^{-\kappa - \boldsymbol{\alpha}(\boldsymbol{C_X} + \boldsymbol{C_{\bar{X}}})\boldsymbol{X} - \boldsymbol{\beta}(\boldsymbol{W_X} + \boldsymbol{W_{\bar{X}}})\boldsymbol{X} - \boldsymbol{\lambda}((\boldsymbol{F_X} + \boldsymbol{F_{\bar{X}}})\boldsymbol{X} - \boldsymbol{Y_F}) - \boldsymbol{\eta}((\boldsymbol{T_X} + \boldsymbol{T_{\bar{X}}})\boldsymbol{X} - \boldsymbol{Y_T})} \tag{35}$$

where κ, (scalar) $\boldsymbol{\alpha}$, $\boldsymbol{\beta}$, $\boldsymbol{\lambda}$ and $\boldsymbol{\eta}$ (row vectors) are the Lagrange multipliers for the normalisation, Kirchhoff's first and second laws, flow rates and the head loss constraints, respectively. The variation of $\mathfrak{L}$ is given by $\delta \mathfrak{L} = 0$. This can be solved, in

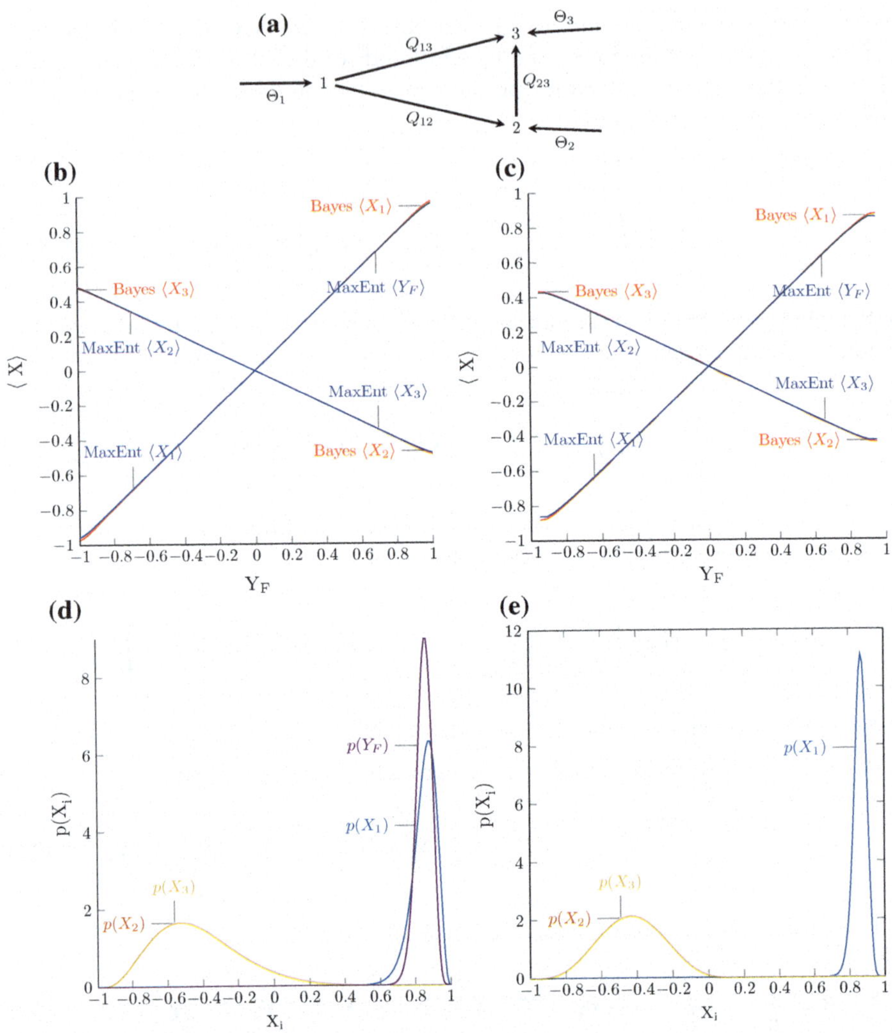

Fig. 1 **a** Network, **b** Truncated normal means **c** Beta means **d** Beta case MaxEnt marginal pdfs **e** Beta case Bayes marginal pdfs

conjunction with the constraints (30)–(34), to give $p^*(\boldsymbol{X}, \boldsymbol{Y_F}, \boldsymbol{Y_T})$ and the Lagrange multipliers κ, $\boldsymbol{\alpha}$, $\boldsymbol{\beta}$, $\boldsymbol{\lambda}$ and $\boldsymbol{\eta}$.

4 Case Study

To investigate the similarities and differences between the predictions of Bayesian and MaxEnt methods, a single loop network was analysed as presented in Fig. 1a. The basis set was chosen as the three inflows. The bounds on each flow rate were constrained to be $\boldsymbol{l} = -\mathbf{1}$ and $\boldsymbol{u} = \mathbf{1}$. The resistances were given by $\boldsymbol{K} = \mathbf{1}$. All link flow priors were selected to have a mean of zero. For the truncated normal case the prior scale parameter was chosen as $\boldsymbol{\Sigma} = 0.1 \times \boldsymbol{I}$ and the beta distribution was assigned a covariance of $0.1 \times \boldsymbol{I}$. To allow numerical integration with the beta prior using the Bayesian method the delta likelihood functions where chosen as normal distributions with a variance of 1×10^{-4}. Figure 1b, c show the mean flow rates when the flow observation Y_{F1} was varied from -1 to 1 with a scale parameter of $\Sigma_F = 0.001$ or variance of 0.001 for the normal and beta distributions, respectively. The marginal distributions for the Beta case of the posterior when $Y_{F1} = 0.9$ using the MaxEnt and Bayesian methods are respectively presented in Figure 1d, e.

The results presented show that the MaxEnt and Bayesian methods obtain similar means with the greatest difference near the integration limits although the pdfs are different. The Bayesian pdf contain cross terms, whereas the MaxEnt pdfs do not although they have more dimensions. The MaxEnt pdfs are able to incorporate cross terms if the second or higher order constraints were applied.

5 Conclusions

The soft constraints applied in the MaxEnt method show that the MaxEnt method with soft prior constraints has a numerical advantage over the Bayesian method, since the interaction terms are represented as Lagrange multipliers rather than covariances in the pdf. This bilinearisation allows the partition function integrations in MaxEnt to be executed as products of separable one-dimensional integrals, rather than as a non-separable multidimensional integrals created by the presence of cross-terms. The examples presented show only small differences between the mean values predicted by the MaxEnt and Bayesian methods.

Acknowledgements This project acknowledges funding support from the Australian Research Council Discovery Projects Grant DP140104402, Go8/DAAD Australia-Germany Joint Research Cooperation Scheme RG123832 and the French Agence Nationale de la Recherche Chair of Excellence (TUCOROM) and the Institute Prime, Poitiers, France.

References

1. Boltzmann, L.: Über die Beziehung zwischen dem zweiten Hauptsatz der mechanischen Wärmetheorie und der Wahrscheinlichkeitsrechnung respektive den Sätzen über das

Wärmegleichgewicht (On the Relationship between the Second Fundamental Theorem of the Mechanical Theory of H. Wiener Berichte **2**(76), 373–435 (1877)
2. Caticha, A.: Relative entropy and inductive inference. Bayesian Inference Maximum Entropy Methods Sci. Eng. **707**, 75–96 (2004). https://doi.org/10.1063/1.1751358
3. Caticha, A., Giffin, A.: Updating probabilities. In: AIP Conference Proceedings, vol. 872, pp. 31–42. AIP (2006). https://doi.org/10.1063/1.2423258
4. Giffin, A.: Maximum entropy: the universal method for inference. Ph.D. thesis, University at Albany, State University of New York (2008)
5. Giffin, A., Caticha, A.: Updating probabilities with data and moments. In: AIP Conference Proceedings, vol. 954, pp. 74–84. AIP (2007). https://doi.org/10.1063/1.2821302
6. Hennig, P., Kiefel, M.: Quasi-Newton methods: a new direction. In: Proceedings of the 29th International Conference on Machine Learning (ICML-12), vol. 29, pp. 25–32 (2012)
7. Hennig, P., Kiefel, M.: Quasi-Newton methods: a new direction. J. Mach. Learn. Res. **14**, 843–865 (2013)
8. Jaynes, E.T.: Information theory and statistical mechanics. Phys. Rev. **106**(4), 620–630 (1957)
9. Jaynes, E.T.: Prior probabilities. IEEE Trans. Syst. Sci. Cybern. **4**(3), 227–241 (1968). https://doi.org/10.1109/TSSC.1968.300117
10. Jaynes, E.T.: Probability Theory: The Logic of Science. Cambridge University Press, Cambridge (2003)
11. Kapur, J.N., Kesavan, H.K.: Entropy Optimization Principles with Applications. Academic Press, Boston (1992)
12. Kullback, S., Leibler, R.A.: On information and sufficiency. Ann. Math. Stat. **22**(1), 79–86 (1951). https://doi.org/10.2307/2236703
13. Manjunath, B.G., Stefan, W.: Moments calculation for the double truncated multivariate normal density. SSRN Electron. J. **1963**, 1–11 (2009). https://doi.org/10.2139/ssrn.1472153
14. Niven, R.K.: Combinatorial entropies and statistics. Eur. Phys. J. B **70**(1), 49–63 (2009). https://doi.org/10.1140/epjb/e2009-00168-5
15. Shannon, C.E.: A mathematical theory of communication. Bell Syst. Tech. J. **27**(4), 623–656 (1948). https://doi.org/10.1002/j.1538-7305.1948.tb00917.x
16. Sharp, K., Matschinsky, F.: Translation of Ludwig Boltzmann's Paper On the Relationship between the Second Fundamental Theorem of the Mechanical Theory of Heat and Probability Calculations Regarding the Conditions for Thermal Equilibrium Sitzungberichte der Kaiserlichen Akademie d. Entropy **17**(4), 1971–2009 (2015). https://doi.org/10.3390/e17041971
17. Shore, J., Johnson, R.: Axiomatic derivation of the principle of maximum entropy and the principle of minimum cross-entropy. IEEE Trans. Inf. Theory **26**(1), 26–37 (1980). https://doi.org/10.1109/TIT.1980.1056144
18. Waldrip, S., Niven, R.: Comparison between Bayesian and maximum entropy analyses of flow networks. Entropy **19**(2), 58 (2017). https://doi.org/10.3390/e19020058
19. Waldrip, S.H.: Probabilistic analysis of flow networks using the maximum entropy method. Ph.D. thesis, The University of New South Wales, Canberra Australia (2017)
20. Waldrip, S.H., Niven, R.K.: Maximum entropy derivation of Quasi-Newton methods. SIAM J. Optim. **26**(4), 2495–2511 (2016). https://doi.org/10.1137/15M1027668
21. Williams, P.M.: Bayesian conditionalisation and the principle of minimum information. Br. J. Philos. Sci. **31**(2), 131–144 (1980)

Using the Z-Order Curve for Bayesian Model Comparison

R. Wesley Henderson and Paul M. Goggans

Abstract BayeSys is an MCMC-based program that can be used to perform Bayesian model comparison for problems with atomic models. To sample distributions with more than one parameter, BayeSys uses the Hilbert curve to index the multidimensional parameter space using one very large integer. While the Hilbert curve maintains locality well, computations to translate back and forth between parameter coordinates and Hilbert curve indexes are time-consuming. The Z-order curve is an alternative SFC with faster transformation algorithms. This work presents an efficient bitmask-based algorithm for performing the Z-order curve transformations for an arbitrary number of parameter space dimensions and integer bit-lengths. We compare results for an exponential decay separation problem evaluated using BayeSys with both the Hilbert and Z-order curves. We demonstrate that no appreciable precision penalty is incurred by using the Z-order curve, and there is a significant increase in time efficiency.

Keywords MCMC · Space-filing curves · BayeSys

1 Introduction

BayeSys is an MCMC-based program that can be used to perform Bayesian model comparison for problems with atomic models. It uses a combination of methods including jump-diffusion sampling, thermodynamic integration, and binary slice sampling to sample from many models simultaneously, ultimately yielding an

R. W. Henderson (✉) · P. M. Goggans
University of Mississippi, Oxford, Mississippi, USA
e-mail: rwhender@go.olemiss.edu

P. M. Goggans
e-mail: goggans@olemiss.edu

A. Polpo et al. (eds.), *Bayesian Inference and Maximum Entropy Methods in Science and Engineering*, Springer Proceedings in Mathematics & Statistics 239, https://doi.org/10.1007/978-3-319-91143-4_28

approximation of the model posterior distribution. To accommodate multidimensional parameter spaces, BayeSys uses a space-filling curve to index the multidimensional parameter space using one very large integer. The Hilbert curve is a space-filling curve that maintains locality well; i.e., points with consecutive Hilbert curve indexes are adjacent in the parameter space. Computations to translate back and forth between parameter coordinates and Hilbert curve indexes are time-consuming, however, which motivates us to find a space-filling curve with more time-efficient transformation algorithms.

The Z-order curve is a space-filling curve with somewhat poorer locality properties when compared to the Hilbert curve but with transformation algorithms that are much faster. This work presents an efficient bitmask-based algorithm for performing the Z-order curve transformations for an arbitrary number of parameter space dimensions and integer bit-lengths. We compare results for an exponential decay separation problem evaluated using BayeSys with both the Hilbert and Z-order curves. We demonstrate that no appreciable precision penalty is incurred by using the Z-order curve, and there is a significant increase in time efficiency.

This paper is organized as follows. Section 2 briefly describes how Bayesian model comparison works and the general mechanism behind BayeSys. Section 3 compares the Hilbert and Z-order curves and details algorithms for performing the bitmask-based transformations associated with the Z-order curve. Section 4 describes our test problem and how each space-filling curve method performed with the problem. Finally, Sect. 5 concludes the paper.

2 Model Comparison Using BayeSys

Model comparison comes in two broad flavors: the first comprises one-at-a-time methods that estimate model probabilities for one model at a time, and the second comprises simultaneous methods that sample from a distribution over the joint parameter space formed by combining the parameter spaces of each model under consideration, then using the samples to compute moments of the model distribution. Nested sampling [9] and thermodynamic integration [2] are examples of the former, while BayeSys (http://www.inference.org.uk/bayesys/) is an example of the latter.

BayeSys is not a general trans-dimensional sampling method; rather, it is designed to work specifically with atomic models. Atomic models are parameterized models that can be broken into a priori identical parts. Each part, or atom, must have identical structures, and each corresponding parameter among the atoms must have equivalent prior distributions. BayeSys uses jump-diffusion sampling [6] to move between model orders. Its sampling moves consist of birth moves (adding an atom), death moves (removing an atom), and within-model moves (varying the parameters of one atom). These moves are accepted or rejected according to the standard MCMC criteria such that detailed balance is maintained and such that in the limit of many iterations, the method is guaranteed to sample from the desired distribution.

Sampling immediately from the joint posterior over the parameters and model orders is usually impossible. BayeSys tries to solve this problem using thermodynamic integration. While thermodynamic integration is usually used to estimate evidence, that is not the goal here. The goal is to start by sampling from the prior distribution and to use the annealing behavior of thermodynamic integration to gradually introduce the likelihood until we are sampling from the posterior distribution.

BayeSys uses several MCMC engines throughout its sampling process. Each of these engines has its own advantages and disadvantages, and the goal of using them all in concert is to maximize our chances of actually arriving at the posterior distribution. Several of these MCMC engines use sampling techniques that perform best in one dimension. However, each atom will likely have multiple parameters. To deal with this issue, BayeSys uses space-filling curves to perform a one-to-one mapping between $\mathbb{N}_0^1$ and $\mathbb{N}_0^N$.

3 Space-Filling Curves

The MCMC engines in BayeSys make use of space-filling curves to perform their sampling. A space-filling curve (in this application) is a one-to-one function $f : \mathbb{N}_0^1 \to \mathbb{N}_0^N$ that maps the 1-dimensional natural numbers to the N-dimensional natural numbers. The space-filling curve allows multidimensional probability distributions to be sampled using one-dimensional sampling techniques, such as binary slice sampling [4, 10]. The ideal space-filling curve for this application would have the following properties:

- Locality. Points that are nearby in the parameter space should be nearby on the curve as well. The converse should be true as well.
- Time efficiency. The algorithms for performing the mapping between parameter space and curve indexes should be time efficient.
- Bidirectionality. Algorithms should exist for mapping parameter space to curve indexes and from curve indexes to parameter space.

BayeSys uses the Hilbert curve as its space-filling curve. We assert that the Z-order curve is a better choice.

3.1 Hilbert Curve

The Hilbert curve [7, Chap. 2], [8] is a space-filling curve that has good locality properties. If two indexes are consecutive on the Hilbert curve, the points in parameter space that correspond to them are adjacent. There are also bidirectional transform functions available for the Hilbert curve, and the implementations of these functions within BayeSys are time efficient. An example of the Hilbert curve for a two-dimensional parameter space with 4 bits per dimension is shown in Fig. 1.

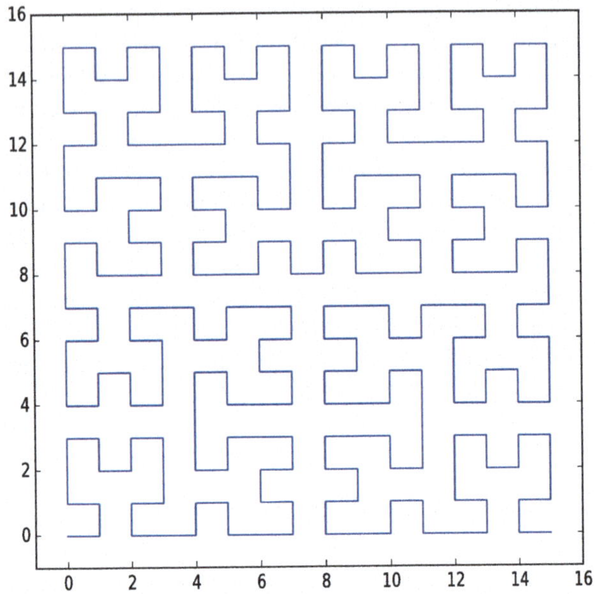

Fig. 1 Hilbert curve for two dimensions with 4 bits per dimension

While the Hilbert curve meets our three criteria for a space-filling curve suitable for sampling with MCMC, its associated transformation algorithms are fairly complex. The Z-order curve is simpler to implement and implementations can be significantly faster than implementations of the Hilbert curve.

3.2 Z-Order Curve

The Z-order curve [7, Chap. 5] (also known as the Lebesgue curve or Morton curve) is a space-filling curve that maintains locality somewhat less well than the Hilbert curve but meets our other two requirements as well. Most importantly, its transformation algorithms are faster than those for the Hilbert curve. In order to transform from the N-dimensional parameter space to the one-dimensional Z-order curve, the bits of the integer coordinates for each dimension are interleaved. If the parameter space has three dimensions and each coordinate axis is represented by a 5-bit integer, the resulting Z-order curve representation will be a 15-bit integer. An example below, in which each letter represents a binary digit, demonstrates the bit interleaving described above.

```
Z-order index        Axes coordinates

                     adgj
abcdefghijkl  <-->   behk
                     cfil
```

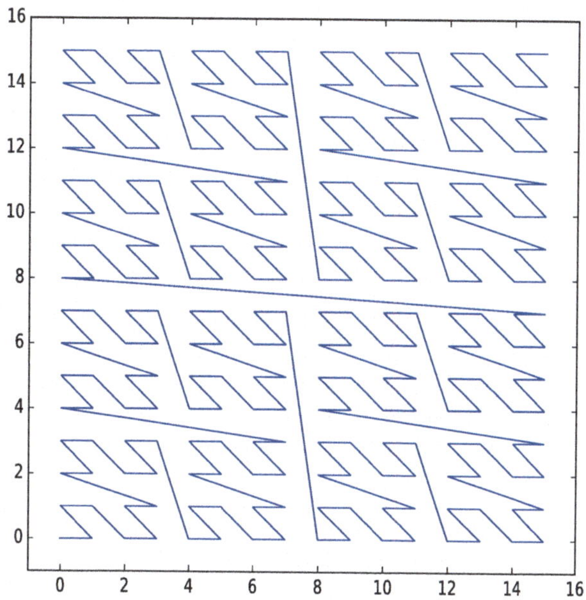

Fig. 2 Z-order curve for two dimensions with 4 bits per dimension

An example of a Z-order curve for two dimensions with 4 bits per dimension is shown in Fig. 2.

3.2.1 The Simple Algorithm

The simple way to perform the Z-order mapping is to loop over the bit and axis indexes and place each bit where it needs to be individually. An example Python 3 function is shown in Listing 1. Note that Python 3 can handle integers with arbitrary precision, so `line` can be as long as you want. NumPy, however, is limited to 64-bit integers, so `b` can be at most 64.

```
import numpy

def line_to_axes(line, b, n):
    axes = numpy.zeros(n, dtype=numpy.int64)
    for i in range(n):
        axes[i] = 0
        for j in range(b):
            axes[i] |= ((line >> (j * n + i)) & 1) << j
    return axes
```

Listing 1: Python code for mapping a Z-order curve index to *n* *b*-bit axes coordinates.

3.2.2 The Mask-Based Algorithm

A cleverer, bitmask-based approach exists, and it is documented in several places online. The most thorough description of an algorithm for generating the necessary bitmasks for arbitrary numbers of dimensions and bits per dimension is given in a Stackoverflow answer by user Gabriel [1]. Gabriel also describes the general method by which the mapping is performed using the bitmasks, but he does not provide algorithms for doing so. The following list outlines the basic procedure for mapping from the Z-order index to axes coordinates:

1. Generate bitmasks based on the number of bits b and number of parameters n.
2. AND the first mask with the Z-order integer to select only every n bits.
3. Loop over each mask. For the ith mask, XOR the Z-order integer with itself shifted to the right by i, then mask the result.
4. Shift the original Z-order integer to the right by 1, then repeat the above from step 2 for each dimension.

Our Python 3 function for computing the bitmasks is adapted from the code presented by Gabriel [1] and shown in Listing 2.

```
def compute_bit_masks(numberOfBits, numberOfEmptyBits):
    bitDistances = [i * numberOfEmptyBits for i in
                    range(numberOfBits)]
    bitDistancesB = [bin(dist)[2:] for dist in bitDistances]
    moveBits = []
    maxLength = len(max(bitDistancesB, key=len))
    for i in range(maxLength):
        moveBits.append([])
        for idx, bits in enumerate(bitDistancesB):
            if not len(bits) - 1 < i:
                if(bits[len(bits)-i-1] == "1"):
                    moveBits[i].append(idx)
    bitPositions = list(range(numberOfBits))
    maskOld = (1 << numberOfBits) - 1
    bitmasks = []
    for idx in range(len(moveBits)-1, -1, -1):
        if len(moveBits[idx]):
            shifted = 0
            for bitIdxToMove in moveBits[idx]:
                shifted |= 1 << bitPositions[bitIdxToMove]
                bitPositions[bitIdxToMove] += 2 ** idx
            nonshifted = ~shifted & maskOld
            shifted = shifted << 2**idx
            maskNew = shifted | nonshifted
            bitmasks.append(maskNew)
            maskOld = maskNew
    return bitmasks
```

Listing 2: Code for computing bitmasks for the more efficient Z-order transform, adapted from [1].

Our Python 3 functions for transforming from axes coordinates to Z-order indexes and back are given in Listings 3 and 4. These functions were inspired by the approach shown in [1], but are entirely original code.

```
from copy import copy
from operator import ior
from functools import reduce
def axes_to_line(alpha, n, b, masks):
    if n == 1:
        return int(alpha)
    alpha = list(alpha)
    alpha = [int(item) for item in alpha]
    max_shift = (n - 1) << len(masks) - 1
    for x, i in zip(alpha, range(n)):
        x &= (1 << b) - 1
        shift = copy(max_shift)
        for mask in masks:
            x = (x ^ (x << shift)) & mask
            shift = shift >> 1
        x <<= i
        alpha[i] = x
    z = reduce(ior, alpha)
    return z
```

Listing 3: Code for transforming from axes coordinates to Z-order indexes using the bitmask method.

```
from numpy import array, int64
from copy import copy
def line_to_axes(z, n, b, masks):
    if n == 1:
        return array(z, dtype=int64)
    masks = copy(masks)
    first = masks.pop()
    masks.reverse()
    masks.append((1 << b) - 1)
    min_shift = n - 1
    alpha = []
    for i in range(n):
        zz = copy(z) >> i
        zz &= first
        shift = copy(min_shift)
        for mask in masks:
            zz = (zz ^ (zz >> shift)) & mask
            shift = shift << 1
        alpha.append(zz)
    return array(alpha, dtype=int64)
```

Listing 4: Code for transforming from Z-order indexes to axes coordinates using the bitmask method.

4 Performance

To test the performance of the Z-order curve space-filling curve in practice, we implemented it in C++ within BayeSys. Both the simple, brute force algorithms and the mask-based algorithms were implemented and tested. Within the framework of BayeSys, the simple implementation of the Z-order curve is fairly straightforward. However, given that BayeSys represents the space-filling curve indexes as n-dimensional arrays of b-bit integers, implementing the bitmask-based approach was less straightforward. Each bit-shift operation requires carries between the array elements to be explicitly computed, reducing the time efficiency.

Table 1 Summary of BayeSys results for Lanczos decay problem using each space-filling curve method over 100 runs

Method	Average atoms		Time (s)		Mean efficiency (%)
	Mean	RMSE	Mean	Stdev	
Hilbert Curve	3.414	0.6032	152.4	18.33	2.191
Z-order, simple	3.325	0.5670	119.9	11.89	2.044
Z-order, mask	3.391	0.6234	135.1	18.27	2.039

This implementation was tested using the Lanczos decay problem originally presented by Lanczos in [3, Chap. 4, Sect. 23] and presented with a Bayesian approach by Ó Ruanaidh and Fitzgerald in [5, Chap. 7]. The specific data used was that presented in Table 7.2 in [5, Chap. 7]. In this problem, the task is to count how many decaying exponentials are present in an observed signal. The signal model for discrete times t_i with J exponential decay components, amplitudes A_j, and decay constants λ_j, takes the form

$$g[t_i] = \sum_{j=1}^{J} A_j \exp(-\lambda_j t_i). \tag{1}$$

For the likelihood, we used a Gaussian distribution with a standard deviation of 0.0001, which corresponds roughly to the amount of error known to exist in our simulated data. For the prior distributions, we used a uniform prior on [0.01, 2.0] for the amplitudes, a uniform prior on [0.1, 1.1] for the inverse of the decay constants, and an unbounded geometric prior for the number of atoms. Regarding the priors for the amplitude and decay constants, we chose these distributions to be broad enough to let the likelihood dominate in the calculation while not being so broad as to allow completely unreasonable parameter values. The likelihood functions were implemented so that only the necessary differential mock data is computed when atoms are added or removed. 100 objects were used in the BayeSys ensemble.

Table 1, Figs. 3, and 4 summarize the results for each space-filling curve method tested, with statistics computed over 100 runs for each method. There was not much difference among the methods observed in precision, i.e., the mean average atoms reported are all close to the correct value of 3, and the RMS error is also similar among the methods. The difference in time was more pronounced, with the simple Z-order method being the fastest and the Hilbert curve method being the slowest.

4.1 Discussion

No significant difference was observed in either precision or sampling efficiency among either method. This is a positive result, as it suggests that the somewhat weakened locality of the Z-order curve does not adversely affect sampling. Both

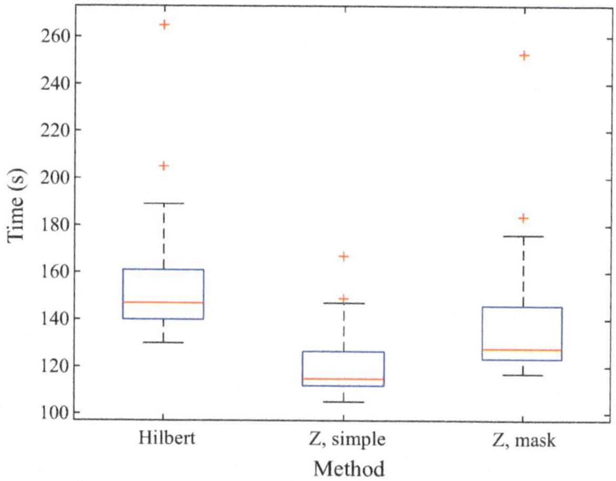

Fig. 3 Box plot of BayeSys run times for each space-filling curve method over 100 runs. One outlier for the Hilbert method at about 500 s was excluded

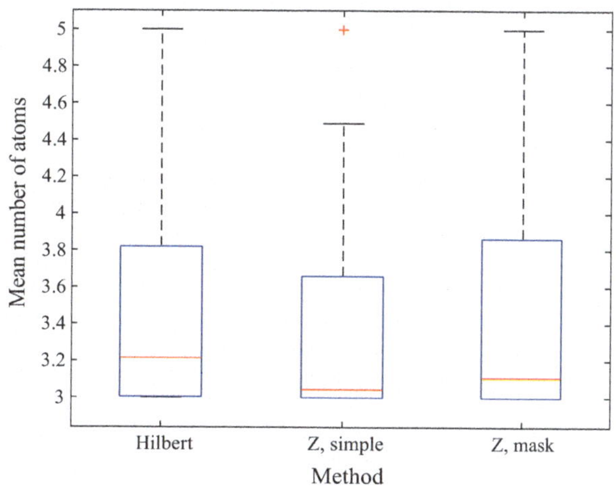

Fig. 4 Box plot of BayeSys mean atom results for each space-filling curve method over 100 runs

Z-order curve implementations were observed to be faster overall than the Hilbert curve implementation, confirming our hypothesis. An unexpected aspect of the time results is that the mask-based Z-order curve method is slower than the simple Z-order curve method. Upon reflection, this result is likely due to the limitations of our implementation of the mask-based method within BayeSys. As previously mentioned, BayeSys stores the space-filling curve indexes as n-dimensional arrays of b-bit integers, requiring us to explicitly compute the carries for each bitshift operation. This likely adds a fair amount of overhead, leading to the unexpectedly slow run time for the mask-based method. In the future, it might be interesting to try something like the GNU Multiprecision Arithmetic Library (https://gmplib.org/) as an alternative way to represent the large integers in the space-filling curve indexes to see if the run time comparisons change.

5 Conclusion

We have presented the Z-order curve as an alternative to the Hilbert curve to perform parameter space mappings within BayeSys. Results obtained from BayeSys for the Lanczos exponential decay separation problem indicate that the Z-order curve can provide a speed increase for some problems at little or no cost to accuracy.

References

1. Gabriel (https://stackoverflow.com/users/293195/gabriel): How to compute a 3d morton number (interleave the bits of 3 ints). Stackoverflow (2013). https://stackoverflow.com/revisions/18528775/2
2. Goggans, P.M., Chi, Y.: Using thermodynamic integration to calculate the posterior probability in Bayesian model selection problems. AIP Conf. Proc. **707**(1), 59–66 (2004). https://doi.org/10.1063/1.1751356
3. Lanczos, C.: Applied Analysis. Prentice Hall, Englewood Cliffs (1956)
4. Neal, R.M.: Slice sampling. Ann. Stat. **31**(3), 705–767 (2003). https://doi.org/10.1214/aos/1056562461
5. Ó Ruanaidh, J.J.K., Fitzgerald, W.J.: Numerical Bayesian Methods Applied to Signal Processing. Springer, New York (1996)
6. Phillips, D.B., Smith, A.F.M.: Bayesian model comparison via jump diffusions. In: Gilks, W., Richardson, S., Spiegelhalter, D. (eds.) Markov Chain Monte Carlo in Practice, Chap. 13. Chapman and Hall, London (1996)
7. Sagan, H.: Space-Filling Curves. Springer, New York (1994)
8. Skilling, J.: Programming the Hilbert curve. In: Erickson, G., Zhai, Y. (eds.) Bayesian Inference and Maximum Entropy Methods in Science and Engineering: 23rd International Workshop, pp. 381–387. American Institute of Physics (2004)
9. Skilling, J.: Nested sampling for general Bayesian computation. Bayesian Anal. **1**(4), 833–859 (2006)
10. Skilling, J., MacKay, D.J.C.: [slice sampling]: Discussion. Ann. Stat. **31**(3), 753–755 (2003). http://www.jstor.org/stable/3448417

The manufacturer's authorised representative in the EU is Springer Nature Customer Service Centre GmbH, Europaplatz 3, 69115 Heidelberg, Germany. If you have any concerns regarding our products, please contact ProductSafety@springernature.com

Printed and bound by CPI Group (UK) Ltd, Croydon, CR0 4YY
22/09/2026
02221059-0008